Computation in Cellular and Molecular Biological Systems

Computation in Cellular and Molecular Biological Systems

Roy Cuthbertson
Dept Human Anatomy & Cell Biology, Univ. Liverpool

Mike Holcombe
Dept Computer Science, Uni. Sheffield

Ray Paton
Dept Computer Science, Uni. Liverpool

World Scientific
Singapore • New Jersey • London • Hong Kong

Published by

World Scientific Publishing Co. Pte. Ltd.

P O Box 128, Farrer Road, Singapore 912805

USA office: Suite 1B, 1060 Main Street, River Edge, NJ 07661

UK office: 57 Shelton Street, Covent Garden, London WC2H 9HE

COMPUTATION IN CELLULAR AND MOLECULAR BIOLOGICAL SYSTEMS
Selected Papers from IPCAT 95

ISBN 981-02-2878-3

This book is printed on acid-free paper.

Printed in Singapore by Uto-Print

This book is dedicated to the memory of

Roy Cuthbertson

(1954–1996)

PREFACE

The First International Workshop on Information Processing in Cells and Tissues (IPCAT) was held in Liverpool between 6th - 8th September 1995. The purpose as advertised to the scientific community was

> "to bring together a multidisciplinary group of scientists working in the general area of modelling cells and tissues".

In the end it attracted 85 delegates from 19 different countries. The discipline specialists attending included: mathematicians, biochemists, cell biologists, neuroscientists, physiologists,computer scientists, biophysicists, immunologists, medics and electronic engineers. This book represents a selection from over forty papers which were presented at IPCAT 95. The first chapter provides an introductory overview to the book which is then divided into three general areas:

- Information processing and signalling processes
- Information processing and cellular systems
- Dynamical models of cellular systems and information processing

It is important to note however that because the material is interrelated in many ways it could have been organised differently. The reader is encouraged to view the text as a whole and find many meaningful links outside of the three sections.

There is often a great deal of discussion these days about multi-disciplinary research and the value of exchanging ideas and methods across traditional discipline boundaries. Indeed, it could be justifiably argued that many of the advances in science and engineering take place because the ideas, methods and the tools of thought from one discipline become re-applied in others. Sadly, it is also the case that many subject areas develop specialised vocabularies and concepts and indeed may also approach more general problems in fairly narrow subject-specific ways. As a result barriers develop between disciplines that prevent the free flow of ideas and the collaborations that could often bring success. This is particularly true in biology and computing where recent rapid developments in research create a danger that many of the advances made do not get exploited in the other subject. As one participant commented:

> "I would say that IPCAT has opened a forum of a unique nature: both experimentalists and theoreticians from the biosciences and computer science badly needed to join their forces in order to make sense of the contemporary glut of new bio-molecular-informational data. It is a timely and rewarding enterprise."

Now is an excellent time to try to bring the two communities together to evaluate progress, educate and inspire each other in the expectation that significant rewards will follow.

A central theme for IPCAT 95 was to explore the nature of biological information and the ways it is processed in cells and tissues. The workshop sought to provide a forum to report research, discuss emerging topics and gain new insights into biological and computational information processing systems. Comments from one participant reflect this view:

> "I was surprised to encounter such a strong interest in models on a subcellular level. Considering the state of cellular and network models, which have seen an enormous development recently, it may be hoped that the subcellular level will equally well develop. However, this level also holds particular difficulties e.g. the limited spatial and temporal resolution of techniques that could provide data on local concentrations of reactants in vivo. Thus, both better modeling and experimental approaches to subcellular information processing are required."

The multidisciplinary and international nature of IPCAT meant that a variety of viewpoints were exposed and discussed. One participant commented:

> "I found the conference very stimulating in many respects. Especially, what I like are as follows.
>
> (1) The conference covered many different topics relevant to biology.
>
> (2) Most of the scientists who participated seem very knowledgeable in a wide variety of topics. Experimentalists had interests in theory and were quite knowledgeable in theory. Theoreticians dealt with biologically relevant problems, not just mathematically oriented problems.
>
> I learned a lot from the talks given by the speakers and also through the discussions during the breaks."

This view was widely shared and most participants felt that the conference was a uniquely stimulating experience. We hope to repeat the event in 1997.

One of the key motivations underlying the first IPCAT Workshop and this book was to attempt to provide a common ground for dialogue and reporting research without emphasising one particular research constituency or way of modelling or singular issue in this area. This is quite a challenging agenda when one considers the great variety of techniques and perspectives involved and the great variety of biological phenomena being studied. There are many exciting and novel avenues that are worth exploring but the understanding of biological phenomena should always be a primary aim.

The IPCAT idea did not take very long to evolve. It started about one year prior to the event when initial discussions between the editors got underway on running a workshop in this area. It soon became clear to us that such an event was timely and this was to some degree endorsed when many of the people we approached happily agreed to become members of the programme committee. However, it was still a risky business and so we are grateful to a number of sponsors (listed below) who provided a much-needed buffer should the event have been unsuccessful.

Part of the challenge of IPCAT and this book is to re-appraise the types of models used to describe and explain the systems of interest and explore complementary viewpoints. Hence, some participants commented:

> "After IPCAT 95 we became more conscious that there is still a lot of work to do concerning the modelling of biological systems. This work concerns not only the sort of models we are working with to represent and quantitate biological systems, but to a real questioning of the implicit built-in assumptions of our models. Many of these implicit assumptions are contained in the experimental evidence we choose as a source of the piece of reality we intend to model ... IPCAT 95 provided us the following question: is information always associated with organization? and if so, how ?"

We hope this issue will be addressed further in future IPCAT events.

It was with sadness that during the production of this book Roy Cuthbertson died. His work on calcium oscillators and positive feedback attracted international recognition and his ideas and enthusiasm contributed greatly to the workshop. As a small token of our appreciation and in recognition of his work we dedicate this book to his memory.

Our thanks to Richard Lim of World Scientific for his interest in IPCAT 95 and encouragement and advice during the production of this book. The editors also acknowledge the help and support of a number of organisations who made the first IPCAT workshop possible: GPT, Unilever, Zeneca, SmithKline Beecham, Merseytravel, The University of Liverpool and The Royal Society.

Finally our thanks to the programme review committee:

Georg Brabant	-	Endocrinology (Hanover)
Michael Conrad	-	Computer Science (Detroit)
Roy Cuthbertson	-	Cell Biology (Liverpool)
Claus Emmeche	-	Philosophy of Science (Copenhagen)
Mike Holcombe	-	Computer Science
George Kampis	-	Philosophy of Science (Budapest)
Douglas Kell	-	Biological Sciences (Aberystwyth)
Gareth Leng	-	Physiology (Edinburgh)
Pedro Marijuan	-	Electronics & Informatics (Zaragoza)
Koichiro Matsuno	-	BioEngineering (Nagaoka)
Ray Paton	-	Computer Science (Liverpool)
Hans-Paul Schwefel	-	Computer Science (Dortmund)
Idan Segev	-	Neurobiology (Jerusalem)
Gordon Shepherd	-	Neurobiology (Yale)
Rene Thomas	-	Molecular Biology (Brussels)
Chris Tofts	-	Computer Science (Manchester)
John Tucker	-	Computer Science (Swansea)
G Rickey Welch	-	Biological Sciences (New Orleans)
Gershom Zajicek	-	Experimental Medicine (Jerusalem)

Ray Paton and Mike Holcombe
June, 1996

Contents

Dedication v

Preface vii

Introduction

1. Information Processing in Cells and Tissues 3
R. Paton and M. Holcombe

Information Processing and Signalling Processes

2. Interplay of Cell-Cell Signalling and Multicellular Morphogenesis during Dictyostelium Aggregation 15
T. Höfer and P. K. Maini

3. Concerted Regulation of Cyclic Adenosine Monophosphate by Calmodulin/Calcium Complex and Dopamine: A Kinetic Modelling Approach 29
R. Kötter, D. Schirok and K. Zilles

4. A Role of Ca^{2+}/Calmodulin-Dependent Protein Kinase II in the Induction of Long-term Potentiation: Switching Characteristics according to Ca^{2+} Signalling and Digitization of Graded Information into Binary Information 43
H. Okamoto and K. Ichikawa

5. Mathematical Aspects of Signal-Processing via G-Proteins 57
J. Nauroschat and U. an der Heiden

6. Oscillations, Positive Feedback and the Flow of Biological Information 69
K. S. R. Cuthbertson

7. A Data and Knowledge Base for Cell Signalling Networks 77
T. Igarashi, T. Kaminuma and Y. Nadaoka

8. Modelling the Emergence of Coregulated Proteins in Biological Regulation Networks 91
E. Chiva and P. Tarroux

9. Specifying Logical Agents in Cellular Hierarchies 105
R. Paton, G. Staniford and G. Kendall

10. Nerve Signal Processing using Artificial Neural Nets 121
M. Bogdan, A. Babanine, J. Kaniecki and W. Rosenstiel

Information Processing and Cellular Systems

11. Long-Range Metabolic Signalling in Heterogeneous Catalytic Systems: Computer Simulation Study 137
P. Mendes, P. Marmillot, J.-F. Hervagault and G. R. Welch

12. Elementary Modes of Functioning in Biochemical Networks 151
S. Schuster, C. Hilgetag, J. H. Woods and D. A. Fell

13. Computational Models of Cellular Processing 167
A. Bell and M. Holcombe

14. The Cell as a Problem-Solving "Engine" 183
P. C. Marijuan

15. Models of Cytoplasmic Architecture and Function 195
M. A. Aon, S. Cortassa and A. Cáceres

16. Models of Information Processing within Inhomogeneous Cellular Structures 209
P. Hartmann

17. Genomic Cellular Automata Transposed on Silicon: Experiments in Synthetic Life 223
P. Marchal, P. Nussbaum, C. Piguet, S. Durand, D. Mange, E. Sanchez, A. Stauffer and G. Tempesti

18. Cohesive Interaction in Biomolecules and their Organizations as Energy Consumers 237
K. Matsuno

19. Bioenergetics and Biocommunication 251
M.-W. Ho

20. Parameter Estimation for a Diffusion Approximation to Stein's Model 265
S. Iyengar

Dynamical Models of Cellular Systems and Information Processing

21. Complex Trajectories from Simple Feedback Circuits 281
R. Thomas

22. Reconstruction of Excitable Tissue Physiology and Pathophysiology from Cellular Models 287
M. R. Boyett, A. V. Holden and H. Zhang

23. Roles of the Ion Channels in Reentrant Cardiac Arrhythmia 301
T. R. Chay

24. The Effects of Random Variation in Stimulation Timing and Magnitude on Excitable and Oscillatory Forms of a Luteinizing Hormone Pulse Generator Model 315
J. P. A. Foweraker and D. Brown

25. A Model of Pituitary Release of Growth Hormone 329
E. A. Stephens, D. Brown, G. Leng and R. G. Smith

26. The Effect of the Embryonic Bottleneck on Vertical Microparasite Transmission 339
M. J. Hatcher, A. M. Dunn and C. Tofts

27. On the Simulation of Health with the Tissue Automat 353
G. Zajicek

Introduction

INFORMATION PROCESSING IN CELLS AND TISSUES

RAY PATON AND MIKE HOLCOMBE

This chapter contains a summary of selected contributions from the proceedings of IPCAT 95. As noted in the preface, the book is collected around three organising themes:

- Information processing and signalling processes
- Information processing and cellular systems
- Dynamical models of cellular systems and information processing

It is important to remember that the material is interrelated in many ways it could have been organised differently. The reader is encouraged to view the text as a whole and find many meaningful links between the many contributions.

The first section of the book looks at information processing and signalling processes in cellular systems. It begins with a study by Thomas Hoeffer and Philip Maini of the interplay of cell-cell signalling and multicellular morphogenesis during *Dictyostelium* aggregation. The development of multicellularity in the life cycle of this microorganism provides a paradigm model system for the study of biological pattern formation. Following starvation, periodic waves of cyclic AMP initiate the aggregation of *Dictyostelium* amoebae. Subsequently, a pattern of branching cell streams is formed which coalesce into an aggregation centre. Thomas and Philip derive a model of the aggregation process which is based on the interplay of the dynamics of the cell distribution, governed by chemotactic movement towards cAMP, with the previously much-studied intercellular cAMP signalling. The model shows how signal relay, chemotactic movement and adaptation orchestrate the collective modes of cell communication and migration in the aggregating cell layer. The chemically mediated cell-cell interaction causes the break-up of the initially uniform cell layer via a chemotaxis-driven instability, leading to the formation of the cell stream morphology which marks the onset of multicellularity in situ. The model provides a unifying framework within which a variety of experimental observations on this unusually complex excitable medium can be explained.

The next few papers report on work concerned with investigations of intracellular signalling. Firstly, Rolf Koetter, Dirk Schirok and K Zilles present a kinetic approach to modelling the concerted regulation of cAMP by calcium, calmodulin and dopamine. They investigated the variation of intracellular cyclic adenosine monophosphate (cAMP) concentrations in response to brief dopamine and calcium stimuli in a kinetic model of adenylate cyclase (AC) and phospodiesterase (PDE) regulation: The model included stimulation of AC by calmodulin/calcium complex ($CaMCa_4$) and dopamine, inhibition of AC by intracellular free calcium, and stimulation of PDE by $CaMCa_4$. The equilibrium concentration of cAMP was in the nanomolar range. Both dopamine and small calcium pulses increased cAMP levels. In contrast, large calcium pulses produced a stimulation of cAMP levels that was initially suppressed by calcium-mediated inhibition. Temporal shifts of large calcium and dopamine pulses in respect to each other revealed an asymmetric interaction: Coincidence of the two pulses abolished the effect of dopamine while a large calcium pulse preceding dopamine by more than 0.5 s produced synergistic increases of cAMP levels. The validity of these findings and their possible significance is discussed.

Following this Hiroshi Okamoto and Kazuhisa Ichikawa present a role of Ca^{2+}/calmodulin-dependent protein kinase II in the induction of long-term potentiation. This work shows relations between the switching characteristics in Ca^{2+} signalling and digitalization of graded information into binary information. Long-term potentiation (LTP) has been extensively studied as an experimental model for usage-dependent change in synaptic efficiency that is thought to be a cellular mechanism underlying memory formation in the brain. A number of studies suggest that Ca^{2+}/calmodulin-dependent protein kinase II (CaMKII) is crucially involved in LTP. However, the way CaMKII participates in LTP is far from elucidated. Hiroshi and Kazuhisa propose a theoretical model for biochemical-reaction networks comprising Ca^{2+}/calmodulin-dependent autophosphorylation of CaMKII and dephosphorylation of the kinase by phosphatase, and investigate a role of CaMKII in LTP by computer simulation of the model. A critical assumption of the model is based on recently reported biochemical evidence indicating that autophosphorylation of threonine 286 residue, the prerequisite process to the generation of Ca^{2+}/calmodulin independent (autonomous) activity of the kinase, is an intersubunit reaction. Results of our simulation show switching characteristics of the model: There is a threshold regarding Ca^{2+} signalling; if the intensity of the Ca^{2+} signal is above this threshold, autonomous activity of CaMKII is elevated maximum (i.e. switched on) if otherwise, it remains near zero (ie switched off). These results suggest a role of CaMKII in the induction phase of LTP: CaMKII digitizes graded information mediated by Ca^{2+} signalling to binary information which may determine whether LTP can subsequently be induced.

Jurgen Nauroschat and Uwe an der Heiden present a mathematical model of transmembrane signalling via G-proteins. Guanine-nucleotide regulatory proteins (G-proteins) are important regulators of various cellular processes such as proliferation, differentiation and protein synthesis. In particular, they are imperative elements of the transmembrane signalling system, processing and conducting external messages from heptahelic receptors to the cytoplasmic domain. Since cellular behaviour crucially depends on such environmental influx, Jurgen and Uwe focus on the omnipresent receptor /G-protein/adenylate cyclase membrane pathway. They outline both its activation and its recently discovered negative feedback regulation by the G-proteins. A mathematical model comprising variables for distinct receptor states, the G-proteins, and the second messenger cAMP is then proposed. The model is based on reaction kinetics. Boundedness of solutions is mathematically shown by specifying a domain (prism) being invariant under the vector field of the model. Referring to the situation of sustained exposure of the membrane to a signal of fixed amplitude, the model exhibits a unique steady-state solution for any value of the amplitude and saturates with large amplitudes. All the steady-states are locally asymptotically stable. They also demonstrate dynamic features of the model such as hypersensitivity, adaption and saturation phenomena.

Roy Cuthberston's chapter is an edited version of some notes for a paper he was working on. Sadly, he died before the paper was completed but we include it here as a tribute. He provides some valuable insights into the nature of information flow in relation to signalling pheneomena. In particular the importance of positive feedback to biological systems. Roy looks at the relations between positive feedback, sigmoidal transfer functions and digital switches in a number of signalling contexts. There are some valuable insights into the role of positive feedback cycles in mouse oocyte fertilisation, CaM Kinase memory and heart disease.

Takako Igarashi, Yoko Nadaoka and Tsuguchika Kaminuma's paper reports on a data and knowledge base for cell signalling networks. They have developed a prototype of the data and knowledge base of the cell signalling networks that consists of interacting extracellular (xenobiotic) chemicals and biomolecules. The system contains structure and functional data and references of these molecules. The system will be useful for modelling cells and their information processing, and to explain important biological phenomena based on these models. The system may also be useful for designing new drugs and for the studies of the so called quantitative structure activity relation (QSAR). The prototype system was implemented on UNIX workstations using an object oriented database management system ACEDB. The system is also available on the Internet.

This first section of the book is concluded with three papers which apply computational techniques to cell signalling systems in distinctive ways. Emmanuel Chiva and Philippe Tarroux report on the modelling the emergence of coregulated proteins in biological regulation networks. The organisation of internal regulation networks enables a cell to differentiate according to environmental stimuli. Since a global experimental approach of both the architecture and the dynamics of such a system is out of range, they design artificial regulation networks. The architecture of the network results from an evolutionary process under environmental constraint. Such a process is mimicked by a genetic algorithm (GA). Emmanuel and Philippe design a fitness function to approximate the constraint to exhibit as many different responses as possible under the influence of the cell environment. They show that the emergence of particular hierarchized structures allows the cell to respond in an adapted way to persistent environmental inducers. Using multivariate analysis techniques, they demonstrate that the model acquires this ability when functionally correlated units appear, an observation that can be related to the coregulated protein groups previously documented in actual regulation systems.

Ray Paton, Geof Staniford and Geoff Kendall discuss the specification of logic-based computational agents in cellular hierarchies with a report on enzymes as information processing agents. The paper begins with a look at some hierarchical and network modelling relations in cells. This leads on to ways of describing enzymes in terms of their switching, pattern recognition and memory capabilities. Two enzymes are used to illustrate these ideas: CaM Kinase II and Cyclin dependent kinase. CaM Kinase II is described from the point of view of a fuzzy system in particular a fuzzy classifier system and the paper concludes with a study of Cyclin dependent kinase from the perspective of Decentralized A.I.

In the final paper of this section Martin Bogdan, Alexei Babanine, Jorg Kaniecki and Wolfgang Rosenstiel discuss the INTER-project (Intelligent Neural InTERface), especially from the point of view of Artificial Neural Nets (ANN). They propose a modus operandi to process real nerve signals using ANNs. They present a method to simulate nerve signals, first experience in separating nerve signals from multi-array recorded data as well as first experience using an artificial neural net. Microfabricated neural interfaces promise to become a powerful tool for applications like the control of motor/sensory limb prostheses for amputees and the direct stimulation of spinal cord injuries. Such interfaces have been successfully implanted and signals of the peripheral nerve system have been recorded. These signals have to be processed for removing limb prostheses. They propose to interpret recorded signals using the Independent Components Analysis (INCA) to separate the incoming signals.

The seccond section of the book deals with information processing and cellular systems. In the first paper Pedro Mendes, Philippe Marmillot, Jean-Francois Hervagault and Rickey Welch look at a computer model of the spatio-temporal behaviour of an enzyme, phosphofructokinase (PFK), in glycolysis and examine the relationship between the aqueous and bound phases. This model emphasises the heterogeneous nature of the system and exposes the wave-like behaviour of the concentration that emerges in the aqueous phase. One interesting outcome is that the front of these waves travel two orders of magnitude faster than its molecules would if diffusion was acting alone. We discuss the relevance of these phenomena in t erms of the physiological conditions of some large mammalian cell structures.

Stefan Schuster, Claus Hilgetag, John Woods and David Fell examine steady state substrate fluxes in a network. A method for detecting all elementary modes in reaction networks of any complexity is presented. The approach sheds new light on the question of properly delimiting metabolic pathways in the living cell, with the implication that an overlap of simple pathways should be catered for. The method is applied to two models of erythrocyte energy metabolism. The elementary modes obtained demonstrate essential functional aspects of the system. Their methodology illustrates another interesting characteristic of biochemical reaction systems, as futile cycles yield as (cyclic) elementary modes. This aspect appears to be helpful in the analysis of cascades of inter-and intracellular signalling.

Alex Bell and Mike Holcombe look at some abstract algebraic models for modelling complex, hierarchical metabolic systems. They present a hierarchical model based on a general theory of computation which sets out to model these systems. Alex and Mike then concentrate on modelling the specific resource-based local metabolic processing that lies at the bottom of this hierarchy.

Pedro Marijuan considers the cell as an information based problem-solving engine. He argues for a view of the cell as a mesoscopic population of microscopic problem solving machines, enzymes and proteins, organized to order to circumvent the overall problems related to the persistence and reproduction of the system. Three abstract properties centre on the biological problem solving organization, self organisation, self modification and self reconfiguration.

Miguel Aon, Sonia Cortassa and A. Caceres address the problem of the global coordination of metabolism by the cytoplasmic architecture. The dynamics of enzymatic reactions coupled to the assembly-disassembly of cytoskeleton components and its significance in the redirection of metabolic fluxes are investigated. A main focus is on the mechanisms of coherence and transduction of signals into a specific cellular response. The tensegrity and fractal percolation models and the molecular crowding view are compared together with their main contributions toward an understanding of the overall organisation of metabolism.

Peter Hartmann also looks at the problems of modelling the dynamics of inhomogeneous cellular structures, in particular tissues. Effects such as cell - cell interactions and morphogenesis are studied. Cellular hypergraphs are discussed and a relative addressing mechanism is presented in this context. A CHG model can be constructed from a biological system. It is demonstrated how state values (like concentrations of chemical substances) can be stored within the CHG model. Then the terms address and component as well as relative addressing are explained in a more general context. Finally it is shown how these addressing mechanisms can be used to formulate the local transition function of the CHG model.

Pierre Marchal and colleagues report on the realisation of VLSI circuits capable of implementing sequential or combinational functions and, like living beings, of repairing themselves both locally (self repair) and globally (self reproduction) without external intervention. the mathematical model underlying this Genomic Cellular Automaton (GCA), as simple proto-cells did evolve through joint venture and finally gather to build the living cell observed nowadays, the GCA cell is a patch work of proto-cells in interaction. Having described both software and hardware implementations of the GCA cell, the emergence of the synthetic counterparts of lifelike properties such as cloning, differentiation, healing and reproduction are discussed.

The next two papaers apply ideas from physics to problems associated with the coherent properties of biosystems. Koichiro Matsuno considers the cohesive interaction exhibited by biomolecules as energy consumers. What is unique to energy consumers is the cohesive capacity exhibited by themselves towards their energy resources. The cohesive interaction latent in energy consumers can be described by the scheme of external description of internal measurement.

MaeWan Ho's contribution is concerned with coherence in organisms. She considers the bioenergetic relationships from a thermodynamic perspective and demonstrates how many of the features of energy mobilisation in the living system can be explained by symmetrically coupled, cyclical flows of stored energy over space-time domains.

Finally Satish Iyengar considers the Ornstein-Uhlenbeck process, which has been proposed as an approximation to Stein's model for the activity of a neuron. According to this model, the firing of the neuron corresponds to the first passage of the process to a constant boundary, or threshold. Statistical inference for this model using interspike interval data has not been developed because the first passage time density is intractable. The author uses the Laplace transform instead to show that maximum likelihood estimation is valid for large samples and illustrates the methods using both simulated data, and spike trains from goldfish retinal ganglion cells.

The final section looks at dynamical aspects of cellular systems in relation to information. Rene Thomas discusses issues associated with how complex trajectories arise from simple feedback circuits. Studies on the biological role of feedback circuits were initially centred on systems with sigmoid or stepwise interactions. It turned out recently that reasoning in terms of feedback circuits (rather than of individual interactions) can be used in a more general context, and, in particular, help understanding the behaviour of weakly non-linear systems "a la Rossler" (with a single non-linear term) known to generate multiple periodicity or deterministic chaos.

The next two papers describe dynamical models of heart tissue. M. R. Boyett, Arun Holden and H. Zhang discuss simulation models concerned with excitable tissue physiology. Pacemaker activity, propagation and reentry in atrial tissue are simulated by reconstructing a particle differential equation model of tissue from biophysically accurate cell models. Isolated cells are represented by nonlinear differential systems, and linear electrical coupling of cells or constructing a continuum model in which voltage interacts diffusively exposes new physiological phenomena - including atrial modulation of pacemaker rate, and illustrates the emergence of spatially coherent structures - travelling and rotating waves that are tissue, not cell behaviours. Cardiac tissue, in which intracellular and cellular dynamics, and the mechanisms and connectivity of cell-to-cell interactions is quantitatively described, provides simple illustrations of the emergence of global dynamics is spatially extended cellular networks.

Teresa Chay seeks to explain why some antiarrhythmic drugs exert proarrhythmic action to patients with a history of myocardial infarction. To examine this, she uses both bifurcation analysis and simulation approaches. Antiarrhythimic drugs are modeled by varying the maximal conductance of the sodium, calcium, time-dependent delayed rectifying potassium, and time-independent inward rectifying potassium channels in the Beeler-Reuter model. The bifurcation analysis predicts that there is a critical ring size (CRS) at which the infinite ring behaviour breaks down. In the neighbourhood of the CRS, the cycle length, the conduction velocity, and the action potential duration exhibit oscillation. The channel blockers affect CRS in different manners - both the sodium and calcium blockers shift the CRS to the left (proarrhythmia), while the K^+ blockers shift the CRS to the right (antiarrhythmia). Based on this criterion, the drugs which block the sodium and calcium channels are, in general, proarrhythmic.

The next two papers deal with aspects of the pulsatile dymanics of certain pituitary hormones. In the first study, Jonathan Foweraker and David Brown describe some investigations on a luteinizing hormone pulse generator. Brown et al. had developed a mathematical model of the luteinizing hormone (LH) pulse generator. This exhibits an experimentally observed non-monotonic relationship between the frequency of LH pulses (fp) and the frequency of discrete adrenergic stimulation (fa). This relationship has previously been shown to be robust to slight random variation in fa. In this paper these analyses are extended to include results spanning the range between regular and Poissonian patterns of stimulation, as well as introducing moderate random variation in the stimulus magnitude. The interactions between these two types of random variation and their effect on the non-monotonic fp/fa relationship using an empirical measure of non-monotonicity are investigated. The boundary between monotonic and non-monotonic relationships statistically are analysed and the interactions between the two types of random variation quantified. Previous work on an oscillator model for the LH pulse generator by introducing random variation is also extended in this paper.

In the second paper Elinor Stephens, David Brown and Gareth Leng describe a model of pituitary release of growth hormone. Growth Hormone (GH) is released from the anterior pituitary in a pulsatile manner, and this patterning of release is essential to the effectiveness of GH in the body tissues. Release is controlled by the coordinated actions of stimulatory growth hormone-releasing hormone (GHRH) and inhibitory somatostatin (SST), which are released by the hypothalamus. In addition to its powerful inhibitory effect, SST acts in a paradoxically positive manner to resensitize the pituitary GH-releasing cells (somatotrophs) to stimulation. In the prolonged absence of SST, a desensitization of the response to GHRH occurs, resulting in smaller GH pulses. After a period of high SST and low GHRH exposure, as is observed in vivo between GH pulses, the somatotroph will respond optimally to a pulse of GHRH. GHRH and SST work together, through their often opposite actions, to regulate the GH output of the anterior pituitary. As a first step to understanding the GH system as a whole, we outline a nonlinear differential equation model of the pituitary response to GHRH and SST which incorporates key stages of the process. We present simulations showing behaviour consistent with in vitro experimental finding.

The final two papers in this section describe the application of two computational techniques namely probabilistic process algebra and cellular automata to two very different biological systems. M. J. Hatcher, A. M. Dunn and Chris Tofts present models of the effect of cell lineage on successful vertical transmission of microparasites. The effects of the relative breeding rates of the parasite and its host are examined. These models have been applied to a microsporidian parasite in the crustacean *Gammarus duebeni* and have demonstrated that the parasite is not segregating unbiasedly. Therefore the parasite may be capable of identifying potential germ tissue in the early embryo.

Finally, Gershom Zajicek introduces the notion of a "Tissue Automat" to models of cancer. Medical systems consist of two kinds of components, the first can be expressed by extensive magnitudes, and the latter, by intensive. Components interact by feedback. Traditional models simulate mainly systems with extensive magnitudes, yet many important properties are intensive, e.g. health. A-life (artificial life) models simulate systems with intensive and extensive components, but usually lack two ingredients that are important to medicine, structured turnover and feedback. The Tissue Automat (TA) is an A-life construct with feedback that implements structured turnover, and simulates extensive and intensive magnitudes, e.g., health and aging. These ideas are illustrated here on the crypt-villus unit of the gut lining. The model simulates extensive properties, e.g., labelling with tritiated thymidine (autoradiography), x-irradiation, regeneration, and first stages of neoplastic progression. Intensive properties constitute, health, and aging. It can be applied also for simulating physiological systems, e.g., anaesthesia monitoring.

Information Processing and Signalling Processes

INTERPLAY OF CELL-CELL SIGNALLING AND MULTICELLULAR MORPHOGENESIS DURING DICTYOSTELIUM AGGREGATION

THOMAS HÖFER, PHILIP K. MAINI
Centre for Mathematical Biology,
Mathematical Institute, University of Oxford
24–29 St Giles', Oxford OX1 3LB, UK

1 Introduction

The cellular slime mould *Dictyostelium discoideum* provides a paradigm model system for the study of multicellular pattern formation. Its life cycle involves a route to primitive multicellular organization which has independently evolved in terrestrial species of at least four groups of microorganisms (myxobacteria, acrasiomycota—the cellular slime moulds, myxomycota, and ciliata [1]). In these species a large number of single cells (nuclei in myxomycota) form, through a process of aggregation and differentiation, a fruiting body.

Of these *Dictyostelium* in particular has been the subject of much experimental and theoretical research; a detailed description of its life cycle can be found for example in [2]. Briefly, under favourable conditions *Dictyostelium* exists in the form of single amoeboid cells which feed on bacteria in the soil and multiply. Eventually this leads to the exhaustion of the bacterial food sources, which induces cells to actively aggregate into streams. The cell streams coalesce into mounds containing typically 10^4–10^5 cells. Passing through a migratory slug-like stage during which cells differentiate in prespore and prestalk cells (and a number of other cell (sub)types), a fruiting body develops. It aids the dispersal of the spores which then develop again into single amoebae.

Here we focus specifically on the aggregation stage; cf. for example Figure 1 in [10]. Cell-cell communication via macroscopic waves of cyclic adenosine 3',5'-monophosphate (cAMP) organizes periodic cell movement towards the aggregation centres [3,4]. These chemical waves take the form of rotating spirals ([10], Figure 1 (a)). The cell layer becomes divided into aggregation territories ([10], Figure 1 (b)), and eventually amoebae form a conspicuous pattern of branching cell streams which radiate from the aggregation centre outwards ([10],Figure 1 (c)). Cell streaming is accompanied by the establishment of direct cell-cell contacts and thus marks the onset of multicellular organization in *Dictyostelium.*

The purpose of the present paper is to integrate the available experimental information at the cellular level into a quantitative model of the aggregation phase in this ensemble of communicating cells. By means of this model we attempt to show how the individual cell properties conspire to produce the sequence of patterning events observed during aggregation. Particular emphasis will be given to the interaction of intercellular signalling and motile cell behaviour which will turn out to be at the heart of the cell streaming phenomenon.

2 Individual cell behaviour

Experimental research on *Dictyostelium* has elucidated many of the cellular and molecular mechanisms involved in morphogenesis and differentiation. Subsequent to starvation cells acquire aggregation competence through the expression of a number of gene products. These equip cells with the ability (1) to synthesize cAMP in an "autocatalytic" feedback loop which plays a central role in establishing long-range cell communication, and (2) to react mechanically to cAMP with directed movement towards increasing cAMP concentration (positive chemotaxis). In particular, elements of the cAMP signalling system are expressed which are responsible for sensing (cAMP receptors), synthesizing (adenylate cyclase, AC) and degrading (cAMP-phosphodiesterases, PDE) cAMP [5]. The major cAMP receptor type, cAR1, also mediates the chemotactic response. In addition, other components of the AC and chemotactic pathways, such as specific G-proteins, accumulate after starvation [6].

A central feature of both pathways is their activation as well as desensitization by cAMP. This is particularly well characterized in the case of the AC pathway. Binding of extracellular cAMP to the cAMP receptors leads to a relatively fast G-protein mediated activation of AC, and the synthesized cAMP is rapidly transported into the extracellular medium (response time of the order of seconds). This provides a self-enhancement mechanism for the cAMP signal. The "autocatalytic" response is terminated by a desensitization (adaptation) of the pathway to further stimulation, brought about by cAMP on a somewhat slower timescale (order of minutes). Multiple desensitization mechanisms have been characterized [7]; most relevant for the timecourse of the cAMP signals during aggregation appears to be desensitization at the receptor/G-protein level.

Activation of the chemotactic pathway follows a very similar pattern [8]. Here inositol 1,4,5-trisphosphate, calcium and cyclic GMP function as intracellular messengers linking the cAMP receptors to the motile machinery of the cell. Again several desensitization mechanisms are known, and it appears that a somewhat faster desensitization reaction than that of the AC pathway is required to achieve accurate cell orientation [8,9].

3 A model of pattern formation in the cellular ensemble

On the basis of this information on individual cell behaviour we can derive a minimal model for pattern formation in the cellular ensemble. The model takes into account three dynamic variables; cell density, $n(x, y, t)$, concentration of extracellular cAMP, $u(x, y, t)$, and fraction of active cAMP receptors per cell (that is, cellular sensitivity towards the cAMP signal), $v(x, y, t)$, all dependent on space, (x, y), and time, t. The evolution equations for these variables take the form [10]

$$\frac{\partial n}{\partial t} = \underbrace{\nabla \cdot (\mu(n) \nabla n)}_{\text{random migration}} \underbrace{- \nabla \cdot (\chi(v) n \nabla u)}_{\text{chemotaxis}}, \tag{1}$$

$$\frac{\partial u}{\partial t} = \lambda[\underbrace{\phi(n) g_+(u, v)}_{\text{synthesis}} - \underbrace{(\phi(n) + \delta) g_-(u)}_{\text{degradation}}] + \underbrace{D \nabla^2 u}_{\text{diffusion}}, \tag{2}$$

$$\frac{\partial v}{\partial t} = \underbrace{-f_+(u) v}_{\text{desensitization}} + \underbrace{f_-(u)(1 - v)}_{\text{resensitization}}, \tag{3}$$

where $\nabla = (\partial/\partial x, \partial/\partial y)$. We briefly discuss each of these equations in turn.

The dynamics of the cell distribution will be a central element of the model. To monitor it, we introduce the cell density variable n which can be crudely viewed as a continuous space-dependent average of the (discrete) cell distribution obtained over a characteristic length of a few cell diameters. It is governed by the advection-diffusion equation (1) which accounts for random cell migration ("diffusion"), directed movement in cAMP gradients, ∇u (chemotaxis), and, in conjunction with equation (3), for a time-dependent adaptation response towards the cAMP signal. This last feature is realized by the dependence of the chemotactic coefficient χ on the fraction of active cAMP receptors, v. The chemotactic coefficient represents a combined measure of sensitivity to the signal and motile response and should therefore record the effect of adaptation. We choose it to be of the form $\chi(v) = \chi_0 v^m/(A^m + v^m)$, $m > 1$, that is, there is a threshold value of active receptors, A, which is required for efficient sensing of the signal. In [11] we have shown that such an extension of the standard chemotaxis model [12] is necessary if the characteristic timescales of changes in the chemoattractant (cAMP) concentration and adaptation towards the chemoattractant are of comparable magnitudes. This is the case in *Dictyostelium* aggregation (cf. discussion in [11]), and probably ensures that cells aggregate by rendering cells unresponsive to the cAMP gradient in the wavebacks of the periodic cAMP waves.

Equations (2)–(3) for the signalling dynamics are adapted from a model of cAMP signalling by Martiel & Goldbeter [13–15]. For computational reasons we take somewhat simpler functional forms for the reaction rates $f_\pm$ and $g_\pm$ in equations (2) and (3) than those derived in [13] which, nevertheless, retain the important characteristics of the latter: $f_+ = k_+u$, $f_- = k_-$, $g_+ = (bv + v^2)(a+u^2)/(1+u^2)$, $g_- = du$. The adaptation and cAMP degradation kinetics follow simple linear or bilinear rate laws, while the rate of the AC pathway (g_+) accounts for stimulation by cAMP depended on the cellular sensitivity. Estimates for the majority of model parameters can be extracted from the experimental literature [10]; see legends to Figures 1 and 3. In contrast to the original model, which was developed for suspensions of constant cell density, we have to account explicitly for the feedback of the cell distribution into the local cAMP dynamics. As both cAMP synthesis and hydrolysis are linked to cells (intracellular AC pathway, PDE carried on cell membranes), the cAMP production and degradation rates at a certain point in the medium depend on the cell density at this point. This gives rise to the cell density factor $\phi(n)$ in (2). In general it will be an increasing function of the cell density. Specifically, we take the form $\phi(n) = n/((1 - \rho n/(K + n)))$; see [10] for details. A small background activity of secreted PDE independent of cell density is added (δ).

Equations (2)–(3) with the cell density "clamped" at a constant value essentially reduce to the system investigated by Tyson *et al.* [14] as a model of cAMP signalling at the onset of aggregation. Alternative models for this situation have been developed [16,17]. The models differ in particular with respect to the actual mechanism of cAMP-induced desensitization of the AC pathway. While more elaborate models can perform better when simulating certain experimental procedures, the two-variable model incorporated here certainly captures the essence of the cAMP signalling system in *Dictyostelium*: fast cAMP production in a positive feedback loop and slow cAMP-induced desensitization. In a spatially extended (2D) medium with cAMP diffusion as the only transport process, the system has the typical properties of an excitable medium, and in particular supports stable spiral wave solutions (Figure 1) [14,18]. In addition, a slight shift in parameters (for example an increase in AC activity [13]) can cause amoebae to produce cAMP in a periodic fashion. This is what normally happens at the beginning of aggregation when the establishment of a number of autonomously oscillating centres (pacemakers) can be observed. From these the cAMP signal is relayed as a series of expanding concentric waves. Inhomogeneities in the medium can break these waves, and the resulting free wave tips give rise to cAMP spirals.

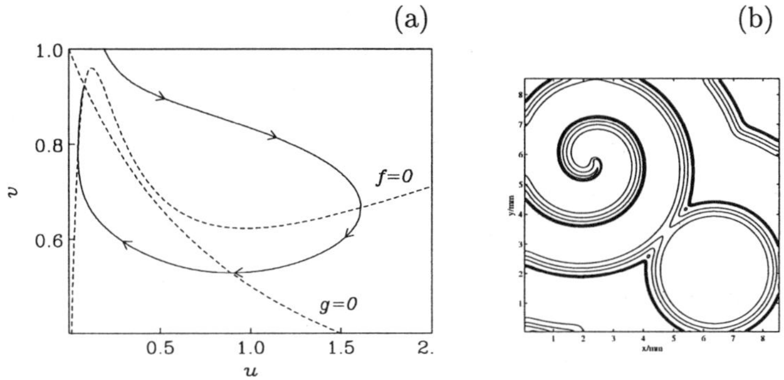

Figure 1: Behaviour of system (2)–(3) with cell density "clamped" at a constant value ($n = 1.0$); (a) phase plane with typical excitation trajectory, and (b) contour plot of spiral wave solution (upper left of domain) and concentric wave (lower right; the pacemaker region in the centre is silent at the time of the snapshot) on a two-dimensional spatial domain. Nondimenional parameters: $\lambda = 70.0$, $a = 0.0014$, $b = 0.2$, $d = 0.0234$, $\delta = 0.11$, $k_+ = k_- = 2.5$, $D = 1.0$; timescale 2.5 min, length scale 220 μm, cAMP scale 5×10^{-7} M.

However, in previous models these wave solutions have been obtained with the idealization of a stationary homogeneous cell layer. We now turn to the investigation of the full model (1)–(3) which includes the dynamics of the cell distribution.

4 Cell streaming—the patterning mechanism

As discussed above, equations (2)–(3), with cell density viewed as a constant and spatially homogeneous parameter, exhibit travelling wave solutions and, in particular, spiral waves. For the full model (1)–(3) it can be shown that travelling wave solutions again exist, which are almost identical to the wave solutions obtained for cell density "clamped" at a constant value. The reason for this is essentially the very small ratio of typical cAMP wavespeed (300 μm/min) to chemotactic (advective) cell velocity (20 μm/min). From equation (1) an estimate for the disturbance of a homogeneous cell distribution, $n_0 =$ constant, in the front of a cAMP wave can be obtained as $n \approx n_0/(1 - w/c)$, where c and w denote the cAMP wavespeed and the average chemotactic cell velocity, respectively. Thus wave-induced cell movement leaves the homogeneous cell layer practically undisturbed—an observation also reported for the *in situ* situation [19].

Therefore the basic ideas on the initiation of pattern formation by cAMP waves carry over to the full model incorporating cell movement. However, we shall show that there is a crucial difference between the density-clamped case and the full model: Whereas the cAMP spiral waves and periodically forced target patterns constitute stable dynamical attractors in the previous models, they will function as a slowly evolving "chemical prepattern" for cellular morphogenesis when cell chemotaxis is included.

This can be best understood by considering the stability of the wave solutions in the full model. For this purpose one looks at the evolution of a small perturbation of the wave solution with time. Such perturbations are naturally supplied by small random inhomogeneities in the cell distribution. The linear stability analysis of the model around two-dimensional periodic wave solutions is carried out in detail in [20]. It essentially reduces to considering a linear problem of the form

$$\frac{d\boldsymbol{a}(z;q^2)}{dz} = \boldsymbol{A}(z;q^2)\ \boldsymbol{a}(z;q^2); \quad z = x + ct. \tag{4}$$

Here periodic plane waves are assumed to propagate in x-direction with speed c, hence z denotes a coordinate frame moving with the waves; $\boldsymbol{a}(\cdot;q^2)$ denotes the amplitude of the q-th Fourier mode of a spatial perturbation of the system variables in the y-direction, that is *perpendicular* to the direction of wave propagation. Hence the total perturbation in the y-direction is

$$\text{perturbation}(y,z) = \int \boldsymbol{a}(z;q^2)\exp\{iqy\}dq \tag{5}$$

Whether a perturbation with a certain spatial wavelength $2\pi/q$ will grow or decay with time depends on the evolution of its amplitude which is determined by (6). The system matrix $\boldsymbol{A}(z;q^2)$, the actual form of which is derived in [20], is periodic in z through a dependence on the unperturbed periodic wave solutions of (1)–(3), that is, the evolution of the perturbation is *forced by the periodic cAMP waves*. This is further emphasized by the fact that the time-like variable in (6) and (7) is the travelling wave variable z: the perturbation essentially evolves in the travelling wave frame. Standard Floquet theory can be employed to obtain the *dispersion relation* from (6), that is the graph of the characteristic linear growth rates of the Fourier modes [20]. The result is depicted in Figure 2. One can clearly see that there is a band of unstable spatial wavenumbers. Thus cAMP wave propagation in the initially homogeneous cell layer is unstable from the outset, and the waves force the slow growth of a pattern parallel to the wavefronts with a characteristic length scale roughly an order of magnitude larger than the typical cell diameter. As we shall see in the next section, it is this patterning instability which gives rise to cell streams observed *in situ*. A more detailed investigation of thc instability mechanism

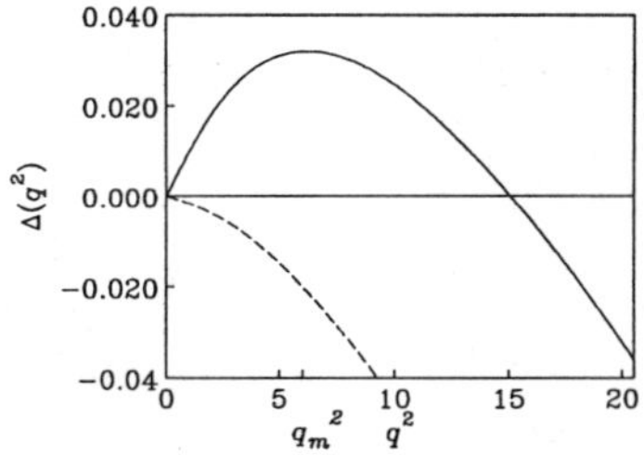

Figure 2: Dispersion relation for patterning parallel to cAMP wavefronts resulting from (6). The ordinata measures the linear growth rate of the spatial Fourier modes given on the abscissa. The solid line is obtained with the parameters given in Figures 3 and 5; there is a band of unstable wavenumbers (small positive growth rates). The dashed line corresponds to a weaker chemotactic response $\chi_0 = 0.2$; here the uniform cell layer is stable.

uncovers the typical features of a *chemotaxis-driven instability* [12], in the present case forced by periodic chemoattractant waves [20]. We conjecture from the linear analysis that cAMP wave propagation gives rise to a slow break-up of the initially homogeneous cell layer perpendicular to the direction of wave propagation, that will eventually lead to the formation of a cell stream pattern.

5 Numerical explorations

To investigate the instability mechanism in the fully nonlinear model, we solve (1)–(3) numerically. A standard finite difference discretization on a rectangular domain is employed, with an alternating direction implicit scheme for the diffusion operator and an upwind scheme for the chemotaxis term.

A typical simulation result is shown in Figure 3. One clearly sees the development of a counter-rotating spiral pair from a disrupted wavefront, the partitioning of the domain in two distinct aggregation territories roughly along the diagonal, and finally the break-up of the cell layer into a pattern of branching cell streams, that is, the onset of multicellularity. During cell stream development the cAMP waves retain their principal spiral geometry, though with an increasingly rugged and irregular concentration profile (Figure 3 (b)).

Thus the model (1)–(3) captures the essential dynamics of aggregation, and we can now proceed to compare different model predictions with the results of the linear analysis and experimental observations.

(i) Cell streams arise on a characteristic spatial scale

(a)

Figure 3: Spatio-temporal evolution of (a) cell density, and (b) cAMP concentration in a numerical simulation of system (1)–(3). The dimensional domain size is as in Figure 3 (b); snapshots are taken in intervals of 15 min (from top left downwards, to bottom right). Initially a small random perturbation between −0.075 and 0.075 was added to the homogeneous cell density (1.0) at every mesh point. Boundary conditions are zero-flux. Parameters as in Figure 3, and $\mu = 0.012$, $\chi_0 = 0.5$, $A = 0.72$, $m = 10.0$, $\rho = 0.7$ and $K = 0.8$; reference cell density $1.5 \times 10^{-4}/\text{cm}^2$

(b)

Figure 3: continued.

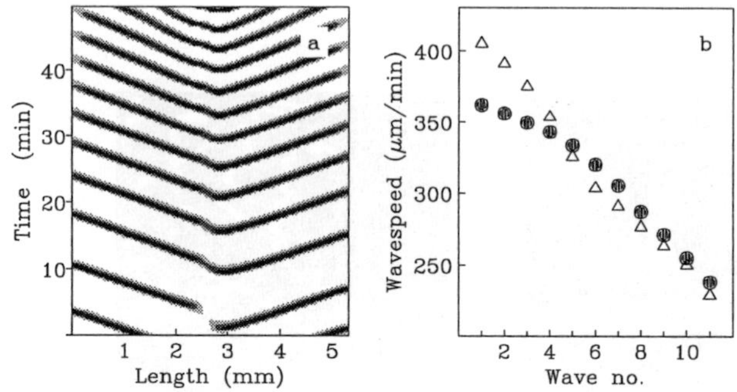

Figure 4: (a) Evolution of cAMP concentration in a spatial slice through an aggregation territory with a cAMP spiral; (b) corresponding wave velocities (•), compared to the experimental data in [21], Figure 3 (△). The spatial slice is taken parallel to the horizontal boundary through the spiral wave core.

The power spectrum of the evolving small-amplitude pattern in cell distribution clearly shows a distribution of modes around a dominant spatial frequency ([20], Figure 7). This agrees with the analytical prediction of a dispersion relation with a single maximum. Moreover, the numerical values for the dominant modes are in good agreement [20]. Obviously, such a characteristic spatial scale together with the spiral or concentric geometry of the cAMP waves provides a recipe for the formation of a branching stream pattern. Quantitative *in situ* data do not exist, but within an aggregation territory streams usually appear to be distributed around a characteristic width. Both a certain "freezing" of initial inhomogeneities and nonlinear competition are familiar phenomena in chemotaxis systems which can contribute to such a width distribution, which is also seen in the simulations.

(ii) Spiral period and wavespeed depend on "wave time"

It has long been recognized that the basic geometry of the cAMP spiral waves changes as aggregation proceeds: the wave frequency increases while the propagation speed drops [21]. This behaviour, which has not been observed in any other excitable medium, has been attributed to somewhat speculative biochemical changes which are thought to be stimulated directly by the cAMP waves. However, our minimal model exhibits this phenomenon with constant parameter values (Figure 4). This certainly does not rule out slow biochemical modifications which may enhance it [5]. On the other hand, the model (1)–(3) offers an explanation of the basic phenomenon within the framework of

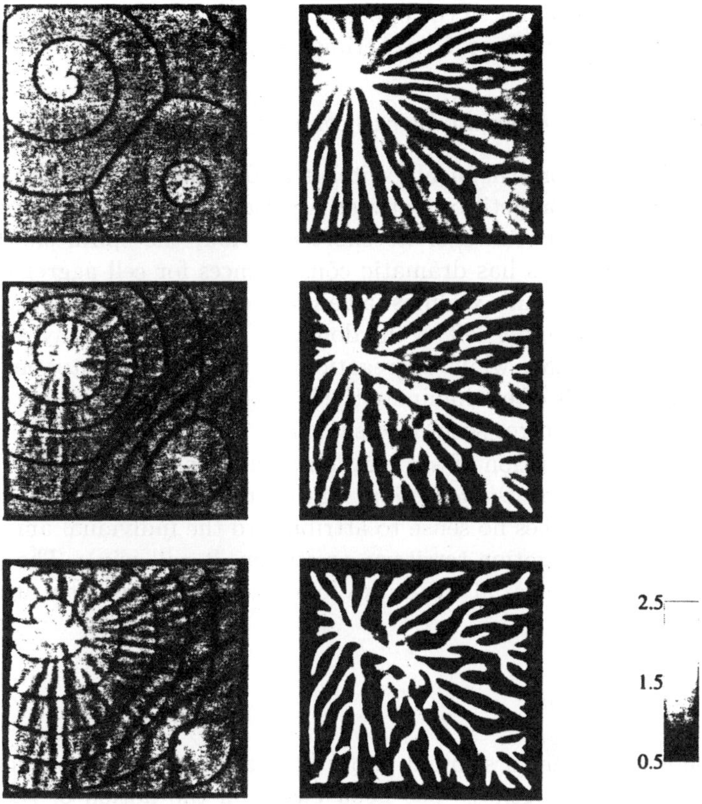

Figure 5: Competition of two aggregation territories, organized by a spiral wave of cAMP and a periodic pacemaker. Shown is the cell density, and in the first three snapshots also the cAMP wave contours. Details of simulation as in Figure 3.

excitable media in terms of slowly evolving inhomogeneous excitability properties of the cell layer, caused by aggregation and streaming [10,20].

(iii) Caffeine induces rotating cell loops

Caffeine is known as a pharmacological agent to affect both aggregation and subsequent development. It inhibits the AC pathway [22], and applied in large quantities stops cAMP wave propagation (P. Newell, personal communication). A simulation of moderate caffeine application, reflected in a decrease of the activity of the AC pathway by about 25 %, shows the formation of a cell loop in the centre of the aggregation territory; cf. [10], Figure 5. The core of the cAMP spiral continuously cycles around this cell loop and consequently induces cell rotation in the opposite direction. Caffeine-induced loop formation has indeed been observed experimentally [4].

(iv) Aggregation territories can compete

It is known from chemical excitable media that wave structures can compete [23]. Such a competition between a periodic pacemaker and a spiral is shown in Figure 5. Because of its shorter period, the spiral slowly overrules the pacemaker. This has dramatic consequences for cell aggregation. Unlike the simulation of Figure 3, almost all amoebae are now attracted to the aggregation centre on the upper left of the domain. This scenario may explain the *in situ* observation that, while aggregation is initiated by autonomously oscillating centres, spirals often dominate the later aggregation process to the extent that the periodic centres seem to disappear.

The competition of aggregation territories seen in Figure again underlines the character of *Dictyostelium* aggregation as a *collective phenomenon* (for example, it makes no sense to attribute to the individual amoebae a choice as to which aggregation territory, or stream, it will join). The potential role of such collective spatial structures for evolution has recently attracted attention [24].

6 Conclusion

Following Turing's seminal paper [25], a variety of theoretical models of biological morphogenesis has been based on the notion of the formation of a global collective pattern from local interactions in a developmental system [26]. However, direct quantitative evidence for developmental mechanisms of this kind remains scarce. Here we have shown that the aggregation patterns of the microorganism *Dictyostelium discoideum* can be rationalized on a physico-chemical basis within this framework. The cell stream morphology is the result of a sequence of patterning instabilities, triggering first cAMP wave propagation, which then induces movement of amoebae and streaming with a distinct polarity parallel to the wavefronts. In this sense cAMP functions as a "dynamic morphogen". Interestingly, there are a number of other developmental systems in which travelling waves appear to act as "precursors" of patterning events, for example the formation of skin organ primordia [27] or somite formation [28].

Recent experimental work points to the central role of periodic cAMP signals and chemotaxis also in the subsequent "tissue-like" stages of *Dictyostelium* morphogenesis [29]. Our results clearly indicate that the dynamic interaction of biochemical signalling and cell mechanics is crucial for the understanding of pattern formation along this primitive route to multicellularity.

Acknowledgements

We thank P. Newell and J. Sherratt for stimulating discussions. T.H. acknowledges support from the Boehringer Ingelheim Fonds, the Physical Sciences and Engineering Research Council of Great Britain, and a Jowett Senior Scholarship at Balliol College Oxford.

1. L. Margulis and Schwartz, *Five Kingdoms*, (Freeman, New York, 1988).
2. J.T. Bonner, *The Cellular Slime Moulds*, (Princeton University Press, 1967).
3. K.J. Tomchik and P.N. Devreotes, *Science* **212** (1981) 443.
4. F. Siegert and C.J. Weijer, in: *Oscillations and Morphogenesis*, ed. L. Rensing, (Marcel Dekker, New York, 1993) p. 133.
5. P.N. Devreotes, in *Advances in Nucleotide Research*, eds. P. Greengard and G.A. Robinson, (Raven Press, New York, 1983) p. 55.
6. M. Pupillo, A. Kumagai, G.S. Pitt, P.N. Devreotes, *Proc. Natl. Acad. Sci. USA* **86** (1989) 4892.
7. P.J. Van Haastert, M. Wang, A.A. Bominaar, P.N. Devreotes, P.Schaap, Mol. Biol. *Cell* **3** (1992) 603.
8. P.C. Newell, G.N. Europe-Finner, G. Liu, B. Gammon and C.A. Wood, in *The biology of the chemotactic response*, eds. J.P. Armitage and J. Lackie (CUP, Cambridge, 1990), p. 273.
9. P.J. Van Haastert and P.R. Van der Heijden, *J. Cell Biol.* **96** (1983) 347.
10. T. Höfer, J.A. Sherratt, P.K. Maini, *Proc. Roy. Soc. Lond. B* **259** (1995) 249.
11. T. Höfer, P.K. Maini, J.A. Sherratt, M.A.J. Chaplain, P. Chauvet, D. Metevier, P.C. Montes and J.D. Murray, *Appl. Math. Lett.* **7** (1994) 1.
12. E.F. Keller and L.E. Segel, *J. Theoret. Biol.* **26** (1970) 399.
13. J.-L. Martiel and A. Goldbeter, *Biophys. J.* **52** (1987) 807.
14. J.J. Tyson, K.A. Alexander, V.S. Manoranjan and J.D. Murray, *Physica D* **34** (1989) 193.
15. J.J. Tyson and J.D. Murray, *Development* **106** (1989) 421.
16. P.B. Monk and H.G. Othmer, *Proc. R. Soc. Lond. B* **240** (1990) 555.
17. Y. Tang and H.G. Othmer, *Math. Biosci.* **120** (1994) 25.
18. J.J. Tyson and J.P. Keener, *Physica D* **32** (1988) 327.
19. F. Alcantara and M. Monk, *J. Gen. Microbiol.* **85** (1974) 321.
20. T. Höfer, J.A. Sherratt, P.K. Maini, *Physica D* **84** (1995), in press.
21. J.D. Gross, M.J. Peacey and D.J. Trevan, *J. Cell Sci.* **22** (1976) 645.

22. M. Brenner and S. Thoms, *Devl. Biol.* **101** (1986) 136.
23. A.S. Mikhailov, Foundations of Synergetics (2nd ed.), (Springer-Verlag, Berlin, 1994).
24. M.C. Boerlijst and P. Hogeweg, *Physica D* **48** (1991) 17.
25. A.M. Turing, *Phil. Trans. Roy. Soc. Lond. B* **237** (1952) 37.
26. J.D. Murray, Mathematical Biology (Springer, Berlin, 1989).
27. G.C. Cruywagen, P.K. Maini, J.D. Murray, *IMA J. Math. Appl. Med. Biol.* **9** (1992) 227.
28. R.J. Keynes and C.D. Stern, *Development* **103** (1988) 413.
29. F. Siegert, C.J. Weijer, A. Nomura, H. Miike, *J. Cell Sci.* **107** (1994) 97.

CONCERTED REGULATION OF CYCLIC ADENOSINE MONOPHOSPHATE BY CALMODULIN/CALCIUM COMPLEX AND DOPAMINE: A KINETIC MODELLING APPROACH

ROLF KÖTTER, DIRK SCHIROK, and KARL ZILLES
C. & O. Vogt Institute for Brain Research, Heinrich Heine University, PO Box 101007, D-40001 Düsseldorf, Germany

Introduction

The complex role of biochemical pathways in the intracellular mediation of synaptically induced effects in neurons is increasingly being recognized. The analysis of such pathways has much to offer, particularly in respect to the non-classical transmitters that exert "modulatory" metabotropic effects in the central nervous system. Such a "modulatory" transmitter is dopamine, which is known to influence intracellular cyclic adenosine monophosphate (cAMP) concentrations in the striatum via G proteins and adenylate cyclase (AC)[1]. However, cAMP levels are also affected by calcium[2,3], a second messenger whose intracellular concentration increases, for example, in response to synaptically released glutamate[4,5]. A previous equilibrium model of dopamine and calcium interactions in the mammalian striatum provided important insights into distinct concentration-dependent modes of postsynaptic signal integration[6]. Here we present a kinetic model of cAMP regulation by calmodulin/calcium complex and dopamine via AC and phosphodiesterase (PDE) in order to further investigate the responses of intraneuronal cAMP concentrations to brief calcium and dopamine pulses.

Methods

An established scheme of intracellular second messenger pathways was adopted in the present study[6-8] (Fig. 1): As a first approximation, AC activity in the striatum is controlled by three factors: It is stimulated by dopamine (DA) and calmodulin/calcium complex (CaMCa4), and inhibited by intracellular free calcium (Ca). Adenylate cyclase forms cAMP from its substrate adenosine triphosphate (ATP). Cyclic AMP is hydrolyzed to AMP by phosphodiesterase (PDE), whose activity depends on CaMCa4.

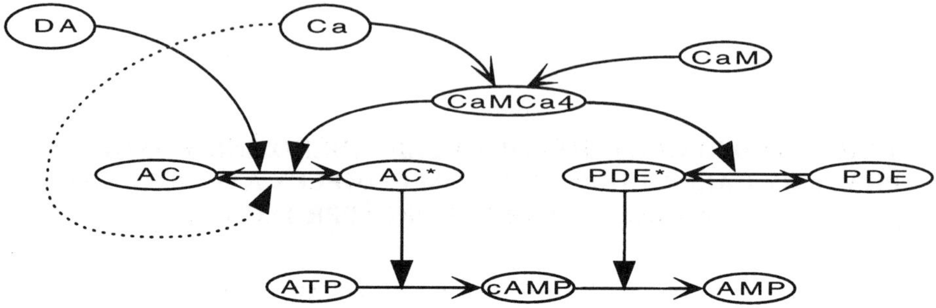

Figure 1: Simplified scheme of cAMP regulation by calcium and dopamine as investigated in this study.

Brief dopamine pulses in the striatum were modeled using data from Kawagoe et al.[9]: The dopamine concentration at the synaptic interface produced by a single pulse and a brief burst (5 pulses at 20 Hz) was described by:

$$d[DA] / dt = (f)\,[DA]_p - (Vmax\,[DA]/Km / (1 + [DA]/Km)), \qquad (1)$$

where f is the stimulation frequency in Hz and $[DA]_p$ is the dopamine concentration increase of 260 nM with stimulus. Vmax and Km are given as 4 μM/s and 0.16 μM, respectively. The dopamine concentrations resulting from the single pulse and the burst stimulation, respectively, are shown in Fig. 2A.

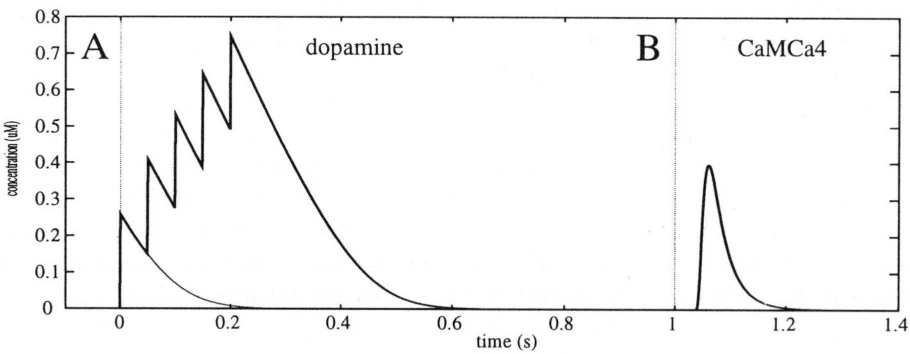

Figure 2: Time course of synaptic dopamine and intracellular CaMCa4 concentrations simulated in response to stimuli starting at times indicated by the dotted lines. A: Synaptic dopamine concentrations following single pulse (thin line) and burst (thick line) stimulation. B: CaMCa4 concentration time course calculated according to Eqns. 3-7 from binding of large pulse of intracellular free calcium to CaM. Another smaller calcium pulse produced increases in CaMCa4 concentrations that are too small to be visible (peak <3 10^{-7} μM).

The time course of intracellular free calcium concentrations in dendritic spines has been modeled in detail by Holmes and Levy[10]. Holmes[11] provides time courses of calcium concentrations in different compartments of a dendritic spine that receives tetanic afferent stimulation via a synapse located at the spine head. The resting calcium concentration is 0.02 μM and stimulation induced an increase peaking at 28 μM in the compartment immediately below the synapse. Taking into account diffusion and elimination processes, the concentration in the compartment at the spine stalk peaked 45 ms after stimulus onset with a concentration of 18 μM. The temporal behaviour of this calcium stimulus was reproduced by curve fitting and extraction of suitable parameters for numerical integration. Another smaller calcium pulse, which approximated the time course of a single stimulus, was described by a mono-exponential alpha function of the form:

$$[Ca]_t = [Ca]_{min} + ([Ca]_{max} - [Ca]_{min})\ t\ e/\tau\ \exp(-t/\tau), \quad (2)$$

with $[Ca]_{min} = 0.02$ μM, $[Ca]_{max} = 0.5$ μM, and $\tau = 5$ ms.

CaMCa4 is formed by four successive steps of binding of calcium ions to calmodulin (CaM) as described by a set of differential equations:

$$d[CaM] / dt = k\_1\ [CaMCa] - k1a\ [CaM]\ [Ca], \quad (3)$$

$$d[CaMCa] / dt = k1a\ [Ca]\ [CaM] + k\_1\ [CaMCa2] - k\_1\ [CaMCa] - k1b\ [Ca]\ [CaMCa], \quad (4)$$

$$d[CaMCa2] / dt = k1b\ [Ca]\ [CaMCa] + k\_2\ [CaMCa3] - k\_1\ [CaMCa2] - k2\ [Ca]\ [CaMCa2], \quad (5)$$

$$d[CaMCa3] / dt = k2\ [Ca]\ [CaMCa2] + k\_2\ [CaMCa4] - k\_2\ [CaMCa3] - k2\ [Ca]\ [CaMCa3], \quad (6)$$

$$d[CaMCa4] / dt = k2\ [Ca]\ [CaMCa3] - k\_2\ [CaMCa4]. \quad (7)$$

Further explanations and values of the parameters are provided in Table 1.

Differential equations for AC and PDE are derived according to the guidelines set out by Segel[12] assuming rapid equilibrium conditions between enzyme and respective activators. All activators and inhibitors are regarded as producing independent, reversible, and non-essential activation or inhibition of the enzyme. The velocity of AC describes the formation of cAMP:

$$d[cAMP] / dt = (pf\ hf\ VmaxAC\ [ATP] / KmAC\ (1 + b\ [CaMCa4] / EC50CaMCa4 + a\ [Ca] / IC50Ca + ab\ [CaMCa4]\ [Ca] / (EC50CaMCa4\ IC50Ca) + DA / EC50DA\ (c + bc\ [CaMCa4] / (0.464\ aEC50CaMCa4) + ac\ [Ca] / IC50Ca\ ))) / denominator. \quad (8)$$

The numerator is the sum of all product forming combinations of substrate, CaMCa4, Ca, and DA, multiplied by the respective maximal velocities. The denominator is simply the sum of all combinations of enzyme, substrate, CaMCa4,

Ca, and DA, each divided by the respective dissociation constants between the complex and free enzyme as described by Segel[12]. The factor 0.464 results from factoring of DA / EC50DA. The concentration of the substrate ATP was saturating at 3 mM. For further details see Table 1.

Table 1: Parameters and values used in the model equations.

Parameter	Description	Value	Reference
[CaM]	Neuronal concentration of calmodulin	10 μM	Klee & Vanaman, 1982
k1a	On-rate for binding of 1st calcium ion to calmodulin	2.67 / s / μM	Klee, 1988
k1b	On-rate for binding of 2nd calcium ion to calmodulin	7.41 / s / μM	Klee, 1988
k_1	Corresponding off-rate	20 / s	Klee, 1988
k2	On-rate for calcium binding to calmodulin (low-affinity sites)	20 / s / μM	Klee, 1988
k_2	Corresponding off-rate	600 / s	Klee, 1988
VmaxAC	Maximal activity of unstimulated adenylate cyclase	1.3 x 10^-6 μmol / s / mg protein	Harrison et al., 1988
KmAC	App. dissociation constant of adenylate cyclase from the substrate ATP	320 μM	Clement-Cormier et al., 1975
EC50CaM	Calmodulin concentration which yields half-maximal calmodulin-dependent stimulation of adenylate cyclase	0.11 μM	Harrison et al., 1988
EC50CaMCa4	App. CaMCa4 concentration which yields half-maximal calmodulin-dependent stimulation of adenylate cyclase	1.32 x 10^-9 μM	*
EC50DA	Dopamine concentration which yields half-maximal dopamine-dependent stimulation of adenylate cyclase	2.8 μM	Harrison et al., 1988
aEC50DA	EC50DA lowered in the presence of calmodulin	1.3 μM	Harrison et al., 1988
aEC50CaM	Calmodulin concentration which yields half-maximal effect on the response of adenylate cyclase to dopamine	0.035 μM	Harrison et al., 1988
aEC50CaMCa4	App. CaMCa4 concentration which yields half-maximal effect on the response of adenylate cyclase to dopamine	9 x 10^-10 μM	*
IC50Ca	Calcium concentration which yields half-maximal inhibition of adenylate cyclase	0.9 μM	Piascik et al., 1981
hf	Correction factor for low substrate saturation in experiments by Harrison et al. (1988)	1.85	
pf	Conversion factor from μmol / s / mg protein to μM / s assuming the presence of 20 g protein per liter tissue	20000 mg / l	
VmaxPDE	Maximal activity of unstimulated phosphodiesterase	0.028 μmol / s / mg protein	Sharma &Wang , 1986
KmPDE	Dissociation constant of PDE from the substrate cAMP	68 μM	Sharma &Wang , 1986
bVmaxPDE	Phosphodiesterase activity with maximal stimulation by calmodulin	0.168 μmol / s / mg protein	Sharma &Wang , 1986
aKmPDE	KmPDE decreased by calmodulin	12 μM	Sharma &Wang , 1986
bEC50CaM	Concentration of calmodulin which yields half-maximal stimulation of PDE (63 kDa, phosphorylated state)	0.003 μM	Sharma &Wang , 1986
bEC50CaMCa4	App. concentration of CaMCa4 which yields half-maximal stimulation of PDE (63 kDa, phosphorylated state)	0.011 μM	**
<u>Scaling factors for VmaxAC in the presence of the respective effectors</u>			
	<u>Effector:</u>		
a	Calcium	0.046	***
b	CaMCa4	3.55	****
c	Dopamine	2.23	****
ab	Calcium + CaMCa4	0.0034	*****
ac	Calcium + dopamine	0.13	*****
bc	CaMCa4 + dopamine	5.32	****

* Calculated from the equilibrium of calmodulin complexes at a calcium concentration of 0.12 μM according to Harrison et al. (1988).
** Calculated from the equilibrium of calmodulin complexes at a calcium concentration of 100 μM according to Sharma & Wang (1986).
*** Factor taken from Fig.1 in Piascik et al. (1981).
**** Factors calculated from netto influence of each effector on VmacAC as detailed by Harrison et al. (1988).
***** Factors describing calcium influence estimated from Fig.1 in Natsukari et al. (1990).

Cyclic AMP hydrolysis results from PDE activity expressed by the differential equation:

$$-d[cAMP] / dt = (VmaxPDE\,[cAMP] / KmPDE + bVmaxPDE\,[cAMP]\,[CaMCa4] / (aKmPDE\ bEC50CaMCa4)) / denominator. \quad (9)$$

Again the denominator is the sum of the products of the concentrations of all ligands in the complex divided by the product of all dissociation constants between the complex and the free enzyme.

Parameter values were derived from the experimental literature (see Table 1). We used consistent units during all simulations (μmol, second, milligram, and liter), into which the various literature values were converted. Simulations were run on a UNIX workstation using the numerical integration program XPP[13]. The minimum integration time step was 0.1 or 1 ms taking into account the shortest time constants in the biochemical system.

Results

Formation of calmodulin/calcium complexes

All simulations start from an initial equilibrium between Ca, CaM, and the four calmodulin/calcium complexes (CaMCa, CaMCa2, CaMCa3, and CaMCa4), assuming saturating intraneuronal CaM concentrations of 10 μM[14]. Following stimulus onset, additional free calcium binds rapidly to CaM and leads to a successive increase in the concentrations of the calmodulin/calcium complexes (not shown). The subsequent dissociation to baseline occurs in the opposite sequence. As a general rule, the peak concentrations are smaller and occur earlier with increasing numbers of incorporated calcium ions. The peak concentration of CaMCa4 is reached less than 20 ms after the peak of the calcium concentration, and about 60 ms after the onset of the tetanic stimulation that led to calcium influx (Fig. 2B). One second after the onset of the stimulus, the concentrations of calcium and CaMCa4 have almost returned to baseline values. In contrast, the concentration of CaMCa remains well above baseline for several seconds.

Adenylate cyclase activity as a function of calcium, calmodulin, and dopamine concentrations

Eqn. 8 replicates in detail the dependence of AC activity on steady-state Ca, CaM, and DA concentrations as experimentally determined by Harrison et al.[7]. The centre plot in Fig. 3 shows the behaviour of the model system under conditions close to the experimental ones at 0.12 μM calcium.

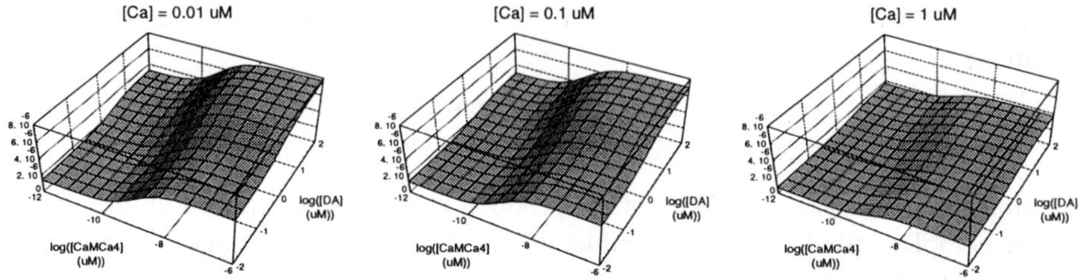

Figure 3: Simulated adenylate cyclase activity in response to varying steady state effector concentrations (Ca, CaMCa4, DA). X and Y axes show the logarithms of CaMCa4 and dopamine concentrations, respectively. Z axis gives adenylate cyclase activity in μmol/s/mg protein. Effect of free intracellular calcium is shown in three successive plots with increasing calcium concentrations from 0.01 μM (left) to 1 μM (right).

As in Harrison et al.[7] the basal velocity of 1.3 10^{-6} μmol/s/mg protein is obtained at low concentrations of CaMCa4 and dopamine. The experimental EC50 of CaM for the stimulation of AC (0.11 μM, see Table 1) is replicated in the model using a EC50 of CaMCa4 of 1.32 10^{-9} μM. At high concentrations of CaMCa4 (10^{-6} μM) the simulated velocity matches the experimentally determined velocity of 4.5 10^{-6} μmol/s/mg protein in the presence of 3 μM calmodulin. In accordance with Harrison et al., dopaminergic stimulation increases the basal velocity to 2.8 10^{-6} μmol/s/mg protein while CaM and DA together have a synergistic effect with a maximal velocity of 6.8 10^{-6} μmol/s/mg protein. The three plots from left to right in Fig. 3 illustrate increasing inhibiton of the reaction velocity by successively increasing calcium concentrations. This effect is in agreement with the IC50 for inhibition of AC by calcium of 0.9 μM determined by Piascik et al.[3]. In the absence of CaM, high calcium concentrations (>1 μM) are assumed to inhibit the basal AC activity to very low values. Calmodulin- and dopamine-mediated stimulation of AC activity is reduced by high calcium levels to basal values as found by Gnegy & Treisman[2].

Responses to brief calcium or dopamine stimulation

The single pulse and the burst release of dopamine (onset at t = 0 s) produce corresponding temporary increases in AC activity from equilibrium (Fig. 4A). The large calcium pulse representing tetanic stimulation with onset at t = 1 s evokes a triphasic pattern of AC regulation: A very brief (<12 ms) initial stimulation is turned into a temporary inhibition (up to 140 ms) followed by a longer lasting activation, which takes several seconds to decay completely. The other calcium pulse characterized by its small amplitude and duration at t = 2 s (time of onset chosen to avoid superposition of curves in Fig. 4A-C) leads to a brief (< 1 s) stimulation of AC activity with a brief inflection half-way up the rising phase, which is just visible in Fig. 4A. This triphasic pattern (stimulation, inhibition, stimulation) of AC regulation by simulated pulses of calcium is caused by the high affinity of AC for CaMCa4 compared to Ca in combination with a high peak calcium concentration. The mechanism is a stimulation of AC by calcium-dependent formation of CaMCa4, that is temporarily suppressed by calcium-mediated inhibition. From a temporal point of view, rising calcium concentrations initially stimulate AC before inhibiting it. The inhibition may prevail (depending on the peak calcium concentration reached) and is followed by a relatively long-lasting CaMCa4-mediated stimulation during the slow return of calcium concentrations to baseline.

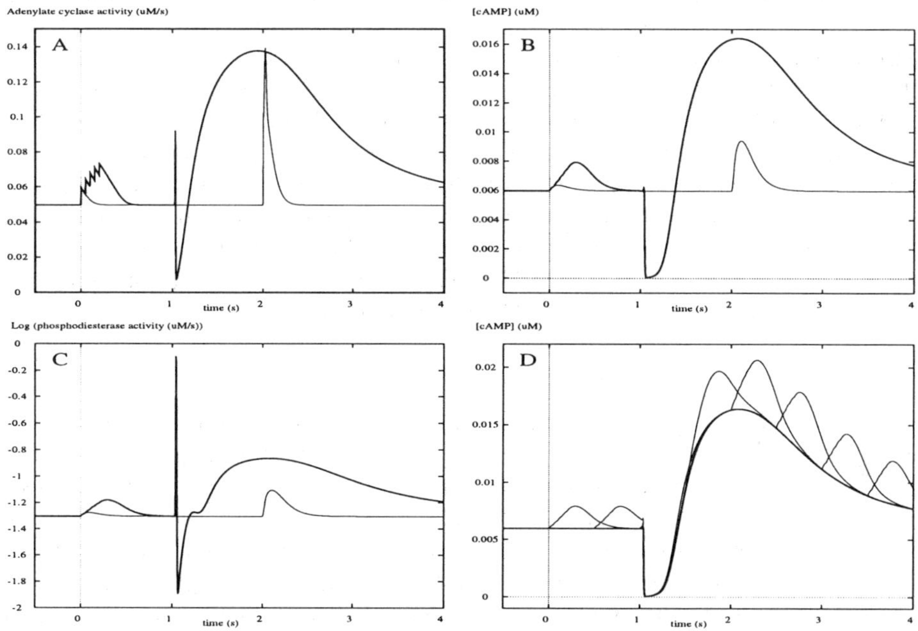

Figure 4: Effects of dopamine and calcium pulses on the time course of adenylate cyclase activity (A), cAMP concentration (B) and phosphodiesterase activity (C). Thick lines in A-C represent the time course in reponse to large pulses of extracellular dopamine (stimulus onset at t = 0 s) and intracellular calcium (t = 1 s) . Thin lines show time course in response to small pulses of dopamine (t = 0 s) and calcium (t = 2 s). D: Time course of cAMP concentrations as a function of temporal separation of large calcium and dopamine pulses. Thick line gives cAMP concentration in response to large calcium pulse only (onset of stimulation at t = 1 s). Thin lines show response of cAMP to dopamine burst superimposed on the calcium stimulus with times of onset of the dopamine burst ranging from t = 0 s to 3.5 s at intervals of 0.5 s.

Fig. 4B illustrates the effect of the four different stimuli on the equilibrium cAMP concentration of 6 nM. Both dopamine stimuli temporarily increase cAMP concentrations. The peaks are delayed in respect to the corresponding peaks in AC activity. The five peaks of the dopamine burst convert into a single peak of cAMP concentrations. The small calcium pulse briefly increases the cAMP concentration without the inflection shown by AC. In contrast, the large calcium pulse eliminates cAMP almost completely preceded by a brief and very small increase. The cAMP concentrations then increase and exceed the starting value after 330 ms. The peak concentration of almost three times steady state is reached after about 1 s. It follows a slow recovery to equilibrium over several seconds.

Phosphodiesterase activity responds to both dopamine stimuli with a comparatively small and slow temporary increase as a consequence of the changing concentrations of its dopamine-sensitive substrate cAMP (Fig. 4C). Similarly, the small calcium pulse produces a small peak in the cAMP concentration and a corresponding peak in the PDE activity. In response to the large calcium pulse PDE shows a brief initial activation similar to AC. This peak limits the formation of cAMP by the initial peak of AC activity. Subsequently, PDE activity falls to very low values as its substrate cAMP expires. Then it recovers fast as a result of increasing concentrations of cAMP, which are caused by continued stimulation of AC activity. As a result of the PDE activity, the triphasic pattern of AC regulation (Fig. 4A) in response to a large calcium stimulus is converted into a biphasic pattern of cAMP regulation (Fig. 4B), which is characterized by a decrease followed by a large and long-lasting increase of cAMP concentrations.

Interactions between large calcium and dopamine pulses

Temporal shifts of the dopamine burst relative to the large calcium pulse have time-dependent effects on the concentration of cAMP (Fig. 4D): If the dopaminergic pulse precedes the calcium pulse, the stimulation of cAMP by dopamine is unaffected until the calcium-induced elimination of cAMP sharply cuts off the dopaminergic stimulation. Simultaneous onset of the dopamine and calcium pulses almost completely abolishes dopamine-induced cAMP formation. This is a combined effect of calcium-induced inhibition of AC and stimulation of PDE. The presence of dopamine is reflected in a minimally accelerated return of cAMP to its steady state concentration. If the dopamine burst follows the onset of the calcium stimulus by 0.5 s and more, the dopamine-dependent stimulation of AC coincides with the CaMCa4-mediated stimulation of AC activity (third phase of the triphasic pattern) and leads to an even stronger formation of cAMP. According to the data of Harrison et al.[7] the combined effect is not just additive but to some extent synergistic as shown in Fig. 4D. This synergism is assumed to depend on the concentrations of both dopamine and CaMCa4: Fig. 4D shows the decay of the synergism as the CaMCa4-induced stimulation of AC wears off. The temporal properties of the CaMCa4- or dopamine-induced stimulation of cAMP are not much influenced by the interaction.

In summary, the model predicts a strong asymmetry of the interactions between dopamine and calcium pulses: 1) dopamine pulses preceding large calcium pulses produce no interaction, 2) temporal coincidence of both pulses abolishes the effect of dopamine, 3) dopamine pulses following large calcium pulses lead to synergistic increases of cAMP concentrations: The strongest synergism occurs if the large calcium pulse precedes the dopamine pulse by about 1 s while longer delays produce a smaller interaction.

Critical parameters of cAMP regulation

The critical parameters of the model system were evaluated by independent variation of all parameters by a factor of ten in both directions. Table 2 lists the effects of parameter changes on the steady state cAMP concentrations. "Increase (%)" denotes the maximal synergistic increase of cAMP levels using a one second delay of the dopamine burst in respect to the large calcium stimulus. "Decrease (%)" gives the maximal decrease of cAMP levels that occurs within 100 ms after the onset of the large calcium pulse. The equilibrium concentration of cAMP had to be re-evaluated for each parameter constellation and was taken as the respective reference value.

Table 2: Effects of parameter variation on cAMP concentrations with synergistically acting dopamine burst and large calcium pulse as shown in Fig. 4D. Parameter values were multiplied or divided by a factor of 10. [cAMP]eq.: Equilibrium concentration of cAMP. Increase (%): Maximal increase of cAMP concentration over equilibrium during synergistic stimulation by dopamine and calcium. Decrease (%): Maximal decrease of cAMP concentration by large calcium pulse.

Parameter	Operation	[cAMP]eq. (μM)	Increase (%)	Decrease (%)
	Default	6 x 10^-3	246.2	99.2
[CaM]	* 10	1.1 x 10^-2	97.1	99.6
	/ 10	5.2 x 10^-3	272.0	98.7
VmaxAC	* 10	6 x 10^-2	245.2	99.1
	/ 10	6 x 10^-4	244.3	99.2
KmAC	* 10	3.2 x 10^-3	244.9	99.2
	/ 10	6.6 x 10^-3	243.0	99.2
[ATP]	* 10	6.6 x 10^-3	245.4	99.2
	/ 10	3.2 x 10^-3	245.5	99.2
EC50CaMCa4	* 10	5.2 x 10^-3	348.8	99.1
	/ 10	1.1 x 10^-2	66.5	99.5
EC50DA	* 10	6 x 10^-3	187.9	99.2
	/ 10	6 x 10^-3	264.4	99.3
aEC50DA	* 10	6 x 10^-3	186.4	99.2
	/ 10	6 x 10^-3	300.7	99.2
aEC50CaMCa4	* 10	6 x 10^-3	186.4	99.2
	/ 10	6 x 10^-3	300.7	99.2
IC50Ca	* 10	6.1 x 10^-3	271.5	95.8
	/ 10	5.1 x 10^-3	127.88	99.9
VmaxPDE	* 10	6 x 10^-4	263.7	93.5
	/ 10	6 x 10^-2	85.3	99.9
KmPDE	* 10	6 x 10^-2	85.2	99.9
	/ 10	6 x 10^-4	255.9	93.5
bVmaxPDE	* 10	6 x 10^-3	242.9	100
	/ 10	6 x 10^-3	244.7	76.5
aKmPDE	* 10	6 x 10^-3	244.7	76.5
	/ 10	6 x 10^-3	242.8	100
bEC50CaMCa4	* 10	6 x 10^-3	244.7	98.8
	/ 10	6 x 10^-3	242.9	99.3

All variations resulting from parameter changes are of a quantitative nature only and do not affect the qualitative features of the results. Proportional effects on the steady state cAMP concentration are noted if the maximal velocities of unstimulated AC (VmaxAC) or PDE (VmaxPDE) or the affinity of PDE for its substrate cAMP (KmPDE) are altered. In addition, the resting level of CaMCa4 limits the equilibrium concentration of cAMP as evidenced by the symmetrical dependence on the CaM concentration and the EC50CaMCa4. The synergistic stimulation of cAMP is critically influenced by those parameters that determine the regulation of AC by the three effectors calcium, CaMCa4, and dopamine. There is also an important effect of parameters controlling the resting activity of PDE (VmaxPDE, KmPDE), which underlines the large capacity of PDE in controlling cAMP levels. Consequently, the size of the cAMP reduction is highly robust: It depends only weakly on factors that affect PDE activity ([b]VmaxPDE, [a]KmPDE) and is invariant to parameters that influence the activity of AC selectively.

Discussion

The current kinetic model of cAMP regulation is based on experimental data and, where specific data are not available, on plausible assumptions. It is shown that the model replicates data on AC regulation by Harrison et al.[7]. Similarly, PDE regulation as reported by Sharma & Wang[15] is reproduced. We chose these specific studies because they provide the most exhaustive lists of kinetic parameters of AC and PDE activation in a series of related experiments. However, comparable data have been provided by these and other groups in a large number of further publications, of which only a few can be cited here[2,3,16,17].

While it is pleasing to see that the present model replicates the experimental results so well, we have to be cautious when moving from the test tube to the living cell. The importance of investigating kinetic responses in dynamic biochemical system in addition to steady-state situations is obvious and will not be discussed here. This issue is reflected in the choice of brief pulsatile stimuli with characteristics that are commonly used in experimental paradigms. However, a consideration is required of the concentrations of the reactants and of interactions that may occur intracellularly in the presence of additional reactants.

There is a general agreement that total CaM concentrations in brain tissues are in the range of 1 to 30 μM[14] but the local concentrations of free CaM may be lower depending on the presence of CaM-binding proteins in addition to calcium. In particular, PDE and protein phosphatase 2B (= calcineurin) show a strong affinity to CaM even at low calcium levels. However, the analysis of parameter variations reveals that the qualitative results of this modelling study are not altered by tenfold variations of the concentration of CaM. Only an increase but not a decrease of the CaM concentration of 10 μM has significant effects on the quantitative data. In the present model, a smaller CaM concentration yields a lower equilibrium concentration and a larger increase of cAMP.

The dynamics of the described variations of AC and PDE activities and cAMP concentrations are similar across a wide range of parameter values. These investigations encompass the possible regulation of PDE by (de-)phosphorylation. As indicated in Table 1 our parameters reflect the phosphorylated state of the 63 kDa isoform of PDE. Dephosphorylation decreases the calcium concentration at 50% activation of PDE by CaM from 1.9 μM to 1.1 μM[15]. The role of this change can be ignored in our model and has been questioned by the original authors themselves[15].

The smoothness of cAMP regulation (low-pass filtering of AC activity, compare synaptic dopamine concentrations in Fig. 2A with their intracellular effects in Fig. 4) is influenced by the equilibrium concentration of cAMP. For example, at tenfold lower cAMP concentrations the individual peaks of the dopamine pulse are relatively larger, while tenfold higher cAMP concentrations show a smooth hump in response to the same burst. Thus, the steady state cAMP concentration is in a position to modulate the responses of cAMP-dependent processes to brief stimuli. For lack of more specific data we assume a homogeneous subcellular distribution of AC and PDE and obtain an equilibrium concentration of cAMP around 6 nM. This reflects the strong PDE activity even in the unstimulated state at 20 nM intracellular free calcium. Although this cAMP concentration is one or two orders of magnitude below whole tissue values[18] it is compatible with recent studies, which show that free cAMP is below 50 nM and constitutes only a small proportion of total cAMP in Aplysia sensory neurons[19]. Since our simulations do not include any cAMP-binding or sequestration, which could prevent hydrolysis of cAMP by PDE, our values are more likely to represent free rather than total cAMP concentrations. Of course, an inhomogeneous distribution of AC and PDE cannot be excluded. A relevant role of anchoring proteins has recently been discussed in respect to the distribution of protein kinase A, which is regulated by cAMP[20]. Concerning AC and PDE, however, the nature of such inhomogeneities remains speculative.

It has not been obvious whether increasing calcium levels should increase or decrease the concentration of cAMP since calcium and CaM affect both AC and PDE[21]. On the one hand, half-maximal activation of AC occurs at lower calcium concentrations than half-maximal activation of PDE in the presence of CaM[6,22]. On the other hand, PDE activity not only exceeds AC activity in the resting state but is also more strongly stimulated than AC by calcium in the presence of CaM[6,15]. In addition, large calcium concentrations inhibit AC[3] and the CaM concentrations required for half-maximal stimulation of AC seem to be higher than those for half-maximal stimulation of PDE. However, as indicated in Table 1 the experimental conditions for AC and PDE differed by a factor of almost 10^3 in the calcium concentrations used. To make the data comparable we calculated the respective EC50 values for stimulation by CaMCa4 and obtained an EC50 of CaMCa4 for AC that is about 7 orders of magnitude lower than that for PDE. This difference would explain the well-known capability of calcium to increase cAMP concentrations. In reality, the situation appears far more complex since PDE may bind CaM in the absence of calcium. This has to be considered in a detailed model of PDE stimulation (manuscript in preparation).

The asymmetry in the response of cAMP to temporal shifts of dopamine pulses relative to calcium pulses puts another aspect to the growing idea of coincidence signalling: we are starting to realize that the investigation of single signals is not sufficient if it is the assimilation of several signals upon which the cell reacts. Specifically, the simulated termporal interactions of dopamine and calcium pulses may be relevant to the role of dopamine in the modulation of striatal functions. During initial learning of a motor task striatal neurons in vivo show increased activity before or during the execution of cue-triggered arm movements. In contrast, dopaminergic neurons respond to the subsequent presentation of a primary liquid reward[23]. Furthermore, there are in vitro slice studies of interactions between neuronal activity and dopamine release in the striatum: Pulsatile dopamine application in conjunction with high-frequency stimulation of corticostriatal pathways and depolarization of striatal principal neurons produces a long-term potentiation in the corticostriatal pathways[24]. Prominent sites for interactions between dopamine and the corticostriatal pathways to take place are the striatal principal neurons, which receive converging dopaminergic and cortical afferent terminals on a proportion of their dendritic spines[25]. In addition, striatal spines possess the intracellular constituents required for interacting biochemical pathways[6]. Thus, concerted regulation of cAMP by dopamine and calcium provides a possible mechanism of interaction between corticostriatal and nigrostriatal signals.

Although the specific temporal requirements of dopamine application in respect to calcium influx in the striatum have not been determined experimentally it appears that pulsatile application of dopamine has different effects on corticostriatal signal transmission than bath application[24]. Based on the present investigation it can be proposed that dopamine release combined with striatal activity that increases intracellular free calcium may have synergistic effects on the regulation of cAMP. If intracellular calcium concentrations rise sufficiently high to inhibit cAMP levels, then a strong synergistic interaction on cAMP can occur only if dopamine is released during the subsequent decay of calcium levels. This synergism appears to occur more than 0.5 s after a large intracellular calcium pulse. Experimental evidence of similar temporal requirements for the interaction of two stimuli has been obtained in Aplysia[26]. Thus, the concerted regulation of cAMP may constitute an important element in the mechanisms that lead to the occurrence of synaptic plasticity in the striatum[6,27].

Acknowledgement

We thank Michael J. Hubbard for his helpful comments on a previous version of the manuscript. This work was supported by a Helmholtz-Scholarship of the BMBF to Rolf Kötter.

References

1. J.W. Kebabian and D.B. Calne, *Nature* **277**, 93 (1979).
2. G.M. Gnegy and G. Treisman, *Mol. Pharmacol.* **19**, 256 (1981).
3. M.T. Piascik, *Mol. Pharmacol.* **20**, 319 (1981).
4. S.N. Murphy et al., *J. Neurosci.* **7**, 4145 (1987).
5. W.G. Regehr. and D.W. Tank, *Nature* **345**, 807 (1990).
6. R. Kötter *Prog. Neurobiol.* **44**, 163 (1994).
7. J.K. Harrison et al., *J. Neurochem.* **51**, 345 (1988).
8. R. Barber et al. in *Adv. Second Messenger and Phosphoprotein Res.*, eds. P. Greengard and G.A. Robison. (Raven, New York, 1992) 1.
9. K.T. Kawagoe et al., *Neuroscience* 51, **55** (1992).
10. W.R. Holmes and W.B. Levy, *J. Neurophysiol.* **63**, 1148 (1990).
11. W.R. Holmes, *Brain Res.* **519**, 338 (1990).
12. I.H. Segel, *Enzyme Kinetics* (New York, Wiley, 1975).
13. G.B. Ermentrout, *Meth. Enzymol.* **240**, 198 (1994).
14. C.B. Klee and T.C. Vanaman, *Adv. Protein Chem.* **35**, 213 (1982).
15. R.K. Sharma and J.H. Wang, *J. Biol. Chem.* **261**, 1322 (1986).
16. Y.C. Clement-Cormier et al., *J. Neurochem.* **25**, 143 (1975).
17. R.L. Kincaid and M. Vaughan, *Meth. Enzymol.* **159**, 557 (1988).
18. J.A. Beavo et al., *Proc. Nat. Acad. Sci. U.S.A.* **71**, 3580 (1974).
19. B.J. Bacskai et al., *Science* **260**, 222 (1993).
20. V.M. Coghlan et al., *Mol. Cell. Biochem.* **127/128**, 309 (1993).
21. D. Schild, *Trends Neurosci.* **17**, 366 (1994).
22. R.R.H. Anholt, *Trends Neurosci.* **17**, 37+367 (1994).
23. W. Schultz et al. in *The Basal Ganglia IV*, eds. G. Percheron et al. (Plenum, New York, 1994) 305.
24. J.R. Wickens et al., *Neuroscience* **70**, 1 (1996).
25. T.F. Freund et al., *Neuroscience* **13**, 1189 (1984).
26. Y. Yovell and T.W. Abrams, *Proc. Nat. Acad. Sci. U.S.A.* **89**, 6526 (1992).
27. J. Wickens and R. Kötter in *Models of Information Processing in the Basal Ganglia*, eds. J.C. Houk et al. (MIT, Cambridge, MA, 1990) 187.

A ROLE OF Ca^{2+}/CALMODULIN-DEPENDENT PROTEIN KINASE II IN THE INDUCTION OF LONG-TERM POTENTIATION: SWITCHING CHARACTERISTICS ACCORDING TO Ca^{2+} SIGNALLING AND DIGITIZATION OF GRADED INFORMATION INTO BINARY INFORMATION

H. OKAMOTO†, K. ICHIKAWA
Foundation Research Lab., Fuji Xerox Co., Ltd.
430 Sakai Nakaimachi Ashigarakamigun, Kanagawa 259-01, Japan
†e-mail: okamoto@rfl.crl.fujixerox.co.jp

1. Introduction

Experimentally, activation of synapses by brief high-frequency stimulation results in long lasting enhancement of transmission efficiencies of activated synapses. This phenomenon, long-term potentiation (LTP, reviewed in [1]), has been extensively studied, especially in rat hippocampus, as an experimental model for usage-dependent synaptic change that is thought of as a minimal process of memory formation in the brain.

Although molecular mechanisms underlying LTP are not yet fully understood, it is now generally agreed that transient elevation of the Ca^{2+} concentration in postsynaptic spines via activation of the N-methyl-$_D$-aspatate class of glutamate receptor channels (NMDA-R/Cs) is necessary for the induction of LTP in rat hippocampal CA1 pyramidal cells [2]. There are also several lines of evidence suggesting that protein kinases, enzymes that catalyze protein phosphorylations, substantially participate in LTP [3-5]. Among them, Ca^{2+}/calmodulin-dependent protein kinase II (CaMKII) is highly concentrated in the postsynaptic region ([6-8], but see also [9, 10]) and, accordingly, considered to be a candidate for involvement in subsequent processes triggered by Ca^{2+} influx through NMDA-R/Cs [11-15]. In fact, specific and nonspecific inhibitors of CaMKII prevent LTP [16-18], and mice genetically lacking the α isoform of the kinase are deficient in LTP [19]. These observations strongly suggest that CaMKII crucially participates in LTP [20-22]. However, details of its role in LTP are far from elucidated.

In the present study, a model for putative postsynaptic biochemical-reaction networks involving activation processes of CaMKII is build upon the bases of regulatory properties of CaMKII that have been characterized *in vitro* experiment by the use of purified kinase. Then, to explore the role of CaMKII in LTP, the dynamical behaviour of CaMKII activity in synapses subjected to stimulation protocol is examined by computer simulation of the model.

2. Model building

CaMKII (for general review for its overall biochemical properties, see [23-26]) is an oligomeric enzyme composed of 10-12 subunits of about 50-62 kDa. There are several isoforms of the CaMKII subunit (α, β/β', etc.), but primary structures of them are highly conserved. A number of experiments, using synthetic peptide or proteolytic fragment, or by means of site-directed mutagenesis, have characterized three domains in the amino acid sequences of a subunit; a catalytic domain in the NH_2-terminal half, a regulatory domain in the central portion, and an association domain in the COOH-terminal half (Fig. 1). The catalytic domain contains ATP-binding and substrate-binding motifs and has potential kinase activity. The regulatory domain (281-309) is comprised of two overlapping subdomains; an inhibitory domain (281-302) [27-34] and a CaM-binding domain (296-309) [35-38]. In the absence of Ca^{2+}/CaM, the inhibitory domain interacts with the catalytic domain and blocks its kinase activity (Fig. 2, a). However, if Ca^{2+}/CaM binds to the CaM-binding domain, this inhibitory effect is neutralized and the kinase subunit becomes active (Fig. 2, b). The association domain is thought to be involved in subunit assembly.

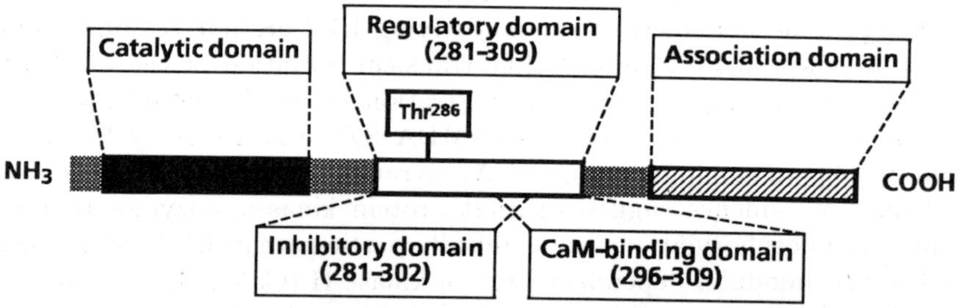

Fig. 1: Domain structure of the α subunit of CaMKII.

Remarkable property of CaMKII that makes it unique from other enzymes is the self-modulation of its activity by autophosphorylation. In the presence of Ca^{2+}/CaM, CaMKII undergoes rapid autophosphorylation. Once autophosphorylated, CaMKII retains its exogenous kinase activity even in the absence of Ca^{2+}/CaM [39-43]. It has been established by now that autophosphorylation of threonine-286 residue (Thr^{286}) located within the inhibitory domain (Fig. 1) is critical to the generation of this Ca^{2+}/CaM-independent (autonomous) activity [44-53]. Autophosphorylation of Thr^{286} (Fig. 2, c) is considered to relieve the inhibitory effect of the inhibitory domain even in the absence of Ca^{2+}/CaM (Fig. 2, d).

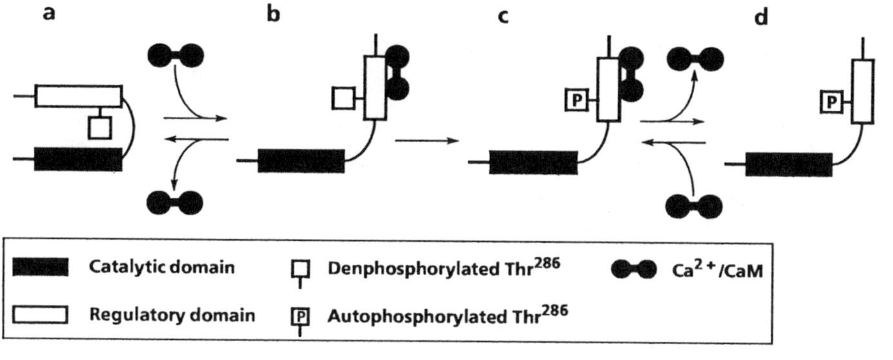

Fig. 2: Regulatory model of CaMKII subunit.

There are also other autophosphorylation sites in a CaMKII subunit. Indeed, $Thr^{305/306}$ within the CaM-binding domain can undergo relatively slow autophosphorylation in a Ca^{2+}/CaM-independent manner, and its phosphorylation blocks subsequent binding of Ca^{2+}/CaM [42, 50, 54-58]. However, it was reported that the major autophosphorylation site of CaMKII in intact nerve cells might be Thr^{286} [59, 60]. It is therefore likely that autophosphorylation of $Thr^{305/306}$ is less important than that of Thr^{286} *in situ*. In the present study, for simplicity, we will assume that Thr^{286} is the only site that can undergo autophosphorylation.

Autophosphorylation of CaMKII is known to be an intramolecular reaction [40, 42, 61]. Recently, autophosphorylation of Thr^{286} has been reported to be predominantly an intersubunit reaction within each holoenzyme [62, 63] and that phosphorylation of Thr^{286} must be preceded by binding of Ca^{2+}/CaM to the "substrate" subunit [62]. These indicate that Ca^{2+}/CaM binding not only neutralizes the inhibitory effect of the inhibitory domain but also makes Thr^{286} an effective substrate for the intersubunit phosphorylation reaction. From these observations, together with the self-modulation mechanisms illustrated in Fig. 2, one can suppose three kinds of Thr^{286}-autophosphorylation processes, as follow:

1) A Thr^{286}-nonphosphorylated (nonautonomous) subunit bound by Ca^{2+}/CaM phosphorylates another nonautonomous subunit bound by Ca^{2+}/CaM (Fig. 3a);

2) an autonomous subunit bound by Ca^{2+}/CaM phosphorylates a nonautonomous subunit bound by Ca^{2+}/CaM (Fig. 3b);

3) an autonomous subunit without Ca^{2+}/CaM phosphorylates a nonautonomous subunit bound by Ca^{2+}/CaM (Fig. 3c).

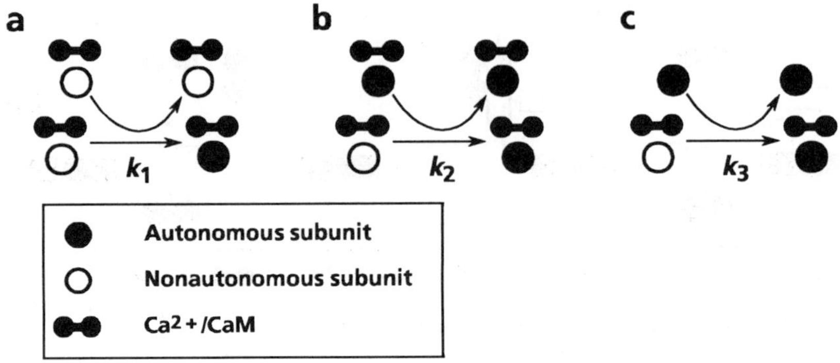

Fig. 3: Intersubunit autophosphorylation of CaMKII.

Thus, autophosphorylation of CaMKII consisting of N subunits can be described by the following scheme:

$$\mathrm{K}_n \xrightarrow{g(n)} \mathrm{K}_{n+1} \quad (n = 0, \cdots, N-1) , \tag{1}$$

where

$$g(n) = g_1(n) + g_2(n) + g_3(n) , \tag{2}$$

$$g_1(n) = k_1\gamma^2(N-n)(N-n-1), \tag{3}$$

$$g_2(n) = k_2\gamma^2 n(N-n), \tag{4}$$

$$g_3(n) = k_3\gamma(1-\gamma)n(N-n). \tag{5}$$

In the above, K_n symbolizes a CaMKII holoenzyme consisting of n autonomous and $N-n$ nonautonomous subunits; $g(n)$ is a rate constant for $\mathrm{K}_n \rightarrow \mathrm{K}_{n+1}$; k_1, k_2, and k_3 are rate constants for intersubunit autophosphorylation reactions 1), 2), and 3), respectively; γ is a probability that each subunit is bound by Ca^{2+}/CaM. Under the assumption that a CaM molecule can bind to a CaMKII subunit only when four Ca^{2+}-binding sites of the CaM molecule are fully occupied by Ca^{2+}, simple kinematical manipulation gives

$$\gamma = \frac{[\mathrm{Ca}^{2+}/\mathrm{CaM}]}{[\mathrm{Ca}^{2+}/\mathrm{CaM}] + K_4} , \tag{6}$$

$$[\mathrm{Ca}^{2+}/\mathrm{CaM}] = \frac{C_{CaM}[\mathrm{Ca}^{2+}]^4}{[\mathrm{Ca}^{2+}]^4 + K_5^4} , \tag{7}$$

where $[Ca^{2+}]$ and $[Ca^{2+}/CaM]$ are the concentrations of Ca^{2+} and Ca^{2+}/CaM in a postsynaptic spine, respectively; K_4 and K_5 are affinities of a CaMKII subunit for Ca^{2+}/CaM and that of a Ca^{2+}-binding site of CaM for Ca^{2+}, respectively; C_{CaM} is the total of the concentrations of CaM and Ca^{2+}/CaM.

It is generally believed that phosphorylation processes in cells and tissues are regulated by counter processes catalyzed by dephospho-enzymes, protein phosphatases. In fact, several reports suggest that CaMKII activity in nerve cells is regulated by phosphatase activity [11, 14, 59, 60]. In fact, phosphatases 1, 2A, and C can erase autonomous activity of CaMKII by dephosphorylation [40, 43, 45, 54, 64-67]. In the present study, dephosphorylation of CaMKII will be described by the standard Michaelis-Menten scheme:

$$\mathrm{P} + \mathrm{K}_n \underset{k_{-6}}{\overset{nk_6}{\rightleftharpoons}} \mathrm{PK}_n \overset{k_7}{\longrightarrow} \mathrm{P} + \mathrm{K}_{n-1} \quad (n = 1, \cdots, N) , \tag{8}$$

where P symbolizes phosphatase; k_6, k_{-6}, and k_7 are rate constants; in the above reaction scheme, k_6 is multiplied by n because K_n has n dephosphorylatable sites.

We have thus constructed a model for putative biochemical-reaction networks in a postsynaptic spine comprising autophosphorylation of CaMKII driven by Ca^{2+} influx through NMDA-R/Cs and dephosphorylation of the kinase by phosphatase. A schematic drawing of the model is given in Fig. 4.

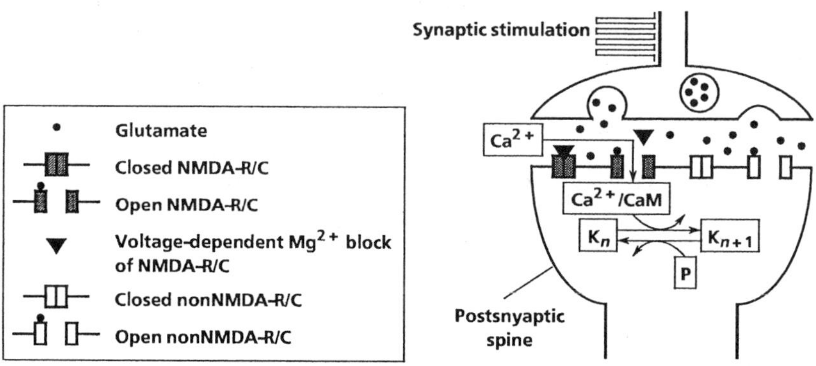

Fig. 4: Schematic drawing of the model.

To simplify analyses, we postulate steady-state assumption for the intermediary metabolites, PK_n ($n = 1, \cdots, N$). Thereby, we have the following set of equations describing the time evolutions of the enzyme concentrations,

$$\frac{\mathrm{d}[\mathrm{K}_0]}{\mathrm{d}t} = -g(0)[\mathrm{K}_0] + \frac{V_D[\mathrm{K}_1]}{K_D + \sum_{n=0}^{N} n[\mathrm{K}_n]} , \tag{9}$$

$$\frac{d[K_n]}{dt} = -g(n)[K_n] + g(n-1)[K_{n-1}]$$

$$-\frac{nV_D[K_n]}{K_D + \sum_{n=0}^{N} n[K_n]} + \frac{(n+1)V_D[K_{n+1}]}{K_D + \sum_{n=0}^{N} n[K_n]} \quad (n = 1, \cdots, N-1) , \qquad (10)$$

$$\frac{d[K_N]}{dt} = g(N-1)[K_{N-1}] - N\frac{NV_D[K_N]}{K_D + \sum_{n=0}^{N} n[K_n]} , \qquad (11)$$

where $[K_n]$ is the concentration of K_n; K_D and V_D are the Michaelis constant and maximal velocity for reaction (8), respectively, and they are defined by $K_D = (k_{-6} + k_7)/k_6$ and $V_D = k_7 C_P$ with C_P being the phosphatase concentration. From these equations, one finds that $\Sigma_{n=0}^{N}[K_n]$ is held constant with time. Hence, one can set

$$\sum_{n=0}^{N} [K_n] = C_K , \qquad (12)$$

where C_K is a constant. This equation describes conservation law for CaMKII in a spine and, accordingly, C_K stands for the kinase concentration.

3. Computer simulation

We examined the dynamical evolution of autonomous activity of CaMKII during synaptic stimulation by computer simulation based on equations (9)-(11). For simplicity, we briefly approximated $[Ca^{2+}]$ to be constant with time, and the amplitude of which was regarded as the stimulus intensity. To investigate the dynamical evolution as a function of the stimulus intensity, we performed computer simulation for various values of $[Ca^{2+}]$. In our simulation, the following parameter values were employed,

$$N = 10 , \quad k_1 = k_2 = k_3 = 1 \ s^{-1} , \quad K_4 = 30 \text{ nM} , \quad K_5 = 10 \ \mu\text{M} ,$$

$$C_K = 10 \ \mu\text{M} , \quad C_{CaM} = 50 \ \mu\text{M} , \quad K_D = 10 \text{ nM} , \quad V_D = 10 \ \mu\text{M} \ s^{-1} . \qquad (13)$$

The initial condition was defined by

$$[K_0] = C_K , \quad [K_n] = 0 \ (n = 1, \cdots, N) . \qquad (14)$$

Fig. 5 displays typical results of our simulation. These results clearly show switching characteristics of the model: There is a threshold with respect to $[Ca^{2+}]$ between 0.91 μM and 0.92 μM; if $[Ca^{2+}]$ is above this threshold, CaMKII is almost maximally phosphorylated, that is, autonomous activity of the kinase is switched on; if otherwise, autonomous activity of the kinase remains near zero (i.e. switched off).

These switching characteristics were also confirmed by numerical calculation of $[Ca^{2+}]$-dependency of autonomous activity in chemical equilibrium.

$[Ca^{2+}]$ was carefully increased so as not to disturb equilibrium, then after $[Ca^{2+}]$ sufficiently exceeds the threshold, it was in turn decreased carefully. Changes in equilibrium autonomous activity along the increase and decrease in γ (note that γ is a monotonic increasing function of $[Ca^{2+}]$, see (6) and (7)) describe a hysteresis loop (Fig. 6), and there are a jump and a falling at the threshold and subthreshold points, respectively (A and B in Fig. 6).

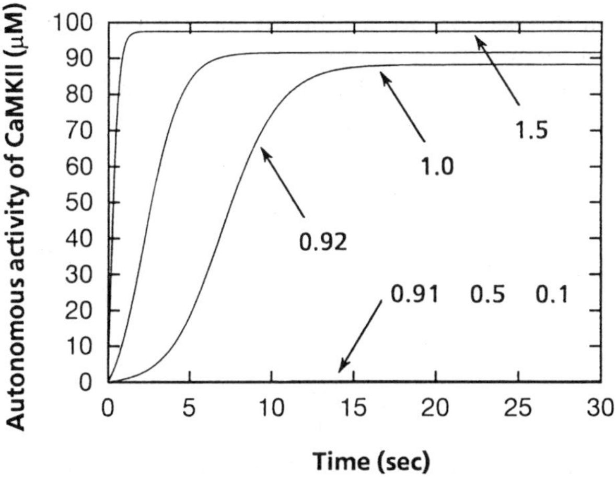

Fig. 5: The time evolution of autonomous activity of CaMKII. Here, "autonomous activity" is defined by $\sum_{n=0}^{N} n[K_n]$, and each curve represents the time evolution of this quantity for $[Ca^{2+}]$=0.1, 0.5, 0.91, 0.92, 1.0, or 1.5 μM.

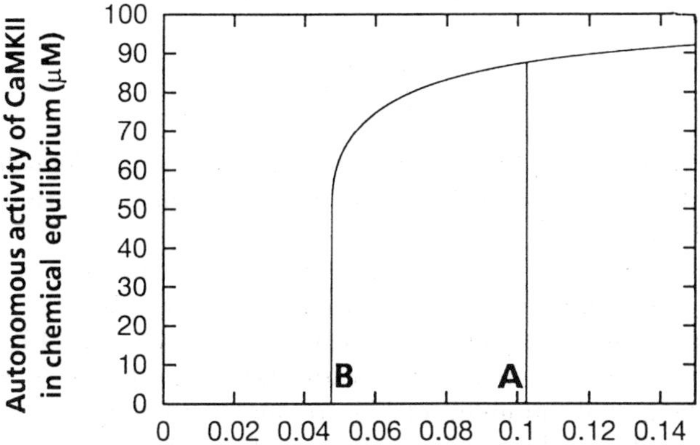

Fig. 6: $[Ca^{2+}]$-dependency of autonomous activity of CaMKII in chemical equilibrium. Here, γ is a monotonic increasing function of $[Ca^{2+}]$ (see Eqs. (6) and (7)).

4. Analytical study

The hysteresis loop in Fig. 6 indicates that the system can be bistable. Several types of theoretical models for bistable chemical switch have been investigated, but in most of them intermolecular autocatalysis is postulated and critical for the generation of bistability [68-73]. It is therefore a novel and interesting finding of the present study that bistability can be generated from intramolecular autocatalysis. In this section, we will briefly discuss a mathematical reason for the generation of bistability in analytical way.

From equations (9)-(11), the following equation can be derived (see [74] for calculations below):

$$\frac{\mathrm{d} < n >}{\mathrm{d}t} =< F(n) > , \tag{15}$$

where

$$< n >= \frac{\sum_{n=0}^{N} n[\mathrm{K}_n]}{C_K} , \quad < F(n) >= \frac{\sum_{n=0}^{N} F(n)[\mathrm{K}_n]}{C_K} , \tag{16}$$

$$F(n) = \gamma^2(N-n)(N-n-1) + \gamma^2 n(N-n) + \gamma(1-\gamma)n(N-n) - \frac{10n}{0.001+ < n >} . \tag{17}$$

For the derivation of (17), parameter values in (13) have been used. Expanding $F(n)$ in Taylor series around $< n >$, and assuming that higher order contributions in the right-hand side of (15) are negligible, we have

$$\frac{\mathrm{d} < n >}{\mathrm{d}t} = -\frac{\partial V(< n >)}{\partial < n >} , \tag{18}$$

where

$$V(< n >) = -\int^{<n>} F(< n >)\big|_{<n>\to x} \mathrm{d}x . \tag{19}$$

Thus, equations (9)-(11) can be approximated as the gradient-descent system described with the potential, V. Therefore, $< n >$ in equilibrium is given as a minimum of $V(< n >)$.

As γ increases from 0 up to 1, the shape of V changes as follows: we first obtain only one minimum (Fig. 7, a), and then two (Fig. 7, b), and finally only one again (Fig. 7, c). The jump at the threshold point (Fig. 6, A) can be explained as a translocation of equilibrium state from the left local minimum of Fig. 7b to the global minimum of Fig. 7c, and the falling at the subthreshold point (Fig. 6, B) can be explained as that from the right local minimum of Fig. 7b to the global minimum of Fig. 7a. These discontinuous translocations are reminiscent of first order phase transition in statistical physics.

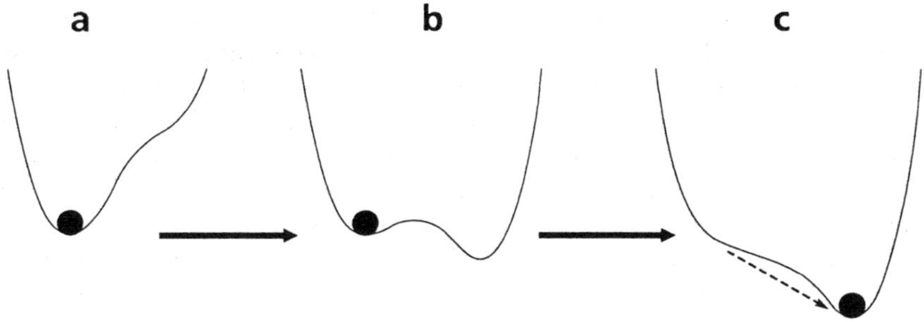

Fig. 7: Change of the shape of V with respect to γ.

5. Summary and discussion

We have build a model for putative postsynaptic biochemical-reaction networks comprised of autophosphorylation of CaMKII driven by Ca^{2+} signalling via activation of the NMDA-R/Cs and dephosphorylation of the kinase by phosphatase. We have then investigated the dynamical behaviour of autonomous activity of CaMKII by computer simulation of the model. Results of our simulation show switching characteristics of the model: There is a threshold regarding synaptic stimulation; if the stimulus intensity is large enough to raise the postsynaptic Ca^{2+} concentration above this threshold, then autonomous activity of CaMKII becomes almost maximum (i.e. switched on); if otherwise, it remains near zero (i.e. switched off). It might be considered that these switching characteristics are due solely to the highly cooperative structure of Ca^{2+} binding to CaM. However, this is not the case because the hysteretic character shown in Fig. 6 cannot be explained by such Ca^{2+}-binding cooperativity of CaM.

Results of our theoretical study suggest that, also in real cells and tissues, CaMKII functions as a switching device that digitizes graded information mediated by Ca^{2+} signalling into binary information. It is therefore probable that whether LTP can be induced or not is determined by this binary information. Thus, we are lead to the following proposal for a role of CaMKII in the induction phase of LTP: If autonomous activity of CaMKII is switched on subsequently to activation of the NMDA-R/Cs, LTP may be induced; if it remains switched off, LTP cannot be triggered.

In contrast to the above proposal, a role of CaMKII in the maintenance phase of LTP was also proposed by several authors [42, 75, 76]. They hypothesized that continual autophosphorylation against phosphatase activity and associating retention of autonomous activity are underlying mechanisms of the maintenance of LTP. To hold this hypothesis, we must assume that autophosphorylation of Thr^{286} still continues even after the termination of synaptic activation and the recovery of the intracellular Ca^{2+} concentration to its resting level (~ 50 nM). However, recent results of biochemical experiments using purified CaMKII suggest that autophosphorylation of Thr^{286} does not occur in the absence of Ca^{2+}/CaM [27, 62, 63, 77]. This implies that autonomous activity cannot be retained against phosphatase activity at a resting level of the intracellular Ca^{2+} concentration. On the other hand, there have been several lines of evidence indicating that a significant portion of CaMKII (2-30 %) is in autonomous state in intact nerve cells at resting state, and this autonomy is responsive to the intracellular Ca^{2+} concentration [11, 15, 59, 60, 78]. These observations imply that autophosphorylation of Thr^{286} can continue even at resting state. Thus, whether CaMKII participates in the maintenance phase of LTP is controversial at the present stage.

To simplify analysis, we have assumed that $k_1 = k_2 = k_3$. However, there are several reports suggesting that autophosphorylation of Thr^{286} is required for full activation of the kinase [79, 80], which implies that k_2 and k_3 are much larger than k_1. There is also a report suggesting the affinity of CaMKII for Ca^{2+}/CaM increases about 1000 times after autophosphorylation of Thr^{286} [81]. Furthermore, Ca^{2+}, CaM, and Ca^{2+}/CaM have been treated kinematically in the present study, but their dynamical behaviour should be included into the model for more realistic simulation. Our model is to be expanded by taking things that might be important but have not been considered here into account in the forthcoming study, and more quantitative discussion will be done there.

References

[1] Bliss,T.V.P. and Collingridge, G.L., Nature **361**, 31(1993).
[2] Malenka, R.C. and Nicoll, R.A., Trends Neurosci. **16**, 521(1993).
[3] Malinow, R., Madison, D.V. and Tsien, R.W., Nature **335**, 820(1988).
[4] Malinow, R., Schulman, H. and Tsien, R.W., Science **245**, 862(1989).
[5] Malenka, R.C. et al., Nature **340**, 554(1989).
[6] Kennedy, M.D., Bennet, M.K. and Erondu, N.E., Proc. Natl. Acad. Sci. USA **80**, 7357(1983).
[7] Kelly, P.T., McGuinness, T.L. and Greengard, P., Proc. Natl. Acad. Sci. USA **81**, 945(1984).
[8] Goldering, J.R., McGuire, J.S. and DeLorenzo, R.L., J. Neurochem. **42**, 1077(1984).
[9] Wu, K., Huang, Y., Adler, J. and Black, I., Proc. Natl. Acad. Sci. USA **89**, 3015(1992).
[10] Suzuki, T., Okumura-Noji, K., Tanaka, R. and Tada, K., J. Neurochem. **63**, 1529(1994).
[11] Fukunaga, K. and Soderling, T.R., Mol. Cell. Neurosci. **1**, 133(1990).

[12] Fukunaga, K., Soderling, T.R. and Miyamoto, E., J. Biol. Chem. **267**, 22527(1992).
[13] Suzuki, T., Okumura-Noji, K., Ogura, A., Kudo, Y. and Tanaka, R., Proc. Natl. Acad. Sci. USA **89**, 109(1992).
[14] Fukunaga, K., Stoppini, L., Miyamoto, E. and Muller, D., J. Biol. Chem. **268**, 7863(1993).
[15] Murphy, T.H., et al., J. Nuerosci. **14**, 1320(1994).
[16] Itoh, I., Hidaka, H. and Sugiyama, H., Nurosci. Lett. **121**, 119(1991).
[17] Funauchi, M., Tsumoto, T., Nishigori, A., Yoshimura, Y. and Hidaka, H., NeuroReport **3**, 173(1992).
[18] Pettit, D.L., Perlman, S. and Malinow, R., Science **266**, 1881(1994).
[19] Silva, A.J., Paylor, R., Wehner, J.M. and Tonegawa, S., Science **257**, 206(1992).
[20] Soderling, T.R., Mol. Cell. Biochem. **127/128**, 93(1993).
[21] Knnedy, M.B., Annu. Rev. Biochem. **63**, 571(1994).
[22] Lisman, J., Trends Neurosci. **17**, 406(1994).
[23] Colbran, R.J., et al., Biochem. J. **258**, 313(1989).
[24] Schulman, H. and Lou, L.L., TIBS **14**, 62(1989).
[25] Colbran, R.J. and Soderling, T.R., Curr. Top. Cell. Regul. **31**, 181(1990).
[26] Hanson, P.I. and Schulman, H., Annu. Rev. Biochem. **61**, 559(1992).
[27] Colbran, R.J., Fong, Y.-L., Schworer, C.M. and Soderling, T.R., J. Biol. Chem. **263**, 18145(1988).
[28] Kelly, P.T., Weinberger, R.P., Waxham, M.N., Proc. Natl. Acad. Sci. USA **85**, 4991(1988).
[29] Payne, M.E. et al., J. Biol. Chem. **263**, 7190(1988).
[30] Colbran, R.J., Smith, M.K., Schworer, C.M., Fong, Y.-L. and Soderling, T.R., J. Biol. Chem. **264**, 4800(1989).
[31] Kwiatkowski, A.P. and King, M.M. Biochemistry **28**, 5380(1989).
[32] Ohsako, S., Watanabe, A., Sekihara, S., Ikai, A. and Yamauchi, T., Biochem. Biophys. Res. Commun. **170**, 705(1990).
[33] Soderling, T.R., J. Biol. Chem. **265**, 1823(1990).
[34] Waldman, R., Hanson, P.I. and Schulman, H., Biochemistry **29**, 1679(1990).
[35] Bennet, M.K. and Kennedy, M.B., Proc. Natl. Acad. Sci. USA **84**, 1794(1987).
[36] Hanley, R.M. et al., Science **237**, 293(1987).
[37] Lin, C.R. et al., Proc. Natl. Acad. Sci. USA **84**, 5962(1987).
[38] Hanley, R.M., Means, A.R., Kemp, B.E. and Shenolikar, S., Biochem. Biophys. Res. Commun. **152**, 122(1988).
[39] Saitoh, T. and Schwartz, J.H., J. Cell Biol. **100**, 835(1985).
[40] Lai, Y., Nairn, A.C. and Greengard, P., Proc. Natl. Acad. Sci. USA **83**, 4253(1986).

[41] Lou, L.L., Lloyd, S.J. and Schulman, H., Proc. Natl. Acad. Sci. USA **83**, 9497(1986).
[42] Miller, S.G. and Kennedy, M.B., Cell **44**, 861(1986).
[43] Schworer, C.M., Colbran, R.J., Soderling, T.R., J. Biol. Chem. **261**, 8581(1986).
[44] Lai, Y., Nairn, A.C, Gorelick, F. and Greengard, P., Proc. Natl. Acad. Sci. USA **84**, 5710(1987).
[45] Miller, S.G., Patton, B.L. and Kennedy, M.B., Neuron **1**, 593(1988).
[46] Schworer, C.M., Colbran, R.J., Keefer, J.R. and Soderling, T.R., J. Biol. Chem. **263**, 13486(1988).
[47] Thiel, G., Czernik, A.J., Gorelick, F., Nairn, A.C. and Greengard, P., Proc. Natl. Acad. Sci. USA **85**, 6337(1988).
[48] Fong, Y.-L., Taylor, W.L., Means, A.R. and Soderling, T.R., J. Biol. Chem. **264**, 16759(1989).
[49] Hanson, P.I., Kapiloff, M.S., Lou, L.L., Rosenfeld, M.G. and Schulman, H., Neuron bf 3, 59(1989).
[50] Lou, L.L. and Schulman, H., J. Neurosci. **9**, 2020(1989).
[51] Waxham, M.N., Aronowski, J., Westgate, S.A. and Kelly, P.T., Proc. Natl. Acad. Sci. USA **87**, 1273(1990).
[52] Ohsako, S., Nakazawa, H., Sekihara, S., Ikai, A. and Yamauchi, T., J. Biochem. **109**, 137(1991).
[53] Ikeda, A., Okuno, S. and Fujisawa, H., J. Biol. Chem. **266**, 11582(1991).
[54] Hashimoto, Y. Schworer, C.M., Colbran, R.J. and Soderling, T.R., J. Biol. Chem. **262**, 8051(1987).
[55] Lickteig, R., Shenolikar, S., Denner, L. and Kelly, P.T., J. Biol. Chem. **263**, 19232(1988).
[56] Colbran, R.J. and Soderling, T.R., J. Biol. Chem. **265**, 11213(1990).
[57] Patton, B.L., Miller, S.G. and Kennedy, M.B., J. Biol. Chem. **265**, 11204(1990).
[58] Hanson,P.I. and Schulman, H., J. Biol. Chem. **267**, 17216(1992).
[59] Fukunaga, K., Rich, D.P. and Soderling, T.R., J. Biol. Chem. **264**, 21830(1989).
[60] Molloy, S.S. and Kennedy, M.B., Proc. Natl. Acad. Sci. USA **88**, 4756(1991).
[61] Kuret, J. and Schulman, H., J. Biol. Chem. **260**, 6427(1985).
[62] Hanson, P.I., Meyer, T., Stryer, L. and Schulman, H., Neuron **12**, 943(1994).
[63] Mukherji, S. and Soderling, T.R., J. Biol. Chem. **269**, 13744(1994).
[64] Shields, S.M., Ingebristen, T.S. and Kelly, P.T., J. Neurosci. **5**, 3414(1985).
[65] Saitoh, Y., Yamamoto, H., Fukunaga, K., Matsukado, Y. and Miyamoto, E., Biomed. Res. **7**, 399(1986).
[66] Saitoh, Y., Yamamoto, H., Fukunaga, K., Matsukado, Y. and Miyamoto, E., J. Neurochem. **49**, 1286(1987).

[67] Dosemeci, A. and Reese, T., J. Neurochem. **61**, 550(1993).
[68] Schlögl, F., Z. Phys. **253**, 147(1972).
[69] Janssen, H.K., Z. Phys. **270**, 67(1974).
[70] Matheson, I., Walls, D.F., Gardiner, C.W., J. Statist. Phys. **12**, 21(1975).
[71] Lisman, J.E., Proc. Natl. Acad. Sci. USA **82**, 3055(1985).
[72] Okamoto, H. and Ichikawa, K., Phys. Rev. E **49**, 3412(1994).
[73] Okamoto, H. and Ichikawa, K., Phys. Rev. E **50**, 1704(1994).
[74] Kampen, N.G.V., *Stochastic processes in physiscs and chemistry* (North-Holland, Amsterdam, 1981).
[75] Lisman, J.E. and Goldring, M., Proc. Natl. Acad. Sci. USA **85**, 5320(1988).
[76] Lisman, J.E. and Goldring, M., J. Physiol. (Paris) **83**, 187(1988).
[77] Mukherji, S., Brickey, D.A. and Soderling, T.R., J. Biol. Chem. **269**, 20733(1994).
[78] Ocorr, K.A. and Schulman, H., Neuron **6**, 907(1991).
[79] Kwiatkowski, A.P., Shell, D.J. and King, M.M., J. Biol. Chem. **263**, 6484(1988).
[80] Katoh, T. and Fujisawa, H., J. Biol. Chem. **266**, 3039(1991).
[81] Mayer, T., Hanson, P.I., Stryer, L. and Schulman, H., Science **256**, 1199(1992).

[illegible]
[illegible]
[illegible], B., [illegible] Soto, G. [illegible]

MATHEMATICAL ASPECTS OF SIGNAL-PROCESSING VIA G-PROTEINS

J. NAUROSCHAT, U. AN DER HEIDEN
Department of Mathematics, University of Witten/Herdecke,
Stockumer Str. 10, D-58448 Witten, Germany

1 Introduction

Regarding operative systems in general, no matter whether stemming from the natural sciences, economy, sociology or any other field, the fundamental question is about the conversion of signal to effect. The mammalian organism as a biological system is permanently targeted by a multitude of endogeneous and exogeneous signals. External messages such as light, smell or sound can be perceived by specific sensory units, which are part of the corresponding sensory subsystems of the organism. For example, rhodopsin is a sensory unit binding photons of light on rod outer segments of the retina. The sensory subsystems, again, are coupled to a complex network of additional organismic subsystems, which can be of neural, immunological, hormonal, or other type, see Fig. 1.

Depending on the aspect of interest, behavior of the organism may be studied at the level of some few, representative subsystems: Referring to the process of hormone-release, for instance, one may concentrate on the network established by the neural/hormonal subsystems *'hypothalamus'*, *'pituitary gland'*, and *'adrenal gland/thyroid'*. For a theoretical analysis, see an der Heiden.[1]

We are by no means restricted to the macroscopic evaluation of such spatially extented organismic networks. In fact, referring to the process of cellular signal utilization, one may study the molecular network established by the subsystems *'membrane'*, *'cytosol'*, and *'nucleus'*, see Fig. 2.

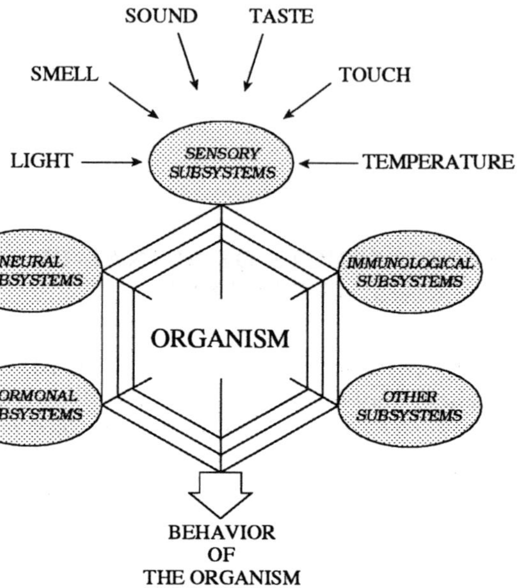

Figure 1: The organism as an overall network of sensory, neural, immunological, hormonal, and other subsystems.

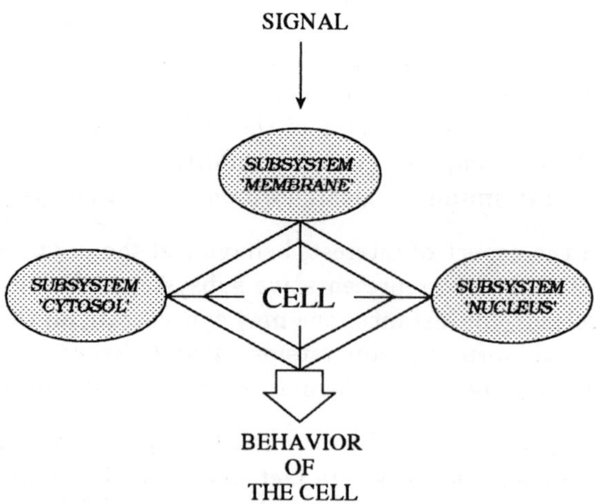

Figure 2: The cell as a molecular network of the subsystems 'membrane', 'cytosol', and 'nucleus'.

There is no universal structure for the network in Fig. 2, because involvement of the different subsystems depends on the type of signal. This is exemplified by the following:

Lipophilic signals, e.g. the steroid hormones, readily traverse the *membrane* to couple to specific cytosolic binding proteins (receptors). The signal-receptor complex then induces enhanced synthesis of mRNA in the *nucleus*. This, in turn, implies enhanced protein-synthesis at the ribosomes. The biological function of these proteins defines the signal- specific behavior of the cell. Obviously, the subsystem *'membrane'* here is of minor importance for signal utilization, whereas the subsystem *'nucleus'* is of predominant importance.

A different situation is found in the networks processing those signals that are unable to traverse the membrane lipid double layer. Here, the signal couples to specific cell-surface receptors. In the majority of these cases, occupation of the receptor initiates a sequence of membrane processes that transform the extracellular ('first') messenger to the intracellular ('second') messenger cAMP. This cyclic nucleotide then activates specific kinases which phosphorylate cellular proteins. The resulting alteration in the functioning of these proteins defines the signal-specific behavior of the cell. Referring to *this* mode of signal utilization it is clear that the network in Fig. 2 crucially depends on the subsystem *'membrane'*. Interestingly, the membrane pathway processing the first messenger to cAMP is of universal structure, whereas the cellular pathway processing the cAMP depends on the type of cell. In order to develop a general mathematical approach to signal utilization, we shall thus focus on the membrane pathway.

2 The receptor/G-protein/adenylate cyclase–pathway in the membrane

Production of cAMP in response to a first messenger is primarily determined by three proteins, namely receptor, guanine-nucleotide regulatory protein ('G-protein') and the catalytic component of the enzyme adenylate cyclase:[2–3] The *receptor* is the unit that specifically detects the external signal. The *G-protein* is the unit that modulates and transports the signal to the enzyme. The *catalyst* is the unit that releases the second messenger cAMP in response to the signal. Communication between the different units is established by sequential coupling, see Fig. 3.

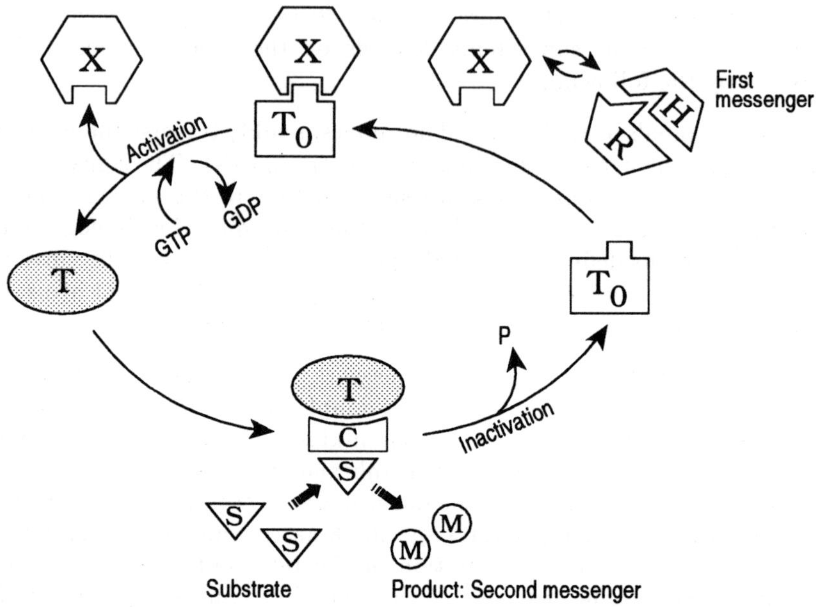

Figure 3: Activation-inactivation cycle of signal processing by the receptor/ G-protein / adenylate cyclase- pathway. H input-signal: extracellular 'first' messenger, R cell surface receptor, $X = H \cdot R$ liganded receptor, T_0 inactive G-protein, T newly activated G-protein, C enzyme adenylate cyclase, S compound $Mg^{2+} \cdot ATP$, M output-signal: intracellular 'second' messenger cAMP, P liberated phosphate.

First, the receptor (R) binds the extracellular signal (H) to the membrane. As a consequence of occupation, the liganded receptor (X) now exhibits a binding-site for a GDP-carrying G-protein (T_0). Interaction between X and T_0 induces exchange of the nucleotide, i.e. dissociation of GDP followed by immediate re-association of GTP. The acquired phosphate destabilizes the link between liganded receptor and G-protein. The consequently liberated GTP-bound G-protein (T) is 'active' in the sense that it couples to the adenylate cyclase (C) and thereby activates enzymic conversion of substrate $Mg^{2+} \cdot ATP$ (S) to product cAMP (M). By virtue of its ability to hydrolyze bound GTP to GDP, *any* functional energetic G-protein ultimately changes back to its original quiescent state, T_0.

Activation of the G-protein negatively controls the functional state of liganded receptor (X):
The inactive G-protein exhibits a heterotrimeric $\alpha/\beta\gamma$-subunit structure, but $\beta\gamma$ dissociates upon activation of the protein. The dimer $\beta\gamma$ recently has been reported to translocate certain receptor- kinase to the membrane.[4–6] However, exclusively the *liganded* membrane-receptor (X) serves as a substrate for the kinase. Phosphorylation by this enzyme induces loss of receptor efficiency – despite of carrying the external signal, the modified receptor cannot couple to the inactive G-protein. Activation of the membrane pathway thus is blocked.

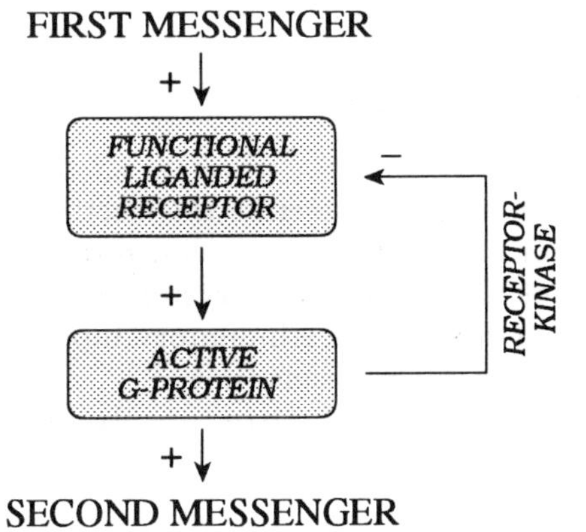

Figure 4: Negative feedback coupling between active G-protein and functional liganded receptor.

Such agonist-specific refractoriness ('homologous desensitization') can be terminated by nullification of the receptor-modification. Results of recent experimental studies indicate that phosphorylated receptors are dephosphorylated in certain vesicle compartments and afterwards recycle to the cell surface as functional proteins again.[7–8]

3 Mathematics of G-protein mediated signal-transduction

Taking the outlined mechanisms as a basis we have developed a mathematical model of G-protein mediated signal-transduction. For details the interested reader is referred to Nauroschat and an der Heiden.[9]

First, we formalized the membrane pathway by proposing a set of biochemical reaction schemes. A corresponding set of rate equations then was derived according to the mass-action law. The number of differential equations could be reduced by analytical evaluation of quasi-steady-state conditions and conservation laws. The final differential system is non- autonomous and of dimension four, with all the variables rescaled:

$$\frac{dr}{d\tau} = -(h(\tau) + c_1)r + c_2x + c_3 \tag{1}$$

$$\frac{dx}{d\tau} = h(\tau)r - c_4x - F(x)(1-a) - Q_1(a)Q_2(x) \tag{2}$$

$$\frac{da}{d\tau} = c_5F(x)(1-a) - c_6a \tag{3}$$

$$\frac{dm}{d\tau} = G(a) - k_m m \tag{4}$$

Here, the variables r, x, and a describe the fractions of free receptor, functional liganded receptor, and active G-protein, respectively. The variable m describes the output of the transduction system (i.e. the second messenger cAMP), whereas $h(\tau)$ describes the stimulatory input (i.e. the first messenger). The independent variable τ means time.
The functions $F(x)$, $Q_2(x)$ and $G(a)$, $Q_1(a)$ are strictly increasing with x and a, respectively, and

$$F(0) = G(0) = Q_1(0) = Q_2(0) = 0. \tag{5}$$

The courses of $F(x)$ and $G(a)$ are saturated and concave in shape.

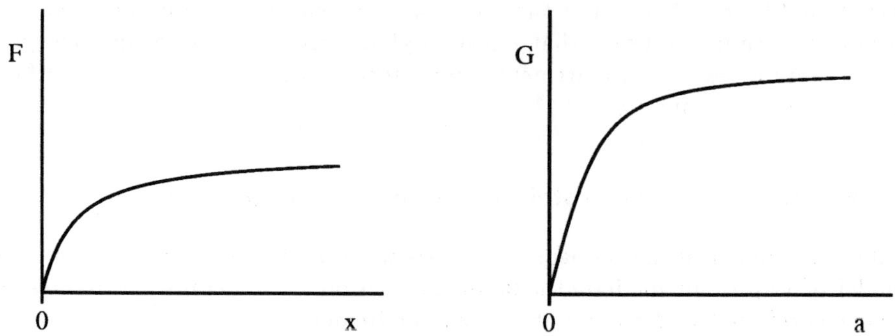

Figure 5: Courses of the functions F(x) and G(a).

Note that $F(x)$ is dynamically weighted by the balance $1-a$ measuring the fraction of inactive G-protein. This expresses the synergism between the 'energetic' receptor and quiescent transducer protein – referring to the notations introduced in Sec. 2, $F(x)(1-a)$ reflects the evolution of the trimolecular intermediate $H{\cdot}R{\cdot}T_0 = X{\cdot}T_0$ (i.e. agonist/receptor/inactive G-protein complex), considered the precursor of newly activated G-protein in Fig. 3.

The model includes yet another type of synergism between the receptor and transducer protein. In fact, the rate function $Q_1(a)Q_2(x)$ in (2) quantifies the negative control of receptor-efficiency induced by the receptor-kinase, see Fig. 4. This process is regarded to depend sigmoidally on the degree of G-protein activation[5] and linearly or hyperbolically on the degree of receptor-occupation, see Fig. 6.

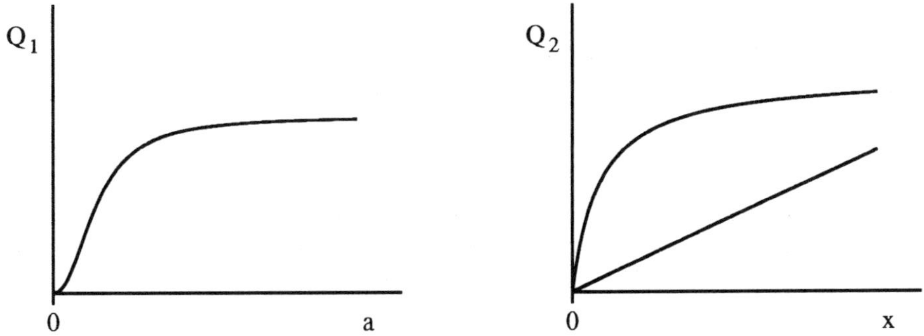

Figure 6: Courses of the functions $Q_1(a)$ and $Q_2(x)$.

The parameters $c_1, \ldots, c_6, k_m$ in the model are positive algebraic expressions of rate-constants and the totalities of receptor and G-protein.[9] It is evident that $c_1, \ldots, c_6$ may be considered the 'main' parameters – in fact, the last equation (4) is decoupled from (1,2,3) so that we may concentrate on the separate (r, x, a)-system.

Analysis of the (r, x, a)-model reveals particular importance of the ratio c_3/c_1. This ratio defines an open prism

$$\Omega := \{(r, x, a) \in (0,1)^3 : r + x < c_3/c_1\} \tag{6}$$

being invariant under the vector field $(\dot{r}, \dot{x}, \dot{a})$.

Simulation of *sustained withdrawal* of the membrane from agonistic input is achieved by defining $h(\tau)$ in (1,2) as the zero function. For this specific situation c_3/c_1 constitutes an equilibrium value for free receptor, since

$$(r_0, x_0, a_0) := (c_3/c_1, 0, 0) \tag{7}$$

is a time-invariant solution of (1,2,3). The values of x_0 and a_0 symbolize relaxation of signal detection and transduction. Note that the equilibrium point (r_0, x_0, a_0) is located on the boundary of Ω.

Simulation of *sustained exposure* of the membrane to a signal of fixed amplitude is achieved by defining $h(\tau)$ in (1,2) as a positive constant function:

$$h(\tau) \equiv h > 0 \qquad \text{for} \quad \tau \geq 0 \tag{8}$$

Contrary to the case of $h(\tau) \equiv 0$, we have no *explicit* steady-state solution of (1,2,3) under the condition of (8). Existence and properties of such solutions yet can be decided by *qualitative* analysis of

$$\begin{aligned} 0 &= -(h + c_1)r + c_2 x + c_3 \\ 0 &= hr - c_4 x - F(x)(1-a) - Q_1(a)Q_2(x) \\ 0 &= c_5 F(x)(1-a) - c_6 a \end{aligned} \tag{9}$$

By recasting (9), $h \geq 0$, into a single equation for the variable r and taking into account the qualitative characteristics of the functions $F(x)$, $Q_1(a)$, and $Q_2(x)$, see Figs. 5 and 6, we obtain the following results:

Existence and uniqueness of steady-states

For any value of $h \geq 0$ the corresponding algebraic system (9) exhibits a unique solution $z_h := (r_h, x_h, a_h)$ in the closure of the prism Ω.

Location of steady-states

For $h = 0$ the steady-state z_0 is located on a corner of $\bar{\Omega}$, see (7).
For $h > 0$ the steady-state z_h is located on the planar surface

$$S_h := \left\{ (r, \chi_h(r), a) : r \in [r_h^-, r_h^+], \chi_h(r) = \frac{r(h+c_1)}{c_2} - \frac{c_3}{c_2},\ a \in [0, a_h^+] \right\} \tag{10}$$

in Ω. Here, the boundaries r_h^-, r_h^+ and a_h^+ are defined by

$$r_h^- := \frac{c_3}{c_1(1 + h/c_1)} \qquad r_h^+ := \frac{c_3}{c_1(1 + h/(c_1 + c_2))} \tag{11}$$

$$a_h^+ := 1 - \frac{1}{c_7 F(x_h^+) + 1} \tag{12}$$

with $x_h^+ := c_3/c_1 - r_h^+ > 0$ and $c_7 := c_5/c_6 > 0$.

Monotonicity of steady-states

The equilibrium value r_h strictly decreases with h, whereas the equilibrium values x_h and a_h strictly increase with h.

The manifold of steady-states z_h, emanating from the corner point $(c_3/c_1, 0, 0)$ of $\bar{\Omega}$, is indicated in Fig. 7.

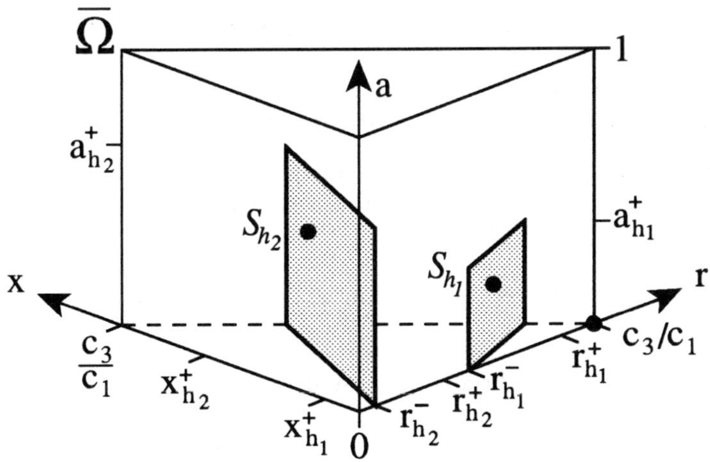

Figure 7: Steady-states of the differential system (1,2,3), $h(\tau) \equiv h \geq 0$, in the compact prism $\bar{\Omega}$. Shown is the situation corresponding to three representative input parameters $h = 0, h_1, h_2$, with $0 < h_1 < h_2$. For $h = 0$ the steady-state z_0 is symbolized by the dot located on the corner $(c_3/c_1, 0, 0)$ of $\bar{\Omega}$. For $h = h_1, h_2$ the steady-states z_{h1} and z_{h2} are symbolized by the dots located on the shaded surfaces S_{h1} and S_{h2}, see (10), respectively.

Stability of steady-states

Linearization of the differential system (1,2,3), $h(\tau) \equiv h \geq 0$, at the steady-state $z_h \in \bar{\Omega}$ takes the form

$$\dot{\xi} = J_h \cdot \xi \tag{13}$$

where J_h denotes the Jacobian matrix of $(\dot{r}, \dot{x}, \dot{a})$, evaluated at z_h.

Since the steady-state z_0 for $h = 0$ is explicitly known, see (7), the solutions λ of the corresponding eigenvalue problem

$$det(\lambda I - J_0) = 0 \tag{14}$$

can be determined:

$$\lambda \in \{-c_1, -c_4 - F'(0), -c_6\} \tag{15}$$

Negativity of these eigenvalues implies local stability of the steady-state for $h = 0$.

Since the steady-state for $h > 0$ remains unknown, we have no explicit formulas for the eigenvalues of the Jacobian $J_h, h > 0$. Yet we can prove local stability of z_h. In fact, taking again advantage of the qualitative characteristics of $F(x)$, $Q_1(a)$, and $Q_2(x)$, we can demonstrate the coefficients of the characteristic polynomial $P(\lambda) = det(\lambda I - J_h)$ to satisfy the conditions of the well-known Hurwitz criterion.[10] Accordingly, all the roots of $P(\lambda)$ have negative real part, yielding the desired result of stability for z_h, $h > 0$.

Since the components r_h, x_h, and a_h of the steady-state $z_h \in \bar{\Omega}$ monotonically depend on h, they will converge towards certain finite limit values as h tends to infinity:

$$r_\infty = \lim_{h \to \infty} r_h, \quad x_\infty = \lim_{h \to \infty} x_h, \quad a_\infty = \lim_{h \to \infty} a_h \tag{16}$$

This saturation of equilibria is illustrated by Fig. 8, resulting from numerical integration of the system (1,2,3).

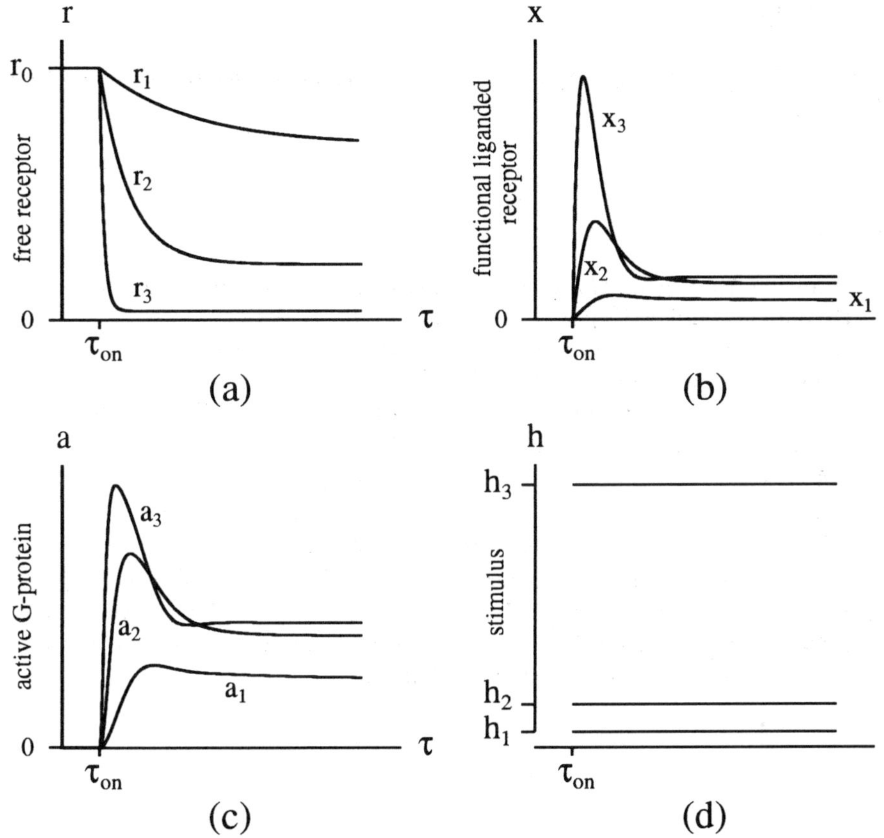

Figure 8: Time-courses of free receptor r (Fig. 8a), functional liganded receptor x (Fig. 8b), and active G-protein a (Fig. 8c), obtained by numerical integration of (1,2,3) for constant input $h(\tau) \equiv h > 0$. Shown are the responses r_i, x_i, and a_i corresponding to three distinct stimulus-amplitudes $h = h_i$, with $i = 1, 2, 3$ and $0 < h_1 < h_2 < h_3$. Stimulation starts at $\tau = \tau_{on}$. For $\tau < \tau_{on}$ the system is at the resting state $(r_0, x_0, a_0) = (c_3/c_1, 0, 0)$ corresponding to $h = 0$. Schemes of stimulation are illustrated by Fig. 8d.

The numerical simulations underlying Fig. 8 revealed that, for a given signal-amplitude h, the variable m (= cAMP) behaves qualitatively similarly to a. The spike-decline behavior of x, a, and m may be interpreted as hypersensitivity of the signalling system at rest.

Acknowledgments

We gratefully acknowledge support by the Deutsche Forschungsgemeinschaft DFG.

References

1. U an der Heiden in *Mathematics applied to biology and medicine*, ed. J Demongeot and V Capasso (Wuerz, Winnipeg, 1993).
2. A G Gilman in *The biology and medicine of signal transduction*, ed. Y. Nishizuka et al. (Raven Press, New York, 1990).
3. J.R. Hepler and A.G. Gilman, *Trends Biochem. Sci.* **17**, 383 (1992).
4. J. Inglese *et al., Nature* **359**, 147 (1992).
5. J.A. Pitcher *et al., Science* **257**, 1264 (1992).
6. S. Müller *et al., Proc. Nat. Acad. Sci. USA* **90**, 10439 (1993).
7. S.S. Yu *et al., J. Biol. Chem.* **268**, 337 (1993).
8. M. von Zastrow and B.K. Kobilka,*J. Biol. Chem.* **267**, 3530 (1992).
9. J. Nauroschat and U. an der Heiden, *to appear in J. Math. Biol.* (1996).
10. L Cesari *Asymptotic behavior and stability problems in ordinary differential equations* (Springer, Berlin Heidelberg, 1971).

OSCILLATIONS, POSITIVE FEEDBACK AND THE FLOW OF BIOLOGICAL INFORMATION

K. S. R. CUTHBERTSON[1]

Department of Human Anatomy and Cell Biology
University of Liverpool
PO Box 147
Liverpool L69 3BX

1 Introduction

Here I propose that positive feedback is a key aspect of all living systems. I will be claiming that biology is digital in key aspects of its behaviour or rather to be slightly more subtle about it there may be the emergence of a new principle combining analogue behaviour with threshold behaviour which in effect leads to digitised or for example pulsative behaviour which would involve threshold phenomena in a major way. Aside: I do recall a time when I was very dubious about Lewis Wolpert's insistence that there were threshold phenomena in biology. I have now completely changed my mind on this. I want to bring in a connection with neural networks. I'll be arguing that there is an analogy between the positive feedback threshold behaviour that I am arguing for and the mathematical basis for the neural network concept. In other words the connection is mathematical and mechanistic rather than anything else. In essence I want to explore the possibility that there is a close connection between the sigmoidal transfer function of neural networks and a positive feedback mechanism which acts as a digital switch.

[1] Editors note: this paper was in draft/note form when Roy died. We have endeavoured to reproduce the material with very little change to his notes.

2 Beta-amyloid aggregation as a positive feedback

A current hypothesis on the origin of Alzheimers Disease holds that it is due to the aggregation of an aberrant peptide called beta-amyloid which is derived from a beta- amyloid protein. In some versions of this hypothesis the aggregation is promoted itself by the presence of aggregated peptide which constitutes a positive feedback mechanism.

3 Multiple inputs to a positive feedback process.

A positive feedback process in biology may have many, indeed very many, positive and negative inputs into a single process. Each of these inputs may be a complex function of the controlling parameters of that part of the process. Shifts in these control parameters may move the system from a stable region into one where possessing positive feedback one should target the control parameters. Note: in this section I should show a figure in which I show the shift produced in a positive feedback system i.e. a system possessing a positive feedback mechanism from a single stable state to a state with three equilibrium points, two of them stable and one unstable, in which the resulting figure looks sigmoidal and I will draw attention perhaps in this figure-legend to the connection between this sigmoidal output of a positive feedback mechanism and the sigmoidal function which is commonly used for the implementation of neural networks.

4 Mouse oocyte fertilisation

During the fertilisation of mammalian oocytes there is a calcium signal which triggers the events of development. This calcium signal consists of complex non-linear oscillations. At the heart of the mechanism of these nonlinear calcium oscillations is at least one positive feedback mechanism. Several such positive feedback mechanisms have been proposed. Firstly, the release of calcium by calcium stores depends on calcium itself. There may be several molecular mechanisms involved in this calcium positive feedback at the site of calcium release from calcium stores involving for example, the IP3 receptor but also other receptors which have yet to be identified. Secondly, there may be a positive effect of calcium on phospholipase C which may be involved in the production of IP3. The positive feedback effect here is that calcium activates phospholipase C. Once could speculate about a possible involvement of other positive feedback mechanisms in this process, especially those involving phosphorylation. One current possibility that is being explored by several laboratories is that protein tyrosine kinase is involved in some way. There is no reason why such an involvement might not actually also be a positive feedback mechanism.

5 Heart disease

Peter Cobbold and colleagues have proposed a mechanism for ischaemic damage during heart attacks which involves a positive feedback mechanism. Cobbold's hypothesis proposes that during a heart attack there is an ATP-catastrophe. Loss of the oxygen supply to the heart results in a loss in the production of ATP, the energy currency of the cardiac cell and indeed all cells. This catastrophe occurs because once the ATP concentration drops below a certain threshold which is currently estimated at 1.2 millimolar, which is substantially below the estimated 5 millimolar ATP. concentration in the normal heart, rigor complexes are activated which activates the actinomycins ATPase in a constitutative fashion. This constitutes a positive feedback process and results in the complete and rapid depletion of ATP from the cell resulting in the death of the cardiac myocite. This insight into the mechanism of irreversible ischaemic heart cell damage has major implications for approaches to the design of new therapeutic agents to protect the heart from ischaemic cell damage during heart attacks.

6 Calmodulin-dependent protein kinase II and memory

Calmodulin-dependent protein kinase II consists of a very complex multi-enzyme system. There are up to twelve sub-units, each of which is a protein kinase. These protein kinases can phosphorylate each other within the complex. When phosphorylated the enzyme has a certain level of intrinsic activity without any additional regulatory input. When unphosphorylated and in the absence of calcium there is an activation of the enzymes. In the presence of calcium-bound calmodulin the enzyme is activated. The auto-phosphorylation within the multi-enzyme complexes of calmodulin-dependent protein kinase II constitutes a positive feedback mechanism. The activation of calmodulin-dependent protein kinase II (CaM-II kinase) by calcium depends very often on the release of calcium from internal stores. The positive feedback behaviour of CaM-II kinase can constitute a molecular switch. This switch in principle could be local to each synapse or any other sub-region of a large multi-synaptic neuron such as for example a hippocampal pyramidal cell. Therefore, again in principle, a single neuron could act as a multi-bit memory with the CaM-II kinase mechanism at the heart of it and positive feedback as the principal mechanism or rather at the heart of this proposal is a positive feedback mechanism. In other words what I am proposing is not in itself a novel suggestion but positive feedback is crucially involved in laying down memory traces.

I have modelled a hypothetical protein kinase closely based on CaM kinase but with some significant differences, notably that it is a monomeric protein. The biochemical data on which it is based is taken from the work of Meyer and colleagues. Meyer and colleagues are dealing with

CaM kinase II monomeric version enzyme is intended to be appropriate for cells such as hepatocytes which do not possess in any significant quantities CaM kinase itself which is largely associated with neurons. In summary:

1. CamK and positive feedback lead to a switch.

2. This switch could be in principle local to each synapse or other region.

3. Therefore one cell could in principle on this basis act as a multi-bit memory perhaps with the same order of magnitude as synapses. This is probably a gross over-estimate. It is probably not a new idea. It would not just be calcium based but involve various levels and types of phosphorylation and other biological means of regulating internal events of a neuron.

7 Concluding Comment

I would argue that in the philosophical conflict relating to the theory of evolution between gradualism and catastrophism there is actually no conflict because both processes can occur together. I am using this as an analogy with my view on positive feedback which philosophically is closely related to catastrophism rather than gradualism. The point is that a mechanism involving positive feedback involves an underlying graduatlistic process which by a process of catastrophism becomes a switch and in terms of the analogy with evolutionary biology that switch would be a change in species or something on a broader front. I suggest that the power of the Darwinian paradigm which is intrinsically gradualistic may have blinded the wider scientific community to the power of catastrophism i.e. to the potential power of catastrophic events. I argue that positive feedback processes may result essential in catastrophic events. Of course, there are other ways in which catastrophic events can arise e.g. through essentially accidental events. In conclusion I have argued that positive feedback processes play a key role in biological systems.

8 Postscript[2]

Over the past decade it has emerged that the calcium second messenger system involves pulsatile non-linear oscillations in which positive feedback plays a central role. In this paper I discuss some ways in which oscillations and positive feedback might facilitate the flow of biological information into tissues and cells, in the context of the calcium second messenger system. Specifically three examples are considered: firstly, the reception of a time-structured agonist signal by an oscillatory calcium signalling system; secondly, a proposed interaction between oscillatory mechanisms at successive levels of the calcium release pathway; and thirdly, the downstream consequences of pulsatile calcium oscillations acting via a non-linear calcium effector system. The first situation might arise from sympathetic innervation of the liver, releasing pulses or nor-adrenalin. Experimental and theoretical evidence is given showing that such pulses could be a more efficient stimulus to the oscillatory calcium signalling system in hepatocytes than a constant input. For the second situation I discuss a model of the calcium oscillator mechanism with two levels of positive feedback, onto phospholipase C and onto the inositol triphosphate-activated calcium channel, which provides an explanation for complex oscillations in hepatocytes. Such a dual mechanism oscillates on two time scales, and provides a richer repertoire of behaviour than either mechanism alone. For the third situation I show how a model calcium-dependant kinase system with positive feedback provided by autophosphorylation can process

[2] Editors note: This section was not part of Roy's original script but was the abstract for his talk at IPCAT 95. It is included here to give some idea of the directions Roy's ideas were taking.

real calcium signals from hepatocytes and selectively respond either to frequency or pulse duration. This shows how information encoded by frequency, which depends on agonist concentration, or pulse shape, which depends on agonist type, might be decoded. These examples illustrate some of the advantages pulsatile oscillators involving positive feedback offer for the transduction of biological information, which include robustness, noise immunity, efficiency, and the potential for discrimination between signals.

A DATA AND KNOWLEDGE BASE FOR CELL SIGNALING NETWORKS

T. IGARASHI, T. KAMINUMA
National Institute of Health Sciences, 1-18-1 Kami-Yoga, Setagaya-ku, Tokyo 158, JAPAN

Y. NADAOKA
The Tokyo Metropolitan Institute of Medical Science, 3-18-22 Hon-Komagome, Bunkyo-ku, Tokyo 113, JAPAN

1 Introduction

Each cell in a multicellular animal is receiving chemical signals that come from other cells. Some of these signal transmitting cells are its neighbors (or even itself) and some are in remote tissues. Hormones and neurotransmitters are well known examples of such chemical signals. Some chemical signals come also from outside of the animal. Drugs and toxic substances are such examples. Cells have receptors both for inter chemical signal agents and for xenobiotic chemical signals. In either case, some of these signal molecules bind to receptors on the cell surface (in the membrane), while some smaller signal molecules penetrate the membrane and directly stimulate or bind to intracellular signal molecules.[1,2]

By analogy each step of cell signal transfer can be considered as a switch, a gate, or a transistor in electric circuits in LSI (Large Scale Integrated Circuit), and the whole set of such signal pathways forms a very complex and sophisticated information processing system.[3] As cells are building blocks for living organisms, so the network is a unit for the information system of the entire organism.

The cell signaling networks change their structure as the multicellular animal develops. As cells proliferate and differentiate new sets of networks emerge, and they form a more complex network of networks. Due to differentiation of cells a signal molecule often has different effects on different target cells. Cell signaling networks respond differently in different metabolic states. Mutated animals have different receptor proteins and different signal molecules. For example the so called oncogenes encode many cell signal proteins which distribute from the surface (membrane) of the cell to the intracellular cytoplasm and the nucleus. When mutated these oncogene products disturb normal signal transductions which may eventually transform the cell into carcinogenic states.

At some stage of development some population of cells in a multicellular animal form cell collectives by direct contacts of cell adhesive molecules on their surfaces. It was hypothesized that cell signal networks coupled with gene regulation networks may control morphogenesis.[4] Cell signaling networks also play very important roles in brain and neural systems. In the so called neural network model one considers the connection of neurons as a network. However in reality subneuronal (intracellular) signal pathways or neuronal transmitters also play essential roles in information processing and in learning.[2]

Before 1980 few such cell signal molecules were known. In 1980s however many signal molecules had been discovered. It was also discovered that many of these molecules belong to a special family of enzyme proteins called protein kinase that phosphorylate other proteins. In 1990s more signal molecules have been found and even three dimensional structures of some of these important kinase proteins or their parts were determined.[5,6] Though the entire set of large and complex cell signaling networks of a cell has not yet completely unveiled, quite a number of characteristic features of such signal molecules and networks have already been unveiled. Moreover the pace of accumulation of data and knowledge in this research field is ever increasing so that it now becomes very difficult for an individual researcher to comprehend and correlate enormous amount of the existing information.

About two years ago we started to develop a computerized data and knowledge base for cell signaling networks. The eventual goal of this project is to catalog all networks and their signal molecules in order to model cells as information processing units, and to make more realistic and quantitative models for development, differentiation, morphogenesis, carcinogenesis, neural information processing, and learning. In this paper we shall introduce the first prototype of the system.

2 Cell Signaling Networks Database (CSNDB)

2.1 The Design Concept

Figure 1 explains over all concept of our Cell Signaling Networks Database (CSNDB). The eventual goal of the system is to simulate responses of a cell to chemical agents which have their origin either in the living system or outside. The key idea is that such a simulation system must be based on sound knowledge on inter molecular interactions among xenobiotic and biotic ligands and their complementary signal molecules in the cell. We also postulate that such

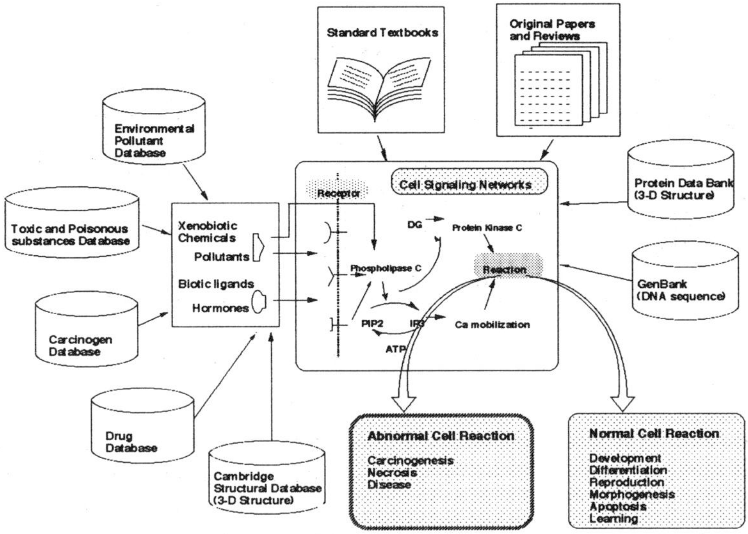

Figure 1: Concept of Cell Signaling Networks Database.

a simulation system should give good insight into many important multicellular biological phenomena such as development, carcinogenesis, apoptosis, and learning.

Although the eventual goal is a remote one, there already exist a large amount of data and knowledge on signal molecules to start considering such a simulator. The simulator basically consists of networks of interacting xenobiotic chemicals and biomolecules.

2.2 Information Contents and the Sources

The computer system is designed to provide available information on :

- xenobiotic chemical agents such as drugs and environmental pollutants
- protein groups such as cell surface receptors and intracellular enzymes that transduce cell signals
- ligands for the receptors on the cell membrane and for the other intracellular signal molecules
- well known signal pathways
- enzyme families
- transcriptional factors.

For each molecule we associate chemical (identification) names, synonyms,

<table>
<tr><th rowspan="2">target tissue</th><th rowspan="2">extracellular signals</th><th colspan="3">membrane signal transduction</th><th colspan="2">inner membrane</th><th rowspan="2">major response</th><th rowspan="2">comment</th></tr>
<tr><th>receptor</th><th>mediating factor</th><th>first effector</th><th>second messenger</th><th>second effector</th></tr>
<tr><td rowspan="2">thyriod gland
adrenal cortex
ovary
muscle
bone
heart
liver
kidney
fat</td><td rowspan="2">thyroid-stimulating hormone(TSH)
adrenocorticotropic hormone(ACTH)
luteinizing hormone(LH)
adrenaline
parathormone
adrenaline
glucagon
vasopressin
adrenaline,ACTH glucagon,TSH</td><td>β-adrenergic receptor</td><td>G protein Gs (αs, $\beta\gamma$)</td><td>adenylyl cyclase</td><td>cAMP ↑</td><td>A kinase chain reaction</td><td rowspan="2">• thyroid hormone synthesis and secretion
• cortisol secretion
• progesterone secretion
• glycogen breakdown
• bone resorption
• increase in heart rate and force of contraction
• glycogen breakdown
• water resorption
• triglyceride breakdown</td><td rowspan="2">cholera toxin</td></tr>
<tr><td>α-adrenergic receptor</td><td>Gi(αi, $\beta\gamma$)</td><td>adenylyl cyclase (inhibition)</td><td>cAMP ↓</td><td>A kinase chain reaction (inhibition)</td></tr>
<tr><td></td><td>acetylcholine</td><td>Muscarinic AChR</td><td>Gi(αs, β, γ)</td><td>ion channel K+ channel</td><td></td><td></td><td></td><td>smell vision</td></tr>
</table>

Figure 2: Cell signal pathways in a table form.

structural data and other necessary information. In case of proteins additional information such as families, functions, sequences, secondary structures, 3-D structures, binding sites, activators, and inhibitors are also included.

Initially knowledge on cell signaling networks particularly signal molecules were taken from standard text books and then organized into tables like Figure 2.[1,2] These tables represent only skeletal frame of the networks. More detailed knowledge were added from reviews and original papers in order to provide the first version of the data and knowledge base for the prototype system.[7–12]

In addition, we referred to various computerized databases for chemicals and biomolecules. Chemical databases were used to search chemical names, synonyms, physicochemical properties, structural data and biological effects including animal tests. Some frequently referred databases were Cambridge Structural Data of X-ray crystallography (CSD), our Structure Database for Carcinogenesis, Mutagens based on IARC monographs, US National Toxicology Program, and our Structural Database for Drugs. As for biomoleculer databases are concerned, standard nucleic acid and protein databases such as PDB or GenBank are accessible by the Internet. Most of these databases were also available at our inhouse system.

2.3 Signal Pathways

A typical signal pathway starts from binding of a ligand to a receptor on a cell. This binding stimulates a signal transduction protein, like G-protein that triggers a series of reactions among intracellular signal molecules in the cytoplasm. These reaction signals are then transferred into the nucleus and activate the transcription factors that regulate gene expressions. These processes control the kind of proteins to be synthesized from the DNA (genes). The product proteins eventually determine the overall cell response to the ligand.

A quite number of signal pathways have been discovered as byproducts of cancer and other disease studies. What is interesting is that though there are wide variety of specific receptors corresponding to chemical ligands there exist fewer common reaction pathways and their associated signal molecules that transfer the signals to gene expression process. In other words different stimulus often utilizes similar pathways that result different responses. We call these important signal pathways as characteristic cell signaling pathways and put high priority on accumulating their data and knowledge, which form the core of the CSNDB.

The CSNDB has a hypertext structure. Figure 3 shows a typical characteristic pathway. The diagram represents the skeletal signal flow. Each arrow corresponds to a signal from one signal molecule to the next one, while each node corresponds to the access point to the window that contains deeper information on this signal molecule. If one clicks an arrow in a signal flow diagram, the modes of molecular interactions are presented by a new window (object), and if one clicks a node in the diagram, then the next window that contains deeper information about this signal molecule is displayed on the screen. One can click some of the items in this window to open further the next window in order to get more information about the item, such as sequences or 3-D structures.

As a prototype system the present CSNDB contains the following characteristic signal pathways: thyroid stimulated hormone, steroid super family, cAMP and PKA, IP3 and PKC, adaptor molecules, ras / raf and MAPK, cytokine, and JAK / STAT.

Some signal pathways can not be represented by a simple one way flow diagram. There are pathways that contain feedback loops and there often exist crosstalks between two or among several pathways. These information are also stored under the arrow objects. (See Figure 4.)

2.4 An Important Signal Molecule, PKC

Usually signal molecules are either small molecules like xenobiotic chemical agents or much larger molecules like biomolecular derivatives, peptides, and proteins. For proteins one may want to know detailed information on their structure and function, which include the gene site on the genome, exon and intron sites, nucleic acid sequence of the gene, sequence homologs to other species, amino acid sequence of the protein, secondary and 3-D structure, domains and functional motifs, the phosphorylation sites, activator and inhibitor binding sites, and mutation sites on the gene.

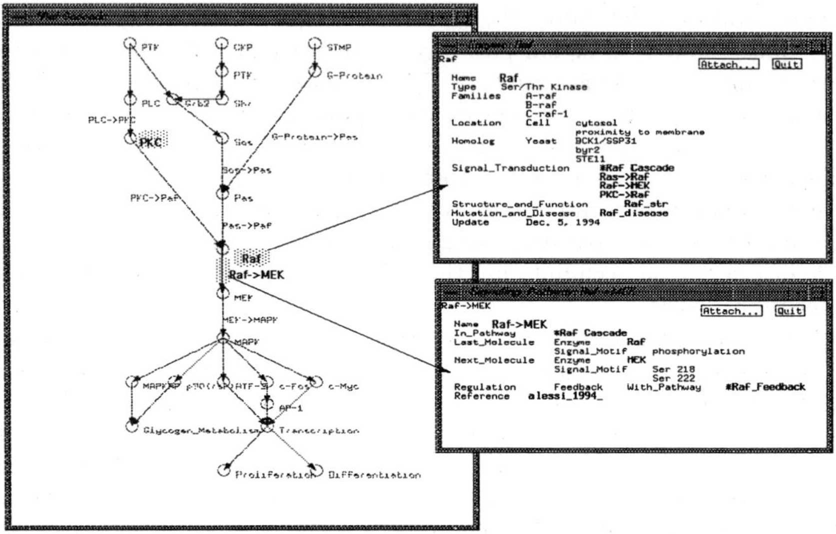

Figure 3: Example of signal pathways diagram (left) and its corresponding deeper information(right).

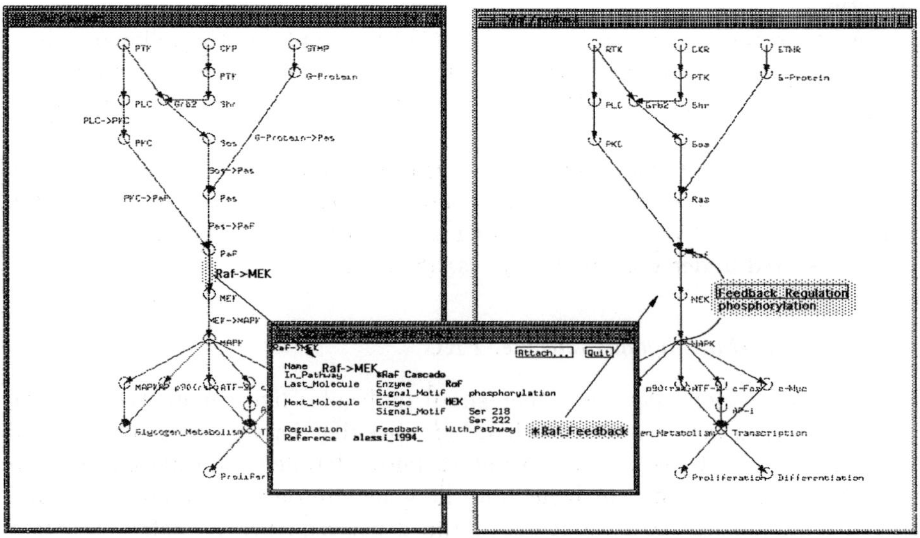

Figure 4: A pathway and its feedback regulation are connected with information under the arrow object.

Figure 5 shows a window of a signal molecule PKC (Protein Kinase C or C-Kinase), which belongs to the class "Enzyme", and which appears also in the signal pathway,"Raf_Cascade"(Figure 3). PKC is one of the most important signal transduction proteins, and enormous amount of experimental data have already been accumulated. We therefore have tried to integrate detailed data and knowledge on PKC for building a comprehensive data and knowledge base model for such an important signal protein.

Some PKC activators are well known tumor promoters, and quite large number of inhibitors to them are also known. It is also known that lipid molecules or membrane molecules are required for PKC activation. Various structure activity relation type studies have been carried out on these collaborative molecules using synthesized chemicals. Knowledge accumulated by these studies are also organized into the CSNDB. Figure 5 shows genetic sequences of PKC and a schematic view of the PKC structure. This picture was drawn by one of the authors and displayed using a share ware picture tool (software) XV. Figure 5 also gives a 3-D view of the catalytic domain structure of PKC. Figure 6 gives the molecular image of staurosporine, which is an inhibitor of PKC. These images which were drawn by using a free molecular graphics software RASMOL could be rotated in real time by using a mouth.

3 System Implementatioan

3.1 Software and Hardware

At the beginning we thought that any standard relational database management system can easily handle data and knowledge represented by tables like Figure 2. However tables like Figure 2 are only very rough sketches of realistic reaction networks. More detailed and deeper information must associate with each item in these tables.

Unfortunately however some of these information may not yet be supplied by experimentalists. Thus if we use a relational database it would become a very sparse one. Moreover structures of data and knowledge may be adjusted or will even drastically be revised when new experimental facts are discovered. Thus we decided to use an object oriented database management system, ACEDB,[16] as the base system for implementing the prototype of the CSNDB.

In ACEDB, the data are organized into objects and classes. Objects are items and classes are sets of similar objects. The hierarchical relations of classes are represented by tree models. A tree model contains two types of entries : tags and data. Tags are labels which define data. The data are numbers,

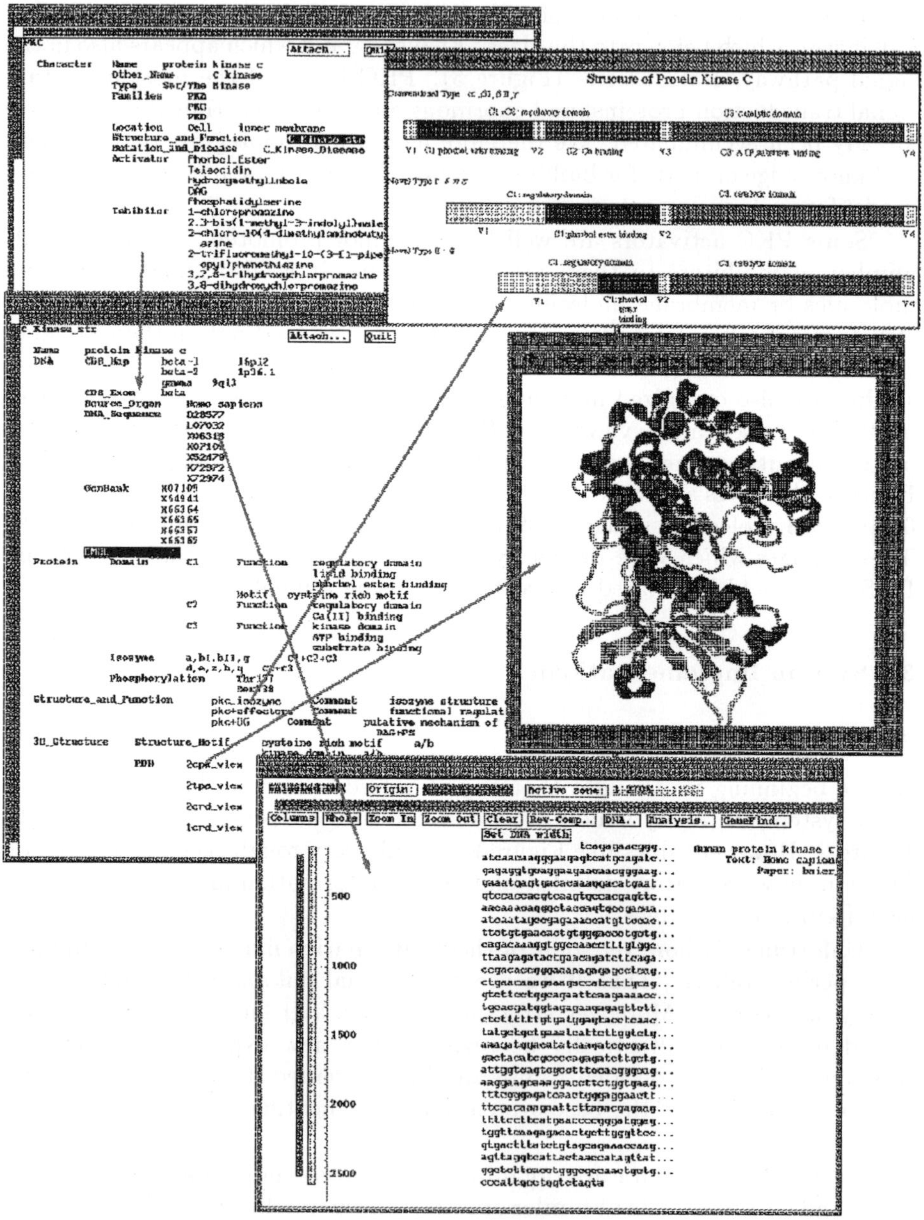

Figure 5: Data and knowledge items are cross referenced in windows.

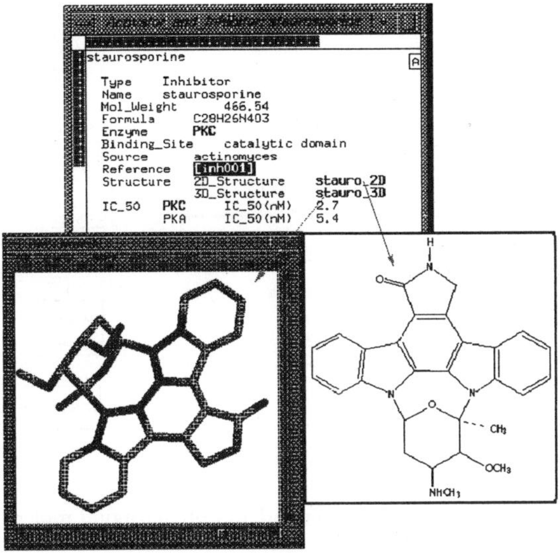

Figure 6: Examples of inhibitor of PKC (staurosporine).

sentences, or pointers to other objects. Pointers are represented by names for the relevant classes. In the case of numbers or sentences their data types are specified by such code words as Int, Float, Text, and Greek. Long texts or raw DNA sequences are handled as arrays of characters. The corresponding annotations are handled in an object with the same name allowing the full flexibility of the tree model structure.

The CSNDB contains seven different types of objects. They are (1) signal transduction molecules, (2) signal transduction pathways associated with these molecules, (3) genomic data of signal molecule, (4) structural data, (5) mutational data, (6) associated chemicals (ligands), and (7) bibliographical data. Objects of signal transduction molecules are grouped into eight classes : hormones, cytokines, receptors, ion channels, effectors, enzymes, messengers, and transcriptional factors (Figure 7A). Each object has its appropriate model format which represents its biological characters. These molecules compose complex networks, and the networks are represented both as signal transduction pathway objects and as pointer type entries in the tree model structure.

ACEDB runs on X-window environment on UNIX workstations. Using ACEDB, we have implemented the prototype system on our SUN Sparc workstations, Silicon Graphics Indigo, and Indy workstations. These workstations are linked to our local area network through ethernet which is linked to the

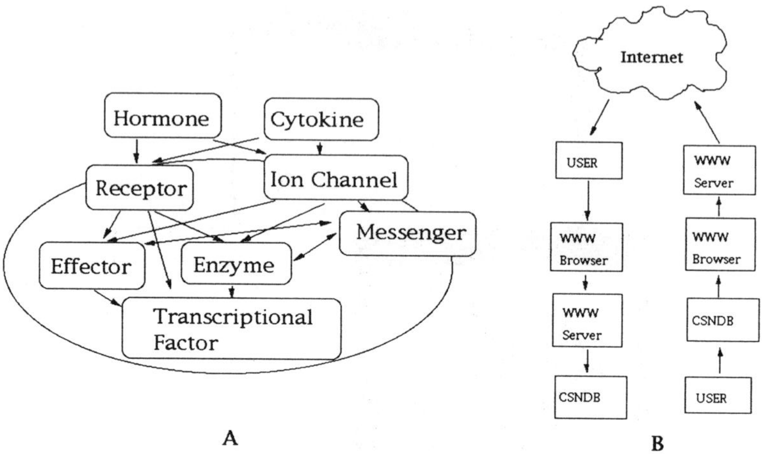

Figure 7: A: Schematic diagram of interdependency of the signal molecules. Each molecule is treated as an object. B: Interfaces between the Internet and the CSNDB.

wide area network, the Internet.

One of the characteristic feature of the CSNDB is that it can handle not only numbers and sentences but also diagrams like signal flow pathways, pictures like secondary structures and structural diagrams of proteins, and 3-D molecular images. We used Metab display.[13] for signal flow diagram, XV[14] for pictures, and RASMOL[15] for molecular graphics. Some of these software were obtained from the Internet.

In general an object oriented database is flexible for changing its data structure, but this strength often makes users difficult to comprehend relations among data items and to comprehend over all data structure. In order to overcome this weakness we utilized the search function that already exists in ACEDB. The data search is carried out by user typed commands that consist of standard and, or, and not operation.

For help function we used (internal) MOSAIC which was originally built in ACEDB. When necessary the user click the help menu in a window to call internal MOSAIC to get explanation and its associated information of the data item.

3.2 Network Environment

From the very beginning the CSNDB has been developed by keeping the Internet environment in mind. ACEDB has a good mailing group that consists of developers and users. We frequently sent questions to this group and got useful information on how to use ACEDB.

The CSNDB can be used in stand alone mode, but can also be used from remote place by accessing to our network WWW server. The CSNDB contains a WWW browser where network addresses (like http) of other useful database are stored. If a user wants to search outside, through the Internet cyberspace he can do so with the internal browser (Figure 7B).

4 Discussion

Development of the prototype system started less than a year ago, and the system is still in its embryonic stage. However we have already learned that our approach is reasonable and ACEDB is very suitable as the base system for the CSNDB. In particular the flexibility of the object oriented database makes us ease to define new data classes as objects, and to implement a hypertext type database. ACEDB also enabled us to handle diagrams, pictures, and 3-D images, so that we consider the CSNDB as a hypermedia database. It should also be emphasized that support from ACEDB mailing group was indispensable for CSNDB.

Right now full function of the CSNDB is only available as stand alone systems, but the Internet (WWW server) version will be ready soon. We are also trying to develop a more sophisticated interface between WWW (browser) and ACEDB in order to improve capabilities of user interfaces.

5 Conclusion

We have developed a prototype of data and knowledge base for cell signaling networks. We are planning to put it on our Internet WWW server for open usage, however before doing it more data on receptors, ligands, inhibitors, and signal pathways should be added to the system. Once the system gets into that phase, we may expect to obtain comments and new data from other researchers through the Internet.

Though the CSNDB was originally designed to be a basic and common tool for wide range of researchers, it could also be used as personal systems by experimentalists who are studying specific receptors, cell signal transduction and other proteins. They may add their own data to the system or fill in their own data to the empty CSNDB.

The system will be useful for modeling receptors and cell signaling mechanisms, cell differentiation, development, carcinogenesis, and learning. The system would be also useful for analyzing bioactivities of xenobiotic chemicals.

Acknowledgments

The authors thank to Dr. J. Thierry-Mieg and Dr. R. Durbin who started the ACEDB project and are very helpful for our work. The authors also thank to their colleagues ; Ms. M. Yukawa and Ms. S. Hasegawa who inputted PKC inhibitor data and drug data respectively, Drs. T. Nakano and K.Nakata who provided the Internet environment and the chemical and biological molecular databases, and Drs. J. Miwa and T. Sano of Fundamental Research Laboratories of NEC Co. and Dr. T. Furuya of Institute for Drug Discovery Research of Yamanouchi Pharmaceutical Co., Ltd. for useful discussion. This work was partly supported by Japanese Human Science Foundation Research Aid Program and by Science Research Promotion Fund from Science and Technology Agency.

References

1. Alberts, B. *et al.*,*Molecular Biology of The Cell*(3rd ed.) (Garland Pub, 1994).
2. Kandel, E. R. *et al.*, *Primciples of Neural Science*(3rd ed.) (Elsevier, Amsterdam, 1992).
3. Hunter, T., *Cell* **50**, 823 (1987).
4. Edelman, G. M., *Topobiology* (Basic Books, NY, 1992).
5. Knighton, D. R. *et al.*,*Science* **253**, 407 (1991).
6. De Bondt, H. L. *et al.*,*Nature* **363**, 595 (1993).
7. Krebs,E. G. *et al. Trends Biochem. Sci.* **19**, 439 (1994).
8. Marshall, C. J. *et al. Cell* **80**, 179 (1995).
9. Pennington, S. R. *et al. PROTEIN PROFILE* **1**, 169 (1994).
10. Vanhoutte, P.M. *et al. Pharmacological Review* **44**, 121 (1992).
11. Edited by Fleeger, C. A.,*USAN and the USP dictionary of drug names* (MD, USA: United States Pharmacopoeial Convention, Inc., 1993).
12. Edited by Society of Japanese Pharmacopoeia, *JAN: Japanese accepted names for pharmacoceuticals* (Tokyo, JAPAN: TAKUJI NIPPO, Ltd., 1992, and 1993).
13. Metab display is a software which runs in ACEDB environment developed by Dr.Stan Letovsky of GDB group of Johns Hopkins University.
14. XV can be obtained from ftp.cis.upenn.edu.
15. RASMOL can be obtained from the following anonymous ftp site: colonsay.dcs.ed.ac.uk.
16. The softwares and documents of ACEDB and related topics were obtained from the following Internet sites:

(a) Thierry-Mieg, J. and Durbin, R. acedbA C.elegans Database: I. Users Guide., II.Installation Guide., III. Configuration Guide.
ftp://ncbi.nlm.nih.gov/repository/acedb, 1991.
(b) Gaeta, B. Tutorial Introduction to ACeDB.
ftp://ncbi.nlm.nih.gov/repository/acedb/ace2, 1993.
(c) Cartinhour, S. Exploring Models Design and Structure.
http://probe.nalusda.gov:8000/acedocs/exploring/toc_modes.html, 1994.
(d) O'Callaghan, M. A Guide to Models and ace Files.
http://probe.nalusda.gov:8000/acedocs/models_guide.html, 1993.
(e) Kirk, J. How to install Guy Ducoux's WWW-ACEDB server.
http://keck.tamu.edu/staff/ace_mosaic_howto.html.

MODELLING THE EMERGENCE OF COREGULATED PROTEINS IN BIOLOGICAL REGULATION NETWORKS

E.CHIVA, P.TARROUX
Ecole Normale Supérieure, Groupe de BioInformatique, URA 686 CNRS, 46 rue d'Ulm, 75230 Paris Cedex 05, FRANCE

1 Introduction

In the past ten to twenty years, the spectacular development of molecular biology techniques and the growing interest for developmental biology, have allowed to investigate very deep aspects of cell functioning. Especially, a large amount of data yielded new insights into the mechanisms responsible for controlling gene expression. However, the questions about the global control mechanisms and dynamics of intracellular regulation networks still remain. Indeed, the experimental study of biological regulation networks is, unfortunately, nearly impossible since known networks appear to be very large. Therefore, an experimental approach integrating a precise analysis of the kinetics of such systems appears to be out of range. In lack of a global comprehensive theory of both transcriptional and post-translational regulation, simulation provides the suitable method to study such complex and highly non-linear systems, for which no analytic approach is available. Indeed, to understand both the cell control mechanisms and their disruption, *e.g.*, in the case of neoplasic growth, makes the research on this topic an unavoidable requirement for modern biology.

Regulation systems have evolved both in prokaryotic and eukaryotic organisms. Considering the prokaryotes, such systems are responsible for the adaptation to complex and unstable environments, in which the food supplies, the pH and the inducers can vary at any time. Considering the eukaryotic cell, the homeostatic stability that characterizes the internal environment of a pluricellular organism leads to a slightly different problem. Yet, such a cell must be able to regulate its growth, to remain stable with respect to unsignificant signals, to respond to pertinent signals such as growth factors, and to differentiate accordingly in a number of different ways. A good example of such a situation is provided by the hematopoietic system, in which a stem cell can give rise to three main lineages, each of them comprising two to four different cell types, under the combined actions of cytokines.[1] In spite of the differences between both types of organisms, the basic requirements of regulation systems appear to be similar: (i) the mechanisms of control should be relatively simple,[15,7] and universal (such as key proteins, ubiquitous relays and feedback loops), (ii) the system should remain stable to small perturbations, thus exhibiting homeostasy, (iii) it should respond in an adapted way to significant signals in its environment and (iv) a given cell type should correspond to a stable state of the system. These properties result from the common mechanisms that have emerged both in prokaryotes and eukaryotes during evolution, both organisms sharing the same genetic code and the same basic gene controls. Mathematical modeling of biological regulation systems has a long history starting with the classical work of Kauffman,[5,6] who described gene regulation networks as autonomous boolean systems. In this connectionnist metaphor, in which the protein classes are defined as nodes of a network of automata, a differentiated cell state is identified as an attractor of the network (limit cycle or fixed point). He showed that such systems could exhibit either a chaotic or an ordered regime. In this last case, he identified order parameters, such as the connectivity between the automata, that are responsible for the emergence of 'frozen cores' of elements that affect the propagation of information (*percolation*) inside the network. He also showed that if such systems can evolve 'at the edge of chaos'[10] then they become preferentiate targets for natural selection. In spite of its simplicity, his model has thus allowed to ponder critical features of biological regulation systems. However, neither Kauffman's model nor more 'traditional' models directly built from experimental results[12] have yet allowed to unveil how the cell environment determines the particular structure of biological regulation networks, and what does this structure look like. Indeed, on one hand, boolean models do not depict the complexity of the regulations that occur, which prevents the model to be directly compared with experimental results. On the otehr hand, models directly derived from experimental data do not allow any generalization for they apply only to the particular situations from which they were built.

To overcome these difficulties, we designed a global model that explicitly accounts for both the evolution of the architecture of protein regulation networks and the emerging dynamics of the resulting systems with respect to their environmental context.[2,3] In this paper, we will focus on the interpretation of this model in terms of metabolism and protein–to–protein as well as protein–to-gene conformational interactions, and, beyond the technical considerations, on the biological interpretations of the results we obtained.

2 The Model

We define a simplified regulation network as a set of interconnected units. Each unit corresponds to an intracellular protein. The interconnections are weighted to account for the influences between the proteins, that is, how a given protein concentration will act on gene promoters to induce or repress other proteins synthesis. The interconnection set defines the weight matrix, W.

2.1 The encoding of the network

Let us consider a discrete-time network of n continuous–state automata $u_i, i \in \{1, ..., n\}$ connected by n^2 non–symmetric couplings $w_{ij} \in [-1; 1]$. A state of the network represents an array of protein concentrations $\{x_1, x_2, ..., x_n\}$. The dynamics of the network is then computed from the equation:

$$x_i(t) = \mathcal{H}[\sum_{j=1}^{n} w_{ij} x_j(t-1)] \tag{1}$$

the function $\mathcal{H}$ being classically defined as: $\mathcal{H}(.) = \frac{1}{1+e^{-(.)/T}}$. Each unit i is assigned a gene G_i, represented by a vector of size ν: $G_i \in \{-1, 1\}^{\nu}$. To integrate as well as possible the biological mechanisms that are responsible for the emergence of adapted regulation systems, the design of the network is achieved through the use of a method based upon an evolutionary approach. In the present implementation, a given regulation network is encoded into a genome (one chromosome) and a population of chromosomes is submitted to evolutionary changes. In contrast to other widely used encoding algorithms,[16] the weight matrix W is not directly computed from the chromosome (which is defined by a family of n genes G_i) but is obtained from an indirect coding. Indeed, each gene G_i is projected onto two linearly independent subspaces, which defines two vectors P_i^+ and P_i^-:

$$P_i^+ = \Phi^+ G_i$$

$$P_i^- = \Phi^- G_i$$

where Φ^+ and Φ^- are two constant $\frac{\nu}{2} \times \nu$ matrices, the structure of which ($\Phi_{(p,q)} \in \{1, -1, 0\}$) defines the contribution of each element of a gene G_i to the resulting 'protein conformation'. Then, the weights between unit i and unit j (resp. between units j and i) are computed from the vectors P_i^+ and P_j^- (resp. from the vectors P_j^+ and P_i^-) as follows:

$$w_{ij} = \frac{1}{\kappa} \sum_{p=1}^{\nu/2} \sum_{q=1}^{\nu/2} (P_i^+(p) - P_j^-(q)) \tag{2}$$

where κ is an appropriate normalization factor. We use a steady-state genetic algorithm (GA) in which only one mutational or recombinational event can occur per timestep; the details are given in Chiva and Tarroux.[2]

The fitness is chosen to account for the ability of biological regulation networks to respond to pertinent signals of the environment. These signals are imposed onto the network by 'environmental units' which do not have back–connections from the network units (their fan–in is zero). The fitness assesses the system stability to transient perturbations and its reactivity to lasting perturbations. An ideal network should show a divergent behavior facing p constant clamps of the environmental units, and a constant behavior facing p transient clamps of the same units. The expression of the fitness $\mathcal{F} \in [-1;1]$ is given by eq.3:

$$\mathcal{F} = \frac{4}{p \cdot n}[\sum_{k=1}^{p} ||x_f^k(p) - x_m(p)||^2 - \sum_{k=1}^{p} ||y_f^k(p) - y_m(p)||^2] \tag{3}$$

where x_f^k represents the final state after the kth continuous clamp exerted, x_m the mean vector of the corresponding p final states, y_f^k the final state after the kth transient clamp exerted, and y_m the mean vector of the corresponding p final states. This criterion allows to account for inducer maintenance, a phenomenon early described.[4]

2.2 *Modelling gene induction*

To simulate the interactions between proteins, we assume that the effects of the other protein concentrations on the evolution of the concentration of a given protein are additive and independent. To study the influence of one given unit $(j, i \neq j)$ on another unit i, let us consider that the other units remain constant at time $t-1$. In these conditions, it is easy to show that the function that gives x_i with respect to x_j is a sigmoid the threshold of which is given by:

$$\Theta_i = -\sum_{k}^{k \neq j} w_{ik} x_k \tag{4}$$

The biological interpretation is straightforward: for one given unit, the activation threshold depends on the network state (that is, the states of the other units). Thus, since the effects of the other proteins determine the inducibility of a given protein, the threshold Θ_i is implicitly coded.

2.3 *Modelling Protein folding*

The non–straigthforward encoding of the weights has been chosen as an abstract model that mimics protein spatial interactions. In this prospect, a 'phenotypic' vector $P_i = (P_i^+, P_i^-)$ represents a given conformation of the polypeptidic chain corresponding to the protein i, in the space of conformations defined by the rules imposed by the two matrices Φ^+ and Φ^- (the protein conformations in this case are fixed by the products Φ^+G_i and Φ^-G_i). The contribution of each element of the gene G_i to the conformation of the protein is defined by the connectivity of the two Φ matrices (*i.e.*, the number of non–null elements per row) [a]. Each gene locus equally contributes to each element of the corresponding P vector. The differences between these elements are obtained through mutual correlation of the effects of different loci according to their informational contents. On the contrary, the information from one locus is potentially spread out over all the elements of P. This model thus truly represents a 'protein folding'. The information the vector P_i bears is twofold: it is both the conformation of protein i and the structure of its gene promoter. Indeed, the weights between unit i and the other units are obtained from the vector P_i. The subvector P_i^+ represents the elements that are responsible for the control of the expression of the protein i by the other proteins: it thus corresponds to the elements of its promoter. The subvector P_i^- represents the elements that are responsible for the control of the expression of the other proteins by the protein i: hence, it corresponds to the conformational sites of protein i that are capable of binding other gene promoters.

The model thus explicitly integrates mechanisms that are part of actual biological regulation networks, though keeping a sufficient level of abstraction to remain computationally tractable. The originality of the weight encoding method allows to account for nonlinear relationships between genotype and phenotype, and for the pleiotropic effects of point mutations and crossing–overs. The strength of gene promoters and protein folding, two basic mechanisms involved in protein mutual regulation, are explicitly represented. At last, the evolutionary approach for the design of the network allows to represent the influence of the extracellular environment on genes and proteins organization.

3 Results

3.1 Evolution of the fitness

We identified the critical parameters that mainly condition the evolution of the fitness, because they influence either the convergence of the GA or the encoding of the weight matrix:

1. the connectivity of the Φ^+ and Φ^- matrices: k_Φ. For high k_Φ values ($k_\Phi = 80\%$), any point change in one given gene will affect every weight in the system. Therefore, we chose to use lower values: $k_\Phi \simeq 35\%$.

---

[a] it was set identical for both matrices and noted k_Φ

2. the length of the genes: ν. For small ν values, there is a strong chance of partial genetic drift: all genes converge roughly toward equivalent structures, preventing the fitness to reach very high values. On the other hand, increasing gene size has two drawbacks: the exponential increase of the search hyperspace and a persistent bias in the weight distribution since the length of the genes directly affects the discretization of weights. In this work, the parameter ν ranges from 20 to 50.

3. the temperature T. The slope of the transition function is $1/4T$. High temperature values make the slope of the transition function $\mathcal{H}$ very soft, which prevents the units to reach high state values for low energy values. Since this directly affects the computation of the fitness (which directly integrates the states of the units), we introduced an amplification factor in the computation of the weights (eq.2) to attenuate this effect. This allows to use high temperature values ($T \simeq 1.0$) and still permits the units to reach high states.

We combined these three critical parameters (as well as the basic genetic rates) to attain higher fitness values than the ones previously obtained.[3] Figure 1 presents the evolution of the fitness in a population of 40 chromosomes, each coding for a 40–units network[b]

3.2 Evolution of the network structure

A basic question is to determine how a fitness increase affects the weights distribution.

Initially, low–fitness matrices exhibit a gaussian weight distribution, as a consequence of the random initialization. This distribution is clearly visible when long genes are used (fig.2–a), because, in this case, the discretization of the weight values is weak. As the networks evolve and become able to categorize the environmental inputs, the weight distribution becomes increasingly heterogeneous and the gaussian pattern disappears, as we see from the figure 2. This truly shows the weight matrix undergoes drastic structural changes, which appear in high–fitness network matrices. This heterogeneous distribution seems to represent a necessary condition to reach high fitness values, because it appears in every simulation. We investigated the modifications of the weight matrix in relation to the fitness of the network. As we observed in

[b]The amplification factor is set to 4.0. The other settings are $\nu = 20$, mutation rate=0.5, evaluation rate=0.8, c.o. swap=0.5, c.o. rate=0.4, p=10, T=1.0, $k_\Phi = 7$. 5 units were clamped and the length of the fitness evaluation test is 20. All these parameters are explained in Chiva and Tarroux.[3]

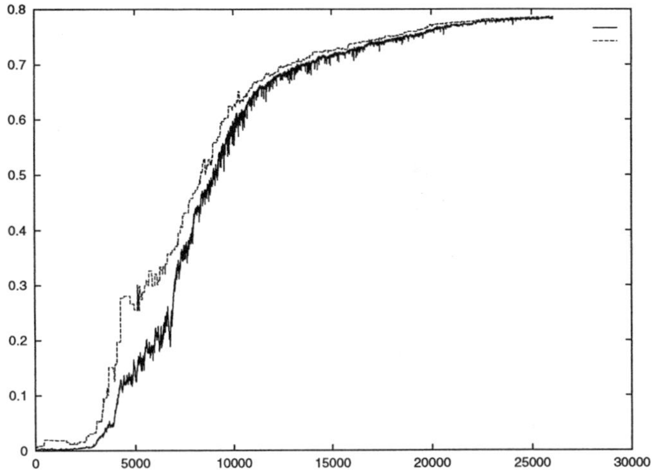

Figure 1: Evolution of the fitness of a 40–units network. The solid line represents the mean fitness and the dashed line the maximal fitness - The vertical axis corresponds to the fitness value and the horizontal axis to the time step

previous work,[3] the network structurates hierarchically (fig.3–b). The system converges towards a situation in which most units present a homogeneous fan–out greater than their fan–in. Then, as the fitness increases, some elements individualize that are strongly connected to the other units of the network. For very high fitness values ($\mathcal{F} = 0.8$), critical units appear, which exhibit both a high fan–in and a high fan–out: these units implement 'receptors' which are extremely sensitive to the signals presented by the other units, and which propagate this information to the entire network. Low–fitness matrices do not exhibit such a structuration (fig.3–a): the units fan–in and fan–out appear to be balanced.

3.3 Principal Component Analysis

Previous studies[14,13] have shown that Principal Component Analysis[8] (PCA) is a powerful tool to examine the structure of the protein expression data in actual biological networks. To investigate the behavior of our present model compared to biological networks, we employed the same kind of analysis. Our use of the PCA is based on the following considerations: each specific environmental condition leads the network to a given state. From a series of such conditions mimicked by clamping a few units, one can record the correspond-

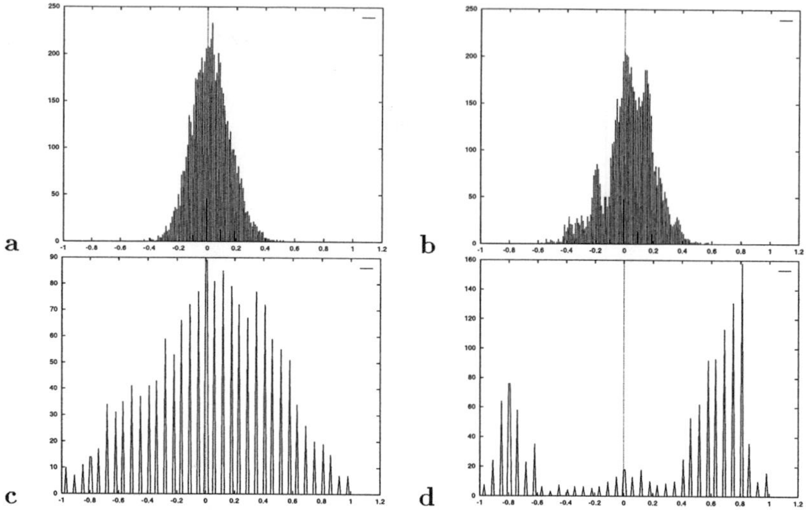

Figure 2: a) gaussian distribution of the weights of a low-fitness 80–units network and b) weights distribution of a 0.7 fitness network. The length of the genes is 50. The same effect is obtained for small networks: c) gaussian distribution of the weights of a low-fitness 40–units network and d) weights distribution of a 0.8 fitness network. The length of the genes is 20.

ing network states in a data matrix, the rows of which represent the units, and the columns the environmental constraints. This matrix carries out the information concerning the way each external constraint acts on the expression of each unit. In biological data, we found that the actual network is organized in such a way that relatively independent groups of proteins are involved in the response to independent constraints. This result led us to propose the existence of a particular structure in expression networks: the regulation group.[13] PCA presents the interest of combining the initial constraints imposed onto the network to define a new set of orthogonal conditions according to which it is easier to study the reaction of the system. Since we successfully applied this technique to the analysis of biological networks, we employed it to investigate the properties of the model.

We performed a PCA test in which the final states corresponding to 10 different continuous clamps of the first 7 units of an 80–units network were stored after 20 iterations[c]. From the results (fig.4) two main conclusions can be drawn.

1. The PCA of a low–fitness network shows a Guttman effect (fig.4–a) which

[c]the clamps we performed were not the ones that were exerted to evaluate the fitness

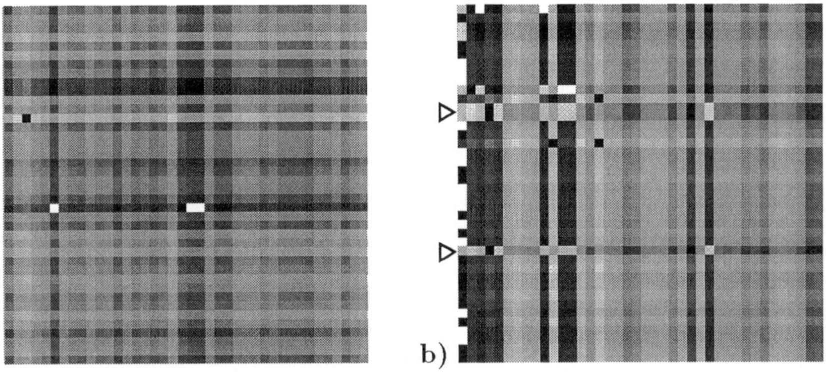

a) b)

Figure 3: Structure of a) a low–fitness matrix ($\mathcal{F} = 10^{-4}$) and b) a high–fitness matrix ($\mathcal{F} = 0.5$). Each square in the image corresponds to a weight in the matrix. Bright (resp. dark) squares correspond to highly positive (resp. negative) weights. Grey squares correspond to low weights. The arrows indicate the weights corresponding to some of the critical units

is usually observed when the responses of a system are functionally correlated. In our case, this effect can be related to the fact that a low–fitness network remains in the same attractor whatever the clamp is. A scattering of the projection of the variables (the 10 different clamp conditions) into the space of the factors would indicate the differences in the reactivity of the network according to the variations of the environment. Fig.4–c clearly shows that low–fitness networks are not capable of presenting different dynamical responses to the perturbations: the ten variables are aggregated. Therefore, the clamps are not discriminant in this case.

2. On the opposite, high–fitness networks seem capable of exhibiting distinct responses to the different clamps: the aggregation disappears, as shown in fig.4–d. Also, from fig.4-b, we can assume the adapted system to be able to react according to at least two major dynamic modes: the functional correlations between the factors tend to disappear. Using the program CONSTEL for cluster analysis,[11] we were able to demonstrate that the units indeed segregate into five separate groups which are subparts of the two major groups.

Thus, from these results, we can postulate that the emergence of primitive groups of coregulated proteins is a direct consequence of the structuration of the network.

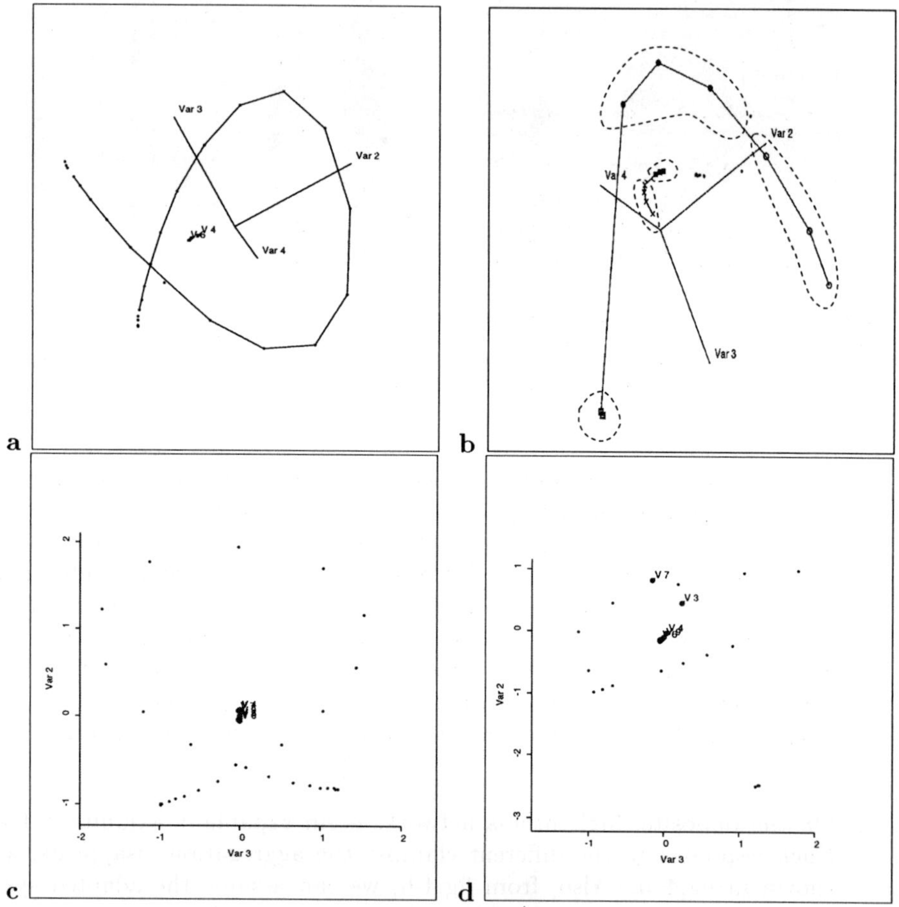

Figure 4: PCA analysis showing a) the Guttman effect observed for a low–fitness network, b) the emergence of two distinct response modes in a high–fitness network, each mode is represented by a trajectory, the five subgroups are visualized using different glyphs and dashed lines c) the clustering of the variables corresponding to the initial clamp set in a low–fitness network and d) the dispersion of the variables observed when the fitness increases ($\mathcal{F} = 0.8$). The size of the network is 80, the length of the genes is 50.

4 Discussion and Conclusion

This model integrates the many mechanisms that are responsible for both transcriptional and post–translational regulation: the existence of an implicit threshold determined by the activities of the other genes, the accounting for cooperative phenomena, protein folding, *cis*–and *trans*–activation, pleiotropy and non–linear relations between genotype and phenotype. Gene induction is modeled from a certain level of abstraction, which allows to get rid of the questions related to the efficacy of the transcription complex stability or to the number of individual gene copies.[9] Using this model, albeit very simplified when compared to the complexity of actual regulation networks, we were still able to study the emergence of specialized structures that enable the cell to react to different combinations of external inducers.

4.1 Interpreting the biased structure of high–fitness networks

Indeed, at the end of the evolutionary process, we observed the individualization of units that both send strong connections to the other units and receive strong back–connections from the entire network (fig.3). Such elements can be related to the *master–switch* genes that possibly control the expression of a number of other genes.[17] This could be a mean to implement internal relays. In the cell, the existence of ubiquitous relays could be a critical factor allowing the signal transmission between two regulation subnetworks. These relays are not related to specific functions: they simply ensure that extra- or intracellular significant signals will indeed affect the dynamics of the entire system, possibly through processes such as cascade propagation (e.g. the G–proteins sytems). The emergence of internal relays such as the key units that have appeared would be a good way to make the system enter very different dynamic regimes according to environmental signals it is confronted to. The weight matrix also exhibits strong weight values on the diagonal for some elements, showing that auto–inductive units are selected.

Thus, we have shown that adapted networks present a strongly biased structure, which might explain why pleiotropic processes such as gene addition are destructive. This structuration has been obtained only under the constraint that the regulation network exhibits as many different responses as possible - we have no information about how a network would be organized if the constraints were different. Still, our results tend to confirm that the efficacy of any genic control system depends on its degree of hierarchy.[17]

4.2 The emergence of modularity

It is likely that a linear independence between regulation groups involved in different functionalities confers to the cell dynamical properties much more efficient to identify different constraints in its environment and react accordingly. Yet, effectors often appear to have correlated effects. It is indeed difficult to select independent conditions to modify the expression of proteins in a biological network. Thus, a superficial examination of the variations in protein expression under the action of a set of effectors is in most cases unable to uncover the *internal* structure of the system. We showed [14] that this structure can be observed only if we plot the changes in protein expression in a system of independent and orthogonal axes.

What is actually done by PCA is, starting from the trivial space in which each axis corresponds to an effector, to build a new representation space. This new space forms a unique system of axes obtained through a linear combination of the former axes, which allows to achieve the best linear separation between the network units. From this projection, we can identify each new axis as a major regulation direction (for it is orthogonal to the others) and, for the same reason, each group as an actual regulation group. The data representation in this new projection space is thus cleared up of the many difficulties blurring our perception of the genuine structure (such as noise in the expression of proteins and potential correlations in the action of different effectors).

We indeed clearly show the problem when we consider the action of different effectors onto the network. The principal component analysis of the results of their actions is expected to unveil the regulation axes and the regulation groups corresponding to the intrinsic structure of the network.

In non–adapted systems, all the units are functionally correlated, which explains the effect observed in figure 4. We have shown that the systems capable of reacting to different clamps can react according to at least two different response modes. This indicates the system evolves towards a situation in which certain units exhibit correlated dynamics with response to different clamps. Therefore, in our case, high–fitness systems are composed of different groups of coregulated units (we identified five such groups). We did not yet clearly demonstrated the emergence of regulation groups as they were observed in actual experiments.[14] However, the results we obtained tend to show that such regulation groups could clearly form, but further work is required to precisely study the conditions that would allow them to appear.

Acknowledgements

This research was supported in part by a grant from the DGA–DRET: 93–070. The authors would like to thank Thomas Vingtrinier for his interest and very helpful discussions.

References

1. L. Borecky. *Acta Virologica*, 37:276–289, 1993.
2. E. Chiva and P. Tarroux in *Lectures Notes in Computer Science* **866**, ed. Y. Davidor, H.-P. Schwefel, and R. Männer, 26 (Springer, Berlin, 1994).
3. E. Chiva and P. Tarroux. *Biol. Cybern.*, **73**,323, 1995.
4. M. Cohn and K. Horibata. *J. Bacteriol.*,**78**,601, 1959.
5. S. Kauffman. *J. Theor. Biol.*,**22**,437, 1969.
6. S. A. Kauffman. in *Origins of Order: Self-Organization and Selection in Evolution.* (Oxford University Press, Oxford), 1987.
7. S. A. Kauffman in *Complexity: Metaphors, Models and Reality* ed. D. G. Cowans and D.Metzler, SFI studies in the Science of Complexity,**19**, 83 (Addison-Wesley, Reading,1994).
8. M. Kendall and H.Stuart. *The advanced theory of statistics*, **2** (Griffin, 1973).
9. M. Ko. *J. Theor. Biol.*, **153**,181, 1991.
10. C. G. Langton. *Physica D*, 42:12, 1990.
11. J.-A. Meyer. *Behav. Res. Met. Instr.*, **6**,561–577, 1974.
12. M.Kraus, P.Lais, and B.Wolf. *Experientia*, **49**(3),245, 1993.
13. T. Rabilloud, P. Vincens, and P. Tarroux. *FEBS lett.*, **189**(2),171, 1985.
14. P. Tarroux. *Electrophoresis*, **4**,63–70, 1983.
15. R. Thomas. *J. Theor. Biol.*, **153**,1, 1991.
16. X. Yao. *Int. J. Neur. Sys.*,**4**,203, 1993.
17. E. Zuckerkandl. *J. Mol. Evol.*,**7**,1, 1975.

SPECIFYING LOGICAL AGENTS IN CELLULAR HIERARCHIES

RAY PATON, GEOF STANIFORD AND GEOFF KENDALL

Department of Computer Science
The University of Liverpool
Liverpool L69 3BX, U.K.

1 Cellular Hierarchies and Networks

The cell is a complex system which can carry out numerous operations in parallel. In order to deal with this complexity scientists have often modelled a cell as if it is like an analogous system. For example, cells and/or their parts have been modelled in a large variety of ways such as: factories and microcosms; circuits and transistors; ecologies and machines; microprocessors and connectionist systems; bags of chemicals and architectural edifices; automata and societies; cognitive and textual systems. Information processing models are also appropriate to ideas associated with parallel, distributed and emergent computation [10]. The idea of the cell as a processor of biochemical symbols has been around for decades.

Cells have a number of cognitive capacities include pattern recognition, memory, adaptability, communication with other cells, problem solving, sense/measurement of external processes and response. These capacities do not only apply to the cell as a whole subsystems within the cell also exhibit these functions in various ways. The logic of cellular information processing can be associated with various hierarchical levels; both their internal workings, how they interact with other components in that level and how interactions occur between levels. There are networks within hierarchies and hierarchies of networks - from molecules to the cellular system as a whole. A simple example of a hierarchy within any cell would be: molecule -> supra-molecular assembly -> trabeculum -> organelle -> supra-organelle system. For our purposes we ask what is the basic unit of information processing in this hierarchy? This is a major problem. Consider another hierarchy in which the same issue persists. Many people view a neurone as the basic unit of computation in the nervous system. Shepherd [12] argues that the basic unit of information processing in the nervous system is the synapse. Neurones are at a more complex level in an organisational hierarchy which includes: synapse -> dendritic spine -> microcircuit -> dendritic tree -> neurone.

A number of complementary models can be associated with each of the organisational levels within the nerve cell. The hierarchical relations presented so far reflect different scales of anatomical size. Hierarchical thinking which focuses on scale alone can be insufficient to take account of the complex information flows in the system. Hierarchical levels do not break down simply into sub-levels. The sublevels are networked in sophisticated ways. Figure 1 summarises some of the ways in which the sublevels of the cellular hierarchy are networked together.

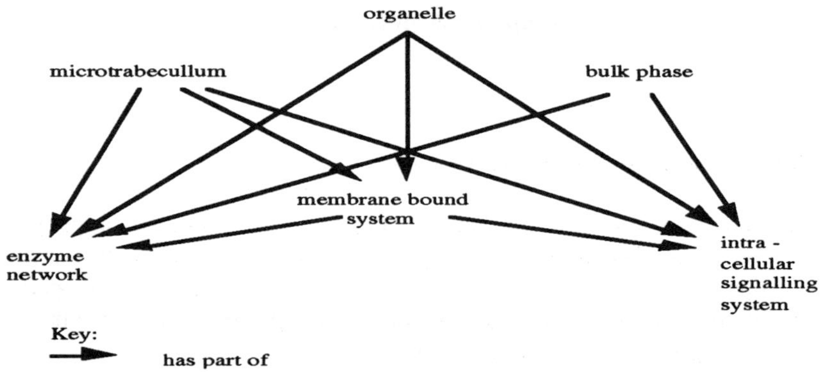

Figure 1 - Networking across Levels

2 Information Processing in Proteins

The organisation of protein structure allows information to be processed in a highly parallel way. If we take this view we treat the molecule as made up of many interacting parts (in 3D) and so in a specification of agency we focus on the atomic, molecular and supra-molecular levels of organisation. Proteins are complex structures in which a number of modes of energy transduction can take place and electrostatic, mechanical and protonic changes occur [16].

The haemoglobin molecule consists of 4 haem-protein subunits which interact to maximise its oxygen binding/releasing capability. It could be argued that in binding/loosing oxygen the tetramer processes information for not only is it sensitive/responsive to its environment, it also exhibits a memory and counting capacity in that it becomes progressively easier for oxygen to bind. Looking at this from certain biochemical or physiological viewpoints it may seem rather strange thinking but maybe not so for people looking at biosensors whether artificial or arterial oxygen chemoreceptors. Suffice to say that even a molecule as subtle and elegant as haemoglobin has an information processing capacity. We argue that this capacity can be seen in two ways viz: internally to the molecule it is a connectionist system and externally the molecule is embedded within a vast parallel distributed network of biochemical processes.

Another group of proteins which exhibit an information processing capacity are those involved in membrane-bound electron transfer systems. For example, cytochrome c acts as an electron shuttle [3] conveying electrons from cytochrome b-c_1 to cytochrome c oxidase. In this case it is not only a shuttle it is also a mobile register, temporarily holding electrons as it moves between b-c_1 and c oxidase. Another example of a shuttle which acts as a register is plastoquinone except this time both protons and electrons are shuttled. Here we have a processor acting on two very distinct data types. Hong [3] discusses the difference between electron shuttles and wires. In the latter case supra-molecular complexes can be said to be "wired" for example, the photosynthetic reaction centre of *R. viridis*.

The examples discussed so far show how the biochemical materials exhibit capacities for memory (e.g., as registers) and that the same structure can be both conduit and processor. Some of the mechanisms which enable these molecules to operate also show cognitive capacities. For example, bacteriorhodopsin is able to make protons flux more effectively through changes to surface protein binding groups. This enables the molecule to actively counteract reverse reactions of protein transfer. Hong [4] calls this an intelligent property of the molecule. The cognitive capacities are intimately related to the structure. Another example of an information processing protein is the nicotinic acetylcholine receptor (nAChR). This is an allosteric protein which can switch between a number of molecular conformational states [6]. These states include a resting state, an open channel state and several desensitised states. Each state has associated distinct binding properties as well as degrees of channel opening. The nAChR molecule has multiple allosteric binding sites for a number of ligands both inside and outside the cell. It also has multiple sites for particular ligands (e.g., approximately 100 for Ca^{2+}). Given this complexity, the nAChR molecule itself exhibits a PDP capacity with the conformational states being arrived at via multiple bindings. Other information processing proteins include multi-enzyme assemblies, microtrabeculum and membrane-based channels and pumps.

A major class of information processing molecules are the enzymes. It will not be possible to look at a lot of examples but we shall consider some simple cases in our efforts to supply some descriptive specifications. On being asked a question "What is an enzyme ?" we may use the idea of cell as a laboratory and reply that it is a biological catalyst. However, in terms of its verbs (i.e., what it does) we may note: catalyse, switch, affect osmotic potential, buffer, regulate, modulate , "glue". In this case the context can determine the meaning that is, what it does. For example, from a thermodynamic point of view an enzyme is a catalyst reducing the free energy of activation of a reaction. From a computational view it is a kind of switch or regulator of control flow allowing one reaction to proceed though not others. So what is an enzyme doing? How does it catalyse, switch and regulate the flow of cellular information and do other things all at the same time?

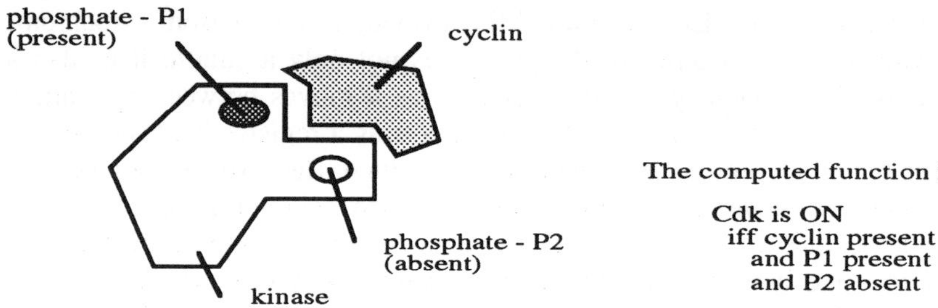

Figure 2 - Cdk (based on [1])

Cyclin dependent kinase (Cdk) can be used to illustrate switch/catalytic capacities (see Figure 2). In this case the kinase is active when cyclin is present and one phosphate site is filled but the other site is empty. Put another way the kinase is a pattern recogniser, only being activated when three elements of the pattern are satisfied. This reflects a further feature of certain enzymes (and other "informational" proteins such as antibodies and receptors) namely, pattern recognition is a key feature of their overall function. Returning to the switch/catalytic capacities, Cdk not only affects a switch/catalytic function, it also undergoes internal switching.

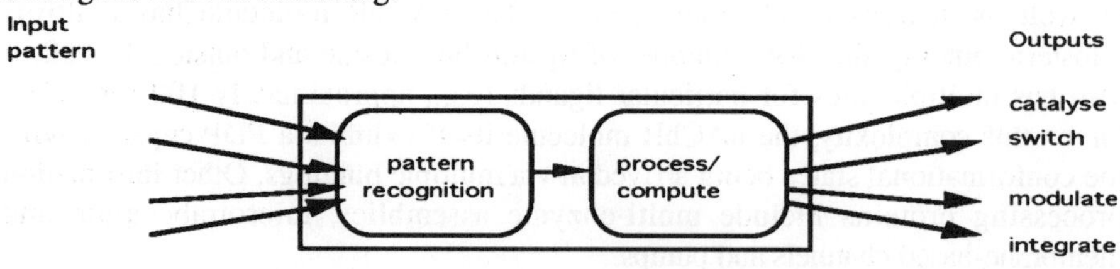

Figure 3 - Simple Information Processing Enzyme

Figure 3 summarises a descriptive information processing specification of Cdk. Note that we separate the pattern recognition action of the molecule from its internal processing in this conceptual model. This is to emphasise the variety of cognitive functions of the molecule rather than suggest that they are independent events. The input pattern will include substrate(s) molecules as well as other factors in the microenvironment. Outputs are defined in general terms but it should be noted that for each output a network of biochemical interactions can be defined across a number of hierarchical levels.

Consider a complex enzyme like CaM-Kinase II. It is a catalyst in its phosphorylating activities and a switch in that it can phosphorylate itself (a catalytic switch) and be activated by Ca^{2+}/Calmodulin. Additionally, it is a "memory molecule" for Ca^{2+} signals may have ceased yet its phosphorylating power can be preserved (see Figure 4). Here we see that what an enzyme does (i.e., "to enzyme") incorporates at least three verbs.

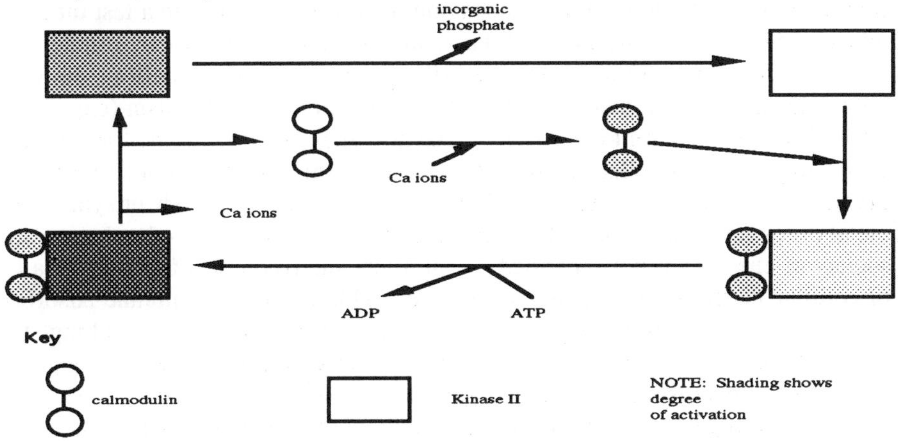

Figure 4 - Simplified Overview of CaM Kinase II Activity (adapted from [1])

We summarise the processing power of a molecule like CaM Kinase II by incorporating a memory module as in Figure 5 (again the simplifying caveats apply).

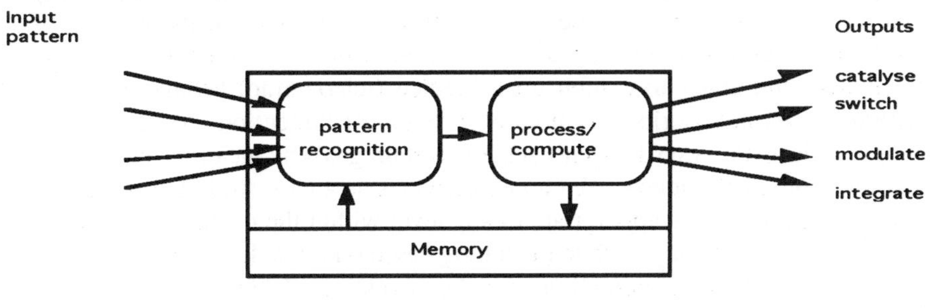

Figure 5 - Enzyme with Memory

Enzymes display a number of cognitive capacities related to: pattern recognition, handling fuzzy input data and fuzzy internal events (e.g., external and self-generated noise), individuality (e.g., whether bound to a membranous subsystem or free), memory capacity (some may act as registers others can remember state changes), internal PDP configuration, communication, counting effects and signal amplification.

So far the descriptive specification has centred on single molecules. We stress this because developing ecologies of interacting molecules is not straight forward. It an be easy to mimic the kinetic behaviour of certain enzymes in a test tube but this will be far from the intracellular reality in terms of microenvironment or potential for interaction. Welch describes two ways of modelling enzyme interactions, as a molecular democracy or a supra-molecular socialism (e.g., [17]). Both views need to employ a society metaphor (see [9], [10]). In the former case, each enzyme is modelled as an autonomous agent and the model of cellular organisation is characterised as a bulk aqueous phase in which enzymes and metabolites are homogeneously dispersed. However, this view of a democratic society of enzymes is not supported by empirical and theoretical findings for a lot of cellular metabolism is spatially organised. This includes membrane adsorbed enzyme clusters, multienzyme complexes and enzyme arrays along the cytoskeleton. Some enzyme systems can reversibly partition between organised states and the bulk aqueous phase depending on metabolic conditions. This emphasis on organisation in Welch's model shifts the focus from a homogeneous bulk reaction-diffusion system to a heterogeneous system which is structured according to the topographical segregation of individual processes. For our purposes we shall stick with small networks of interaction and gradually build up the types, numbers and organisations of the molecules.

The notion of agency now developed shares ideas with cognitive views of biosystems [11]. We apply the following criteria to describe agents: autonomy in that they have a degree of independence from other components in the system, social interaction with other agents, perception of their environment and response to change, and ablity to exhibit goal-directed (teleonomic) behaviour. The last criterion does not ascribe any notion of "will" or "intention" to the agents but simply reflects the ways in agents achieve certain goals by means of their structure and the biological context. We choose enzymes as a case study but would argue that the techniques are apporpriate to any level within the cellular hierarchy. The work emphasises an approach to modelling based on the individuality of objects in hierarchies and how interactions can be investigated "from the bottom up". However, because "the bottom" can be any specified level we are not promoting a totally reductionist approach.

3 Towards a Fuzzy Specification of CAM-Kinase II

We consider one approach to modelling the information processing capacities of CAM-Kinase II. An enzyme such as this: (i) may be affected by a large number of environmental factors, (ii) has many possible internal configurations, and (iii) may catalyse many different proteins. It is clear that a vast number of parameters are involved. Some of these parameters can be thought of as simple discrete 'switches', an example being the presence (on) or absence (off) of a specific second messenger. Specification in standard bivalent logic terms seems satisfactory in such cases. However, other parameters such as the overall level of charge within an enzyme may require a more continuous model if behavioural relevance is to be maintained.

When specifying an enzyme, we cannot hope to accurately incorporate every intricate and subtle effect which small variations in and around the enzyme may introduce. If only a small finite number of discrete-valued environmental influences, internal conformations, and catalytic abilities are considered in an effort to simplify the specification, the danger is that the behaviour of the model as a whole may be lost. A satisfactory first step would seem to be an approximation of certain 'characteristic' enzyme activities, which when combined exhibit dynamics similar to that of the real enzyme. In recent years, fuzzy system theory has been successfully applied in many areas where only approximate models are available or required [5]. We now develop our specification of enzymes within this context.

The specification takes a simplified view of an enzyme's surroundings to avoid the spatial and temporal problems associated with modelling the intracellular structure. Every type of molecule which the enzyme may potentially catalyse is considered to be either present in (1) or absent from (0) the enzyme's local environment. Similarly, the possible 'inputs' of the enzyme (factors affecting its conformation) may also be present or absent. The 'concentration' of each molecule type within some vicinity of the enzyme is used as the basis for determining whether that molecule is present at a given instant. Note that concentration is here a property of the cellular micro-environment, effectively 'a level above' the individual enzyme specification within the hierarchical cellular model.

Concentration is not a bivalent quality. Real concentration levels are likely to fall between the extreme situations of 'none' and 'saturated' respectively. Similarly, any realistic enzyme specification must take account of internal properties such as level of charge, which cannot be adequately described in simple 'on / off' terms. Fuzzy theory provides a way to specify and manipulate such multivalent qualities more realistically.

One might describe concentration as being 'low', 'medium' or 'high'. Fuzzy theory permits the formal representation of such categories, each being defined as a "fuzzy set" over C (the set of all possible concentration levels) via a membership function

$$m_S : C \dashrightarrow [0,1]$$

The value of the membership function indicates the degree of membership in that fuzzy set: the extremes of 0 and 1 are equivalent to the corresponding bivalent logic values. In Figure 7 for example, concentration C1 is 'high', but not as high as C2. Conversely C2 isn't as 'medium' as C1. Fuzzy set operations (see [19]) enable the construction of "fuzzy rules", such as:

"if temperature is high and concentration is high then rate of catalysis is high"

Evaluating - or "firing" - such a rule for specific input values (temperature and concentration in this example) results in a fuzzy value indicating the degree of truth of the rule. A fuzzy system comprises a number of fuzzy rules (a "fuzzy rule base"), each of which are fired simultaneously for a given set of inputs. The fuzzy results are then combined to produce single-valued real outputs, the interpretation of which is dependent on the specific problem. It should be noted that the pattern matching ability of enzymes described earlier is implicit in this approach, being embodied in the firing of fuzzy rules. The combination and subsequent interpretation of the fuzzy outputs may be considered as the second (computational) aspect also mentioned previously.

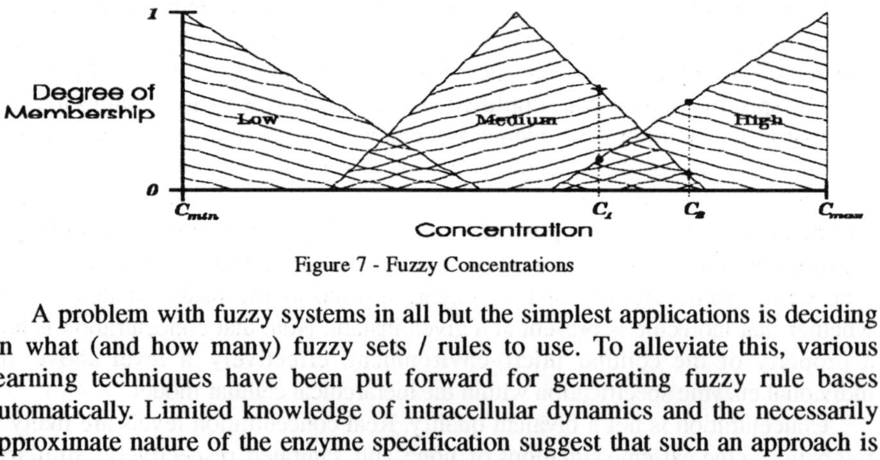

Figure 7 - Fuzzy Concentrations

A problem with fuzzy systems in all but the simplest applications is deciding on what (and how many) fuzzy sets / rules to use. To alleviate this, various learning techniques have been put forward for generating fuzzy rule bases automatically. Limited knowledge of intracellular dynamics and the necessarily approximate nature of the enzyme specification suggest that such an approach is required here. The Fuzzy Classifier System or FCS (eg. [19]) - which evolves a rule base using a genetic algorithm and user-specified fitness measures - is one promising approach. An FCS contains a fuzzy rule base in which each rule - called a classifier - has an associated strength which indicates how large an influence it has on the output. When an "input message" is received and the rule base evaluated, the resulting "output message" is compared to user-specified criteria to determine the fitness of the response. Depending upon how 'good' or 'bad' the response is found to be, the strengths of those classifiers which fired (ie. contributed to the output message) are increased or decreased accordingly. At user-specified intervals, a genetic algorithm (GA) discards weak classifiers and mutates / breeds strong ones. As long as the GA is 'turned on', the FCS constantly adapts to it's environment. The FCS - with a few enhancements - is well suited to modelling an enzyme as an agent for the following reasons:

- The GA can be turned on and off at will - this permits iterative refinement and testing during model development.
- When one possesses a rough idea of a particular aspect of enzyme behaviour (eg. how temperature rise affects a reaction), it should be possible to incorporate this information from the outset to accelerate model refinement. The FCS provides this capability since the user has complete control over the initial contents of the rule base.
- Since the proposed enzyme specification is only an approximation, vast quantities of empirical data relevant for training by example are not available: the appeal of the general 'fitness' criteria required by the FCS is apparent.

The FCS so far described only considers inputs and outputs. An enzyme's internal conformation can have a significant affect on its behaviour, so this too must be incorporated in the model. This led the authors to develop the Enhanced Fuzzy Classifier System, or EFCS. An example EFCS is shown in Figure 8, which also illustrates the facility to incorporate multiple rule bases: this extra functionality is not utilised in the present enzyme specification, so is not discussed further.

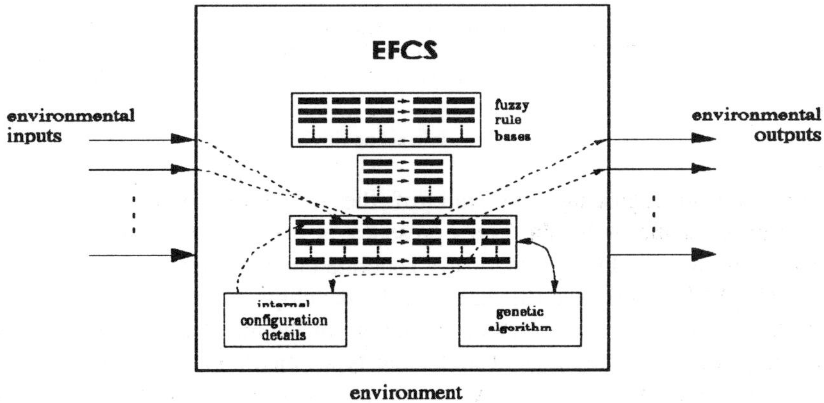

Figure 8 - EFCS in an Environment

The EFCS encodes internal properties as fuzzy values, which are supplied as extra input parameters to the rule base when an environmental input is received. Corresponding output parameters are used to update the properties, as shown in Figure 8. In this way, an enzyme's internal configuration can be specified to any required degree of accuracy. For example, an EFCS CAM-Kinase II model (with an 'activation level' property) would permit classifiers such as:

"if second messenger is present and activation is low then activation is high"

During initial development and training, the EFCS's environmental inputs will in general be determined by the user. When the agent's behaviour is deemed to adequately approximate to the enzyme specification, analysis within a simulated environment can proceed: Figure 9 shows a conceptual scheme of a local intracellular neighbourhood containing several enzymes (each a copy of the EFCS developed).

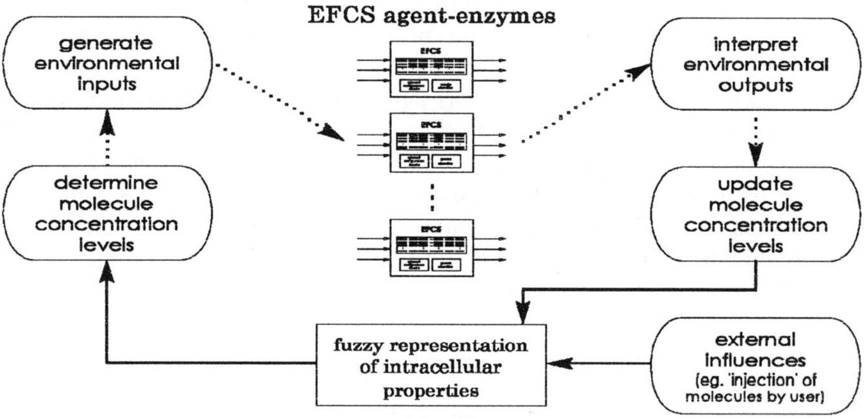

Figure 9 - Interacting EFCSs

Information regarding the state of the neighbourhood - such as the concentration of the molecular types present - is stored as a collection of fuzzy values. These are used to automatically generate environmental inputs for a time step. Each EFCS is activated (in parallel if the development platform allows), then the resulting outputs are interpreted. The nature of the interpretation is very much model dependent and the range of molecular types incorporated in the model will dictate how many fuzzy outputs are required. Intracellular properties are then updated as necessary (eg. concentration of a molecule type may change if catalysis occurs), and the cycle continues. The intracellular properties may be altered at any time (see "external influences" in Figure 9) to simulate a variety of situations such as a rise in temperature or decreased availability of a particular protein. Qualitative analysis of enzyme-agent behaviour against empirical findings is therefore straightforward.

4 An Agent Architecture for Cyclin Dependent Kinase

In this section of the paper some methods for specifying agents which are based on ideas from Distributed A.I. are introduced. We use the example of cyclin dependent kinase (CdK) to develop a partial specification which illustrates a general agent model which can be applied to biological hierarchies. We begin with some basic definitions. So that an agent may receive external stimuli we require the sensory function:

Definition 1:

$$sense: R \dashrightarrow S$$

where the set R is the set of states that represents the agents world and in order to characterise the agents sensory capabilities we partition the set R of external states into a set S of disjoint subsets (associated with the function *sense*) such that the agent is able to distinguish states in different partitions but is unable to distinguish states in the same partition •

The sensory function is used to characterise the way in which an agent receives external stimuli and is used to enable an agent to determine the local state of its environment. In order that an agent's internal state may be modelled we define the function:

Definition 2:

$$internal: N \times \Delta \times S \dashrightarrow \Delta$$

where the set Δ is an arbitrary set of logic databases representing an agent's internal state •

The function *internal* maps an agent's internal state and the received stimuli from the external world into the next internal state. We assume that agent actions cause only local changes to an environment andin order that an agent may make changes we define two more functions:

Definition 3:

$$action : N \times \Delta \dashrightarrow A$$

$$db : A \times R \dashrightarrow R \times W$$

The set N is the set of natural numbers and is used to represent the passage of time internal to an agent based upon the notion of time points. The set A is a set of possible actions that an agent may perform and the set W is a set of words that characterise the informatio that an agent may produce •

The function *action* is a discriminatory function which maps the sets of internal states and observations onto a particular action, thus providing an agent with the ability to choose which action to perform according to its internal condition and perceived stimuli. Finally, the function *do* is an executory function which maps an action and state onto the new state and an output; providing the agent with the ability to change the local state of its environment and to communicate information. We require all the functions defined previously to be computable functions and note that in general there is no simple linear correlation between the internal state changes within an agent and the external state changes of the environment.

Definition 4:

A *logic based agent* (LBA) in an environment is a tuple of the form

$$< N,\ \Delta,\ R,\ S,\ A,\ W, \mathit{sense}, \mathit{internal}, \mathit{action}, \mathit{do} >$$

where the sets and functions are taken to have been defined and described in Definitions: 1-3 Note that the functions *sense, internal, action, do* have a sequential ordering imposed upon them relative to the set N, thus allowing for the detailed functional specification of the agents using a method such as the calculus of communicating systems [7] •

The key idea in defining such agents is the use of an automated inference method such as resolution and the use of the Horn clause subset of predicate calculus to form the basis of a mechanised reasoning ability with a form of deontic logic. In modelling the control of a complex virtual structure in a mathematical space we use two primary mechanisms to impose social structures upon autonomous agents, eg. networks of cooperating equals and hierarchies of the division of responsibilities; the vocabulary of the agent may be used to prescribe which other agents are allowed to meaningfully communicate with it and we can use the computational ability of an agent to implement cooperation strategies that enforce a hierarchy of control.

Consider Figure 10 in which there are four agents with communication paths shown. Each communication channel is a one way medium and this is indicated on the diagram both by the direction of the arrows and by the port labels at each end of the channel. A label with an over-bar is an output port. A further abstraction is that we treat each channel as having in effect infinite bandwidth and zero length meaning that agents can handshake and we always assume that the communication medium can carry the information that an agent needs to transmit.

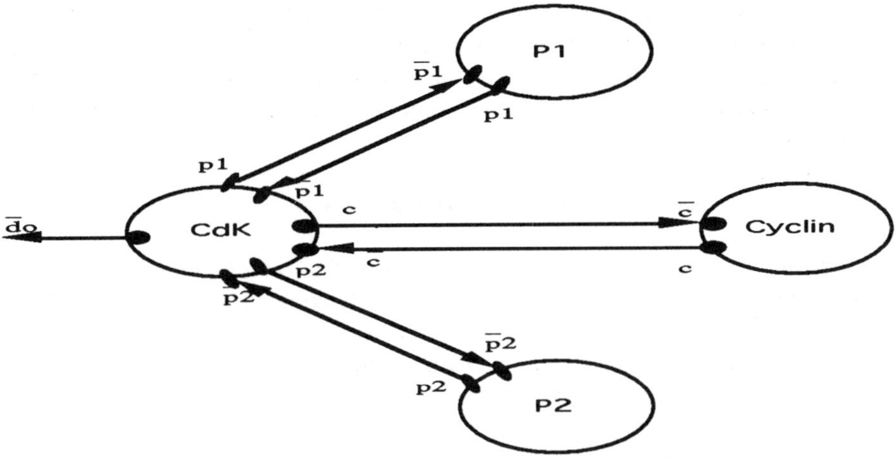

Figure 10 - CdK Agent

In the Figure 10 P1, P2, and Cyclin, all communicate with CdK. In reality they may communicate with other agents but we abstract those communications away for simplicity without loss of generality. CdK communicates with P1, P2, and Cyclin, through the labelled ports and also communicates directly with the environment through the port labelled "over-bar do". In the following model we treat P1, P2, and Cyclin as black boxes and building upon the general architectural specification we provide a fairly informal functional specification and explanation for the CdK agent. In the definition which follows we apply some of Milner's terminology [7] and Staniford's models [13].

Definition 5

We define the Cdk agent as:-

$R = \{hold, break, null\}$

$S = R \times R \times R$

$\Delta = S \times C$ and

$c \in C = n \bmod k$ where $n \in N$ and k is a positive integer determined by experiment.

$A = S$

$W =$ ON, OFF

Cdk = *sense* (r, r',r'').τ.*do* (s,w).Cdk

sense (r,r',r'') = p1(r).p2(r').p3(r'') --> s

τ = *internal* (n,s,s').*action* (n, δ)

do (a,r) = $\overline{p1}$(r). $\overline{p2}$(r'). $\overline{p3}$(r''). $\overline{do}$ (w)

Notice that in the action function we use n to provide the time points and c to provide a division of time into periods thus we have the ability to model internal states at every time point and to have the action function either choose or delay choosing an action as a consequence of some time period after the detection of an external or internal event.

For example given:

s' = (r, r', r") and s = (*hold, null, hold*) and s' ≠ s

then

internal (n, *hold, null, hold,* r, r', r") changes δ to δ (*hold, null, hold,* 0)

if s = (*hold,null,hold*) and s' = s and c = m > 0, (determined by experiment) then

internal (n,*hold,null,hold,hold,null,hold*) changes δ to

δ (break,null,hold,0) causing the P1 agent to break off contact.

for all other values of s internal changes δ to

δ (s,c) where c = c'+1 mod k

When δ = δ (*hold,null, hold,* n) action changes a = (*hold, null, hold*) and *do* outputs p1(*hold*).p2(*null*).p3(*hold*).*do* (ON)

For all other values of a *do* outputs p1(r).p2(r').p3(r").*do* (OFF)

This computational model of CdK provides a flexible tool for experiment that may be formally verified and transformed into a working logic program using the method proposed in [8]. Similar specifications are required for the P1, P2, and Cyclin agents in order to complete the model; we leave them as "black boxes" for reasons of space. More details of the application of this approach to biosystems can be found in [14].

This paper has sought to follow the modelling process from descriptive and explanatory models of the cell as a whole to more detailed computational models of cellular agents based on logic. We propose that this approach, which focusses on an information processing account of intracellular activity based on logics, could be used to complement both dynamical and algebraic models.

References

1. B. Alberts, D. Bray, J. Lewis, M. Raff, K. Roberts & J .D.Watson, *Molecular Biology of the Cell*, (Garland Publishing, New York, 1993)
2. Braun, A.P. & Schulman, H.*Annu. Rev. Physiol.*, **5 7**, 417-445 (1995).
3. F. Hong, *BioSystems*, **2 7**, 189-194 (1992).
4. F. Hong, *Nanobiology*, **1**, 39-60 (1992).
5. B. Kosko, *IEEE Transactions on Computers*, pp. 1153-62 (1993).
6. C. Léna & J-P. Changeux, *TINS*, **1 6**, 5, 181-186 (1993).
7. R. Milner, *Communication and Concurrency*, (Prentice Hall International, 1989).
8. E. R. Olderog, *Terms, and Formulas* (Cambridge University Press, 1991).
9. R.C. Paton, *Biology and Philosophy*, **7**, 279-294 (1992).

10. R.C. Paton, *BioSystems,* **29**, 63-75 (1993).
11. R.C. Paton, "Glue, Verbs and Text in Biology" submitted.
12. G.M. Shepherd in *Computational Neuroscience* ed E.L. Schwartz (Cambridge, MIT/Bradford, 82-96, 1990).
13. G. Staniford, Unpublished PhD Thesis, University of Liverpool, 1993.
14. G. Staniford, G. & R.C. Paton in *Intelligent Agents*, eds M.J. Wooldridge & N.R. Jennings, (Springer Verlag, Berlin, 145-159, 1994).
15. G.R. Welch, *Trends in Ecology and Evolution*, **2**, 10, 305-309 (1987).
16. G.R. Welch in *Computing with Biological Metaphors*, ed. R.C. Paton (London, Chapman and Hall, 1994).
17. G.R. Welch & T. Keleti, T., *Trends Biochem. Sci.*, **12**, 216-217 (1987).
18. M. Valenzuela-Rendon in *Parallel Problem Solving from Nature, Lecture Notes in Computer Science*, eds. Schwefel, H-P. & Maenner, R. (Springer, Berlin, 330-334, 1991).
19. H.J. Zimmermann, *Fuzzy set theory and its applications*, (Kluwer, Boston, 1988).

NERVE SIGNAL PROCESSING USING ARTIFICIAL NEURAL NETS

M. Bogdan, A. Babanine, J. Kaniecki, W. Rosenstiel
Lehrstuhl für Technische Informatik, Universität Tübingen
Sand 13, 72076 Tübingen, Germany
E-mail:
<bogdan|babanine|kaniecki|rosenstiel>@peanuts.informatik.uni-tuebingen.de

1 Introduction

The aim of the INTER[a] project (*I*ntelligent *N*eural In*TER*face) is to investigate fundamental issues related to the design and fabrication of a new generation of microsystems applicable as neural prostheses. A global overview for a PNS-remoted limb prostheses is given in [1] and is shown in figure 1.

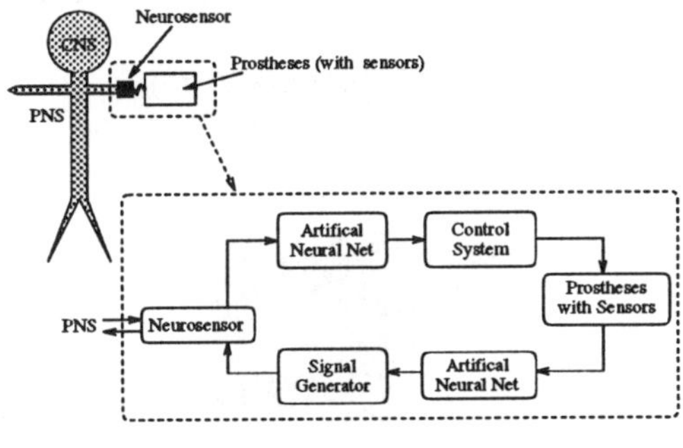

Figure 1: Scheme Configuration of a bio-neural controlled prostheses

Nerve signals will be recorded and amplified by a neurosensor. The neurosensor is a regeneration-type sensor. The principle of the neurosensor is explained below. The signal will first be processed by an artificial neural net (ANN), which eliminates crosstalk and leads to pure single axon signals. We describe the computation of artificial test signals, the implementation of the neural net, and present examples of source separation.

Then, a second neural net is applied which classifies the resulting signals in order to assign certain limb movements to the signal classes. We present first

[a]The INTER project is supported by the European Community under ESPRIT BR project #8897.

experiences with classification of nerve signals and discuss the choice of apropriate parameters of this ANN. A control unit uses the resulting information to regulate the movement of the prostheses.

Ideally, the prostheses is equipped with sensors. Signals from the sensors will be processed by an ANN and transmitted via a signal generator and the neurosensor to the peripheral nervous system (PNS). This means, the prostheses will be completely controlled from the PNS like a natural limb.

2 Neurosensors

What appears to be the earliest (unpublished) work on regeneration-type neurosensors was carried out between 1965 and 1967 by Frishkoff, Goldstein, et al.[2]. In 1969 Marks demonstrated regeneration of bullfrog sciatic nerve axons through porous implants[3]. In 1973 Llinás, Nicholson and Johnson[4] published a proposal for a silicon-based microelectronic neurosensor. They proposed to use on-chip circuitry to address individual microelectrodes in the array of via holes and to transfer the recorded signals to external processing units.

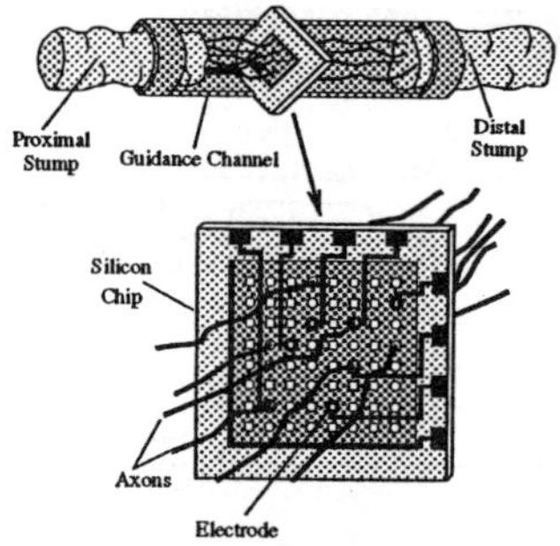

Figure 2: Implementation scheme for the regeneration-type neurosensor

Edell[5] reported first recordings obtained with a silicon substrate regeneration-type array. The recordings with amplitudes on the order of 150 μV were obtained from such devices implanted in the sciatic nerves of rabbits[6]. In 1991 Najafi reported nerve stimulation using a regeneration-type neurosensor[7]. His neurosensor allows stimulating current pulses up to 40 μA.

Kovacs has implanted a microelectrode array, fabricated on a silicon substrate, in the pereonal nerves of rats[2]. He used these arrays to record and to stimulate the nerves up to 13 months postoperatively. Action potentials with peak amplitudes on the order of 100 μV were recorded. He stimulated the nerve with relatively high-current/short-duration stimulus of approximately 500 μA/10 μsec phase.

The principle of the implementation and the neurosensor[8] which will be used in the INTER-project is shown in figure 2.

Peripheral nerves of vertebrates will regenerate if severed. For this reason, the peripheral nerve can be surgically severed in order to insert the proximal and the distal stump into the guidance channel. The guidance channel encloses the chip. The main functions of the guidance channel are to provide a stable physical connection between the chip and the stumps of the nerve, and to promote nerve generation and guide axons growth from the proximal nerve stump towards the distal nerve stump. The chip is fabricated on a silicon substrate perforated by multiple 'via holes'. The axons regenerate through the via holes from the proximal stump towards the distal stump of the nerve. Nerve signals can be recorded by electrodes, which are enclosing some of the via holes. An on-chip circuitry amplifies and preprocesses the nerve signals. The amplified signals are transferred to the units which are controlling the prostheses as shown in figure 1. For more details about the chip, please refer to [9].

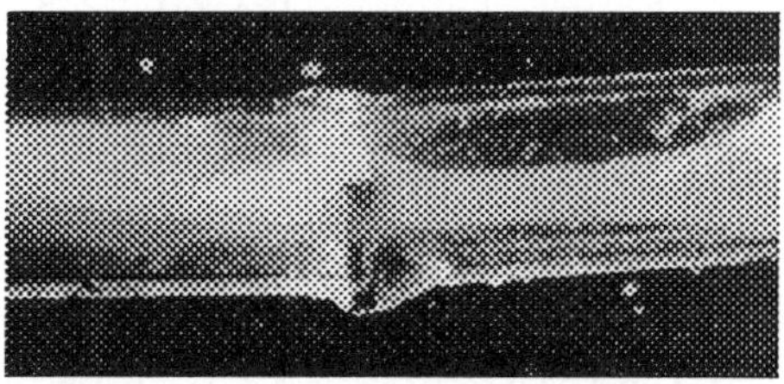

Figure 3: A nerve, regenerated through a dice inside the guidance channel (thermoplastic polyurethane, ChronoFlex^{TM}). On the left hand side the proximal stump, on the right hand side the distal stump. The photograph was provided by our INTER partner CHUV at Lausanne.

The regeneration through these perforated dice has been reported[9] and is shown in figure 3. Myelinated axons were counted proximal and distal to the dice based on histological specimen taken four weeks after implementation of the dices (N = 2). The number of proximal myelinated axons were 2633 and 2385 respectively. 300 and 168 myelinated axons were counted on the distal side.

3 Proposed Signalprocessing using ANN

3.1 Signal Related Problems

Some problems are directly related to the kind of nerve signals and the recording method. Although the information within the signals is pulse-frequency encoded, it can be easily imagined, that the nerve signals are very complex while one nerve bundle contains about 20000 axons. Additionally the axons are interconnected and influencing among themselves and only a small fraction of them can be recorded.

Another problem is due to the recording of the nerve signals. The principle of the recording is shown in figure 4. The signal of one spike is well known. But in our case, we will record the sum of nerve signals from several axons. We are assuming that approximately 5 axons will regenerate through an electrode, so we are recording a weighted sum of 5 signals, without knowing the weights

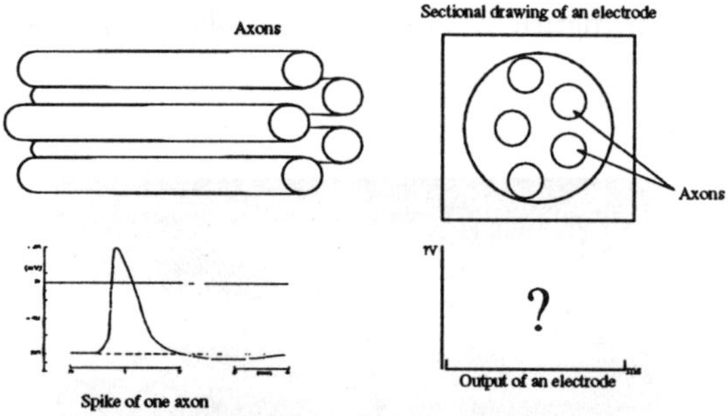

Figure 4: Principle scheme of the signal recording

of the mixture. The weights are depending on several items: the distance between the axons, the distance between the axons and the ring electrode and the distance between the node of Ranvier and the electrode.

The separation of motor and sensor signals is a problem, too. For remoting the prostheses, only the motor nerve signals are required. So it is necessary to separate the motor nerve signals from the sensor nerve signals. This separation cannot be detected by the waveform of the signal using the circumstance, that the motor nerve signals and the sensor nerve signals are transferred in opposite directions, because the signal recorded by the electrode will have the same waveform for both signals.

3.2 Modus Operandi

A global overview of our proposal for the signal processing is shown in figure 5. The preprocessed data (in this case preprocessed means amplified and filtered) will be separated concerning to their sources using Independent Components Analysis (INCA). The separation into motor/sensor signals will be done before the use of INCA by applying different guidance channels favouring either the growth of motor nerves or the growth of sensory nerves.

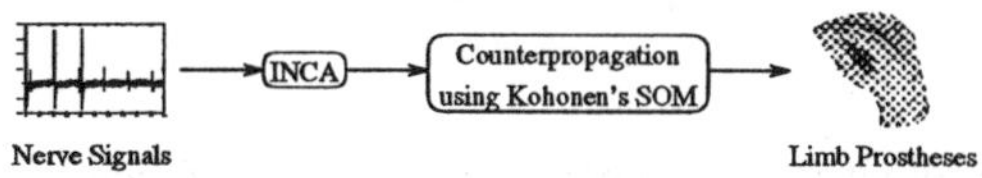

Figure 5: Proposal for signal processing

For separating the different nerve signals from the recorded mixture, we are proposing to use INCA, presented by Jutten and Hérault[11]. The algorithm allows to separate different signals from several independent sources (in our case a source is an axon) without any knowledge on the sources. It is a so called blind separation of sources. The records of the signals are a mixture of several sources. We can use INCA to detect the raw signals within the recorded data. This is a basic request to detect information within the nerve signals.

The classification of the nerve signals must be done by unsupervised learning algorithms because we have to verify our approach using the results of In Vivo tests. In a first approach we are going to apply Kohonen's self-organizing map (SOM). In order to choose the right dimension of the SOM, we determine the information dimension of the recorded signals using the proposal of[12]. A more detailed description of the proposal is given in[13].

4 Simulation of the nerve signals measured with the ring electrode

4.1 Recording problem

Since we have not yet recorded signals from the dice, we have developed a tool which provides the data for computational experiments on interpretation of neural signals by artificial neural net (ANN).

The axons regenerate through the dice of overall dimensions of 1.5 x 1.5 mm^2, according to several evaluations from 5 to 10 through each 'via hole'. The size of the 8-10 μm thin membrane is 0.8 x 0.8 mm^2. Dices containing 10 μm and 40 μm diameter holes, for a total of 400 and 121 holes, respectively. We consider herein a ring electrode, situated on one of the via holes of the active dice (see figure 4).

4.2 Neurophysiological background

The neurophysiological approach we used is based on the work of Meier [10]. A nerve containing some fascicles is considered. The fascicle is assumed to be electrical homogeneous, anisotropic and infinitely long extending cylinder. Longitudinal and radial conductivity of fascicle are

$$\sigma_z = 0.5\ \Omega^{-1}m^{-1}$$
$$\sigma_\rho = 0.1\ \Omega^{-1}m^{-1}$$

respectively.

Thus the ring electrode is situated inside of fascicle. Some axons are growing through the electrode as a result of regeneration (refer figure 6). The

Figure 6: The fibres are growing through the ring electrode. A cross signifies a node of Ranvier

active fibre is simulated as an array of active point current sources along the nerve bundle. Each source represents a node of Ranvier.

4.3 Assumptions and formula

Some simplifications were applied that should be insignificant compared to the additional noise.

So actually we have a superposition of electrical fields generated by a set of active nodes, belonging to some nerve fibres. The assumption of an internodal distance of 1 mm allows us to neglect the influence of a second active node of the same fibre.

Dropping the secondary field part and using it for evaluation of the "lead field" we have finally the following formula for the potential at an electrode at position (x_e,y_e,z_e):

$$\Phi(x_e, y_e, z_e) = \sum_{j=1}^{N} \left\{ \Psi_{x_e,y_e,z_e}(x_j, y_j, z_j)\, I(t) \right\} \tag{1}$$

$$\Phi_i = \frac{I}{4\pi\sqrt{\sigma_\rho \sigma_z}\sqrt{(x-t)^2 + y^2 + z^2\sigma_\rho/\sigma_z}} \tag{2}$$

I(t) in the first formula is the current of the node. According to the reciprocity principle the "lead field" Ψ_{x_e,y_e,z_e} is equal to the potential field that would be generated by a point source at position (x_e,y_e,z_e), divided by the source current I. According to the taken assumption we replace the multiplication under the sum in (1) with Φ_i according to (2).

The singular active node (node of Ranvier) is supposed to generate a sequence of spikes series of certain frequency. The frequency of spike series on the active node (one fibre) are varying in bigger scale of the time. It is also assumed that axons can keep quiet sometimes between transmissions of spike series.

The noise reflecting normal distribution is added in the stage of simulation of singular active node signal.

Lets note nevertheless that absolute values of amplitude and frequency are not so important in terms of ANN implementation. On the contrary relative values have to be introduced as accurately as possible.

4.4 Experiments

A series of experimental simulations was performed. A part of active nodes were assumed to generate the spike series, intermitted by relatively long periods of zero signal. We took the duration of quiet periods to be varied from 5 ms to 70 ms.

Frequency of spike series varies in range of 50-200 Hz. Other active nodes are assumed to generate continues sequences of spikes in the scale of a considered time period (in order of 100 ms). We are taking into account only the closest active node in each fibre. According to the internodal distance of 1000 μm, the second active node must be situated at least 500 μm away. Since Φ_i is proportional to $1/\sqrt{x^2+y^2+z^2}$ it will not influence the measured signal significantly. Signal patterns we use are action potential generated for a distance of 10 μm.

In table 1 an example of signal simulation of 5 active nodes is given. The resulting signal is shown in figure 7.

Table 1: Description of simulated signal in time gap of 100 ms

1 axon	50 Hz, 50 ms	pause 50 ms	
2 axon	pause 6 ms	100 Hz, 94 ms	
3 axon	57 Hz, 100 ms		
4 axon	pause 13 ms	30 Hz, 66 ms	pause 21 ms
5 axon	pause 59 ms	170 Hz, 25 ms	pause 16 ms

Different configurations of relative positions of active nodes and the electrode were simulated due to needs of signal separation using INCA.

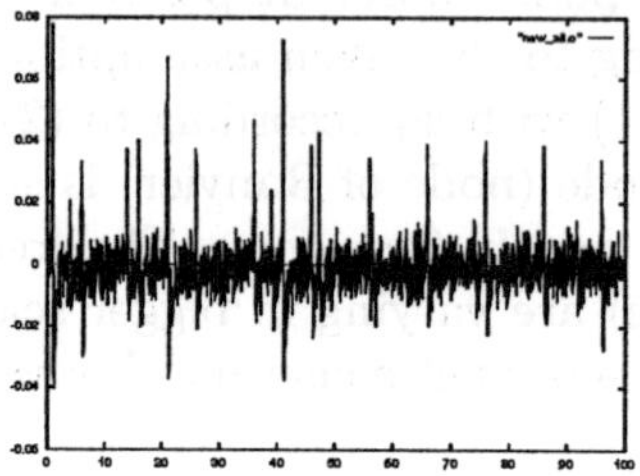

Figure 7: Resulting signal with time windows of 0.05 ms.

The described model allows to simulate signals to be measured with extracellular ring electrode. These signals are shown to depend on the position of the fibres and their active nodes relative to the electrode.

5 Independent Components Analysis

5.1 Origin of signal mixtures

As we mentioned above, we expect that many axons will grow through the 'via holes' of a neurosensor. Also, we expect that every electrode will be passed by more than one single axon. So the data collected by a neurosensor chip will contain many differently composed signal mixtures (one mixture every sensor) which can be separated into their raw signals by Independent Components Analysis, an algorithm developed by Jutten and Hérault[11].

5.2 Theory of source separation

Differently composed mixtures ($\vec{E}(t)$) of unknown sources ($\vec{X}(t)$) are measured at a number of sensors. Crosstalk is eliminated by application of a matrix. Figure 8 shows the functionality of Independent Components Analysis in principle.

The elements of the matrix C represent the amount of one mixture that is to be subtracted from another mixture. The Independent Components Analysis algorithm evaluates these values. The algorithm computes the weights of the neural net sequentially, starting with a pre-trained matrix (C) or a zero-matrix. At first, a preliminary output signal $\vec{S}(t)$ is computed by application of this initial matrix (C) on the input signal $\vec{E}(t)$:

$$\vec{S}(t) = (1 + C)^{-1} \vec{E}(t) \tag{3}$$

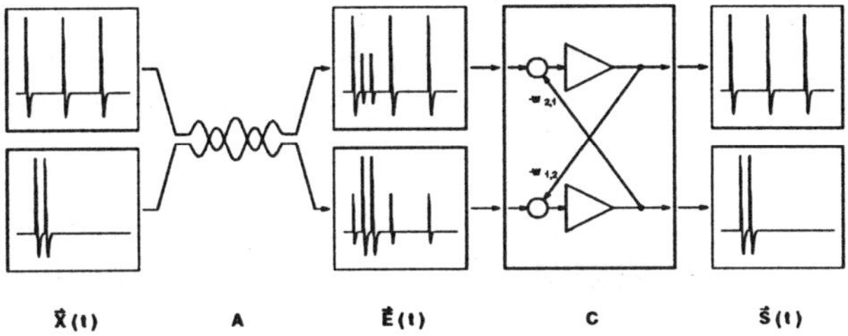

Figure 8: Modification and restoring of signals

Then, an estimation value of the output is evaluated (where '*' means convolution product):

$$< \vec{S}(t) > \; = \; \vec{S}(t) * e^{-t/T} \tag{4}$$

The differences between output values and estimation values is computed according:

$$\vec{s}(t) = \vec{S}(t) \; - < \vec{S}(t) > \tag{5}$$

and used for application of a gradient method in order to modify the net weights:

$$\forall i \neq k : \quad dc_{i,k}/dt = a \; f(s_i(t)) \; g(s_k(t)) \tag{6}$$

With

$$f(s_i(t)) = s_i(t)^3 \tag{7}$$

and

$$g(s_k(t)) = \begin{cases} tan^{-1}(s_k(t)) \\ sign(s_k(t)) \\ s_k(t) \end{cases} \tag{8}$$

After adjustment of the weights, the set of data one time step further is taken for computation.

5.3 Separation of artificial signals

In order to validate the approach to separate nerve signals using INCA, we have used data calculated by the nerve signal simulation tool described above. A set of signal mixtures was analysed in order to realize the effects of blind source separation and signal noise. Since artificial white noise was added to every signal, the separation of two independent sources and additional noise was tested by the computation of three differently composed mixtures. The aim of this noisy input is to separate the noise as a raw signal.

The sampling rate was 20 kHz, the number of samples was 3000. the adaption gain (a) was set to 5.000, the number of learning steps was 30.000, and the binary learning function was selected. The results are presented in figure 9.

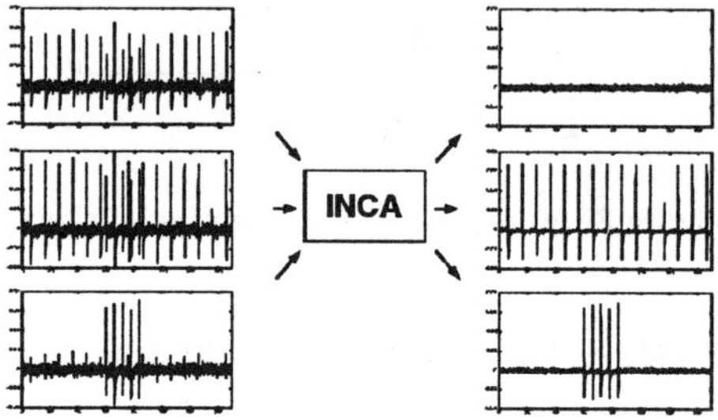

Figure 9: Separation of tho mixed signals and an additional artificial noisy input: On the left hand side the mixtures of the signals are shown, on the right hand side the signals after the separation by INCA

Two different signals and the additional noise are processed by INCA. As figure 9 shows, INCA separates 3 different signals from the mixtures. First, noise was recognised as an isolated source and separated from the other signals. The second signal we obtain is also a raw signal, but encumbered by a reduced noise. Finally, we obtain a nearly pure raw signal. The fact, that the noise cannot be completely rejected is due to the generation of the white noise. We use a mathematical model to generate the noise. For this reason, a difference occurs which is detected by INCA. This difference is distributed to the raw signals. Nevertheless, the noise distortion of the raw signals can be partly rejected using an additional input to INCA containing an artificial noise.

6 Classification of nerve signals

6.1 Nerve Signal processing using Kohonen's SOM

After processing data with INCA, signals are separated into their components. Further processing has to be done in order to assign certain types of signals to a corresponding limb movement. We use a self-organizing map to classify the data. The dimension of the map is determined according to[12].

6.2 The Data Set

Since we have no yet recorded signals from our neurosensor, we are evaluating the ANN with a data set from Barry Lee from the Max-Planck-Institut für biophysikalische Chemie at Göttingen. Unfortunately, these signals have been recorded directly from one axon, so we have no mixture of signals and need not to apply INCA in this case. Lee has recorded signals from ganglion cells of a Macaques retina. The nerve signals have been recorded while presenting pictures with different luminosity (Michelson contrast) under different angles in front of the Macaques eye. The signals have been recorded with a resolution of 0.1 msec for 600 msec from presenting the image in front of the eye.

6.3 Evaluation of Kohonen's SOM

We have trained a two dimensional map with 7x7 neurons with vectors, averaged over 5 samples. A two dimensional map has been chosen to see the topological defects of the map and for the possibility to visualise the map.

Evaluating the received maps for the different components of the vectors, we have found, that some are very folded, others preserve the topology. Obviously some components can be trained preserving the topology of the map. As a component of a vector represents a time window of 20 msec, it seems, that the different time windows contain more efficient information than other time windows. This is a fingerpoint to use only certain time windows for training the map.

The obtained classification map is shown in figure 10. None of the classes

Figure 10: Classification of the data. Black corresponds to 10 % luminosity, light grey to 20 % and white to 40 %.

builds a correct cluster. The cluster for the class of a luminosity of 10 % and 20 % are nearly contiguous. The class of a luminosity of 40 % is splitted into two parts by the class of 10 % luminosity. Assuming the luminosities less then 15 % as gloomy luminosity and luminosities over 15 % as bright luminosity, it is also shown, that the bright luminosity and the gloomy luminosity can extracted from the map, even if the frontiers are not very clear. For further discussion of the results, see [13]. To conclude, the map is orderly, but the clusters have topological defects. This fact may be result of choosing a two dimensional map instead of the optimum four or five dimensional map. Another problem is that the data set is partly equivocal.

Obviously, choosing the correct spiketrains from the data, the classification can be rectified. The errors are due to a small data set, short recording and a great variation if the data within the recording. So, one of further steps must be to identify the beginning and the end of such spiketrains. Also, overlapping time windows might be useful and the influence of the duration of the time windows must be explored.

7 Summary and Conclusion

In this paper we have presented the aim of the INTER-project. The final goal of the project is to build a PNS-remoted limb prostheses, which has the same behaviour as a natural limb. An important package within the project is the signal processing of the recorded nerve signals. We have presented an approach for the solution of this problem. We have proposed to use INCA for separating the individual nerve signals from the recorded mixture of multiple sources. For the classification of the data we propose a SOM because the approach must be verified by using results from In Vivo tests.

We have presented an implementation to simulate nervous signals recorded by a ring electrode. The obtained signals are used to validate the approach to separate nervous signals using INCA. We have shown that INCA is able to separate mixed signals into raw signals, even if the waveform is like spikes of the nervous system. Finally, we have reported about first experiences with biological nerve signals using Kohonen's SOM. We have shown that classifying nerve signals is possible, but there are still problems in detecting the apropriate spiketrains. So the further step is to evaluate more adapted methods for preprocessing of the data. We intend to explore overlapping time windows as well as the duration of the time windows. Another important point is the detection of significant spiketrain packages.

Acknowledge

The authors like to thank Berry Lee from the Max-Planck-Institut für biophysikalische Chemie at Göttingen for the permission to use the Macaque data set and Thomas Wachtler from the Max-Planck-Institut für biologische Kybernetik at Tübingen for his attempts to get this permission.

The INTER project is granted by the European Community under ESPRIT BR project #8897. Thank you to all participating partners of the project for their outstanding effort in this project.

References

1. Dario, P. and Cocco,M.; Technologies and Applications of Microfabricated Implantable Neural Prostheses. In *IARP Workshop on Micromachine & Systems 1993, Tokyo* 1993
2. Kovacs, G.T.A. and Storment, C.W. and Rosen, J.M.; Regeneration Microelectrode Array for Peripheral Nerve Recording and Stimulation. In *IEEE Transactions on Biomedical Engineering, Vol. 39, No. 9* 1992
3. Marks, A.F.; Bullfrog Nerve Regeneration into Porous Implants. In *Anatomical Record, Vol. 163* 1969
4. Llinás, R. and Nichelson, C. and Johnson,K.; Implantable Monolithic Wafer Recording Electrodes for Neurophysiology. In *Brain Unit Activity During Behaviour, Chapter 7* 1973
5. Edell, D.J.; Development of a Chronic Neuroelectronic Interface. In *Doctoral Dissertation U.C. Davis* 1980
6. Edell, D.J.; A Peripheral Nerve Information Transducer for Amputees: Long-Term Multichannel Recordings from Rabbit Peripheral Nerves. In *IEEE Transactions on Biomedical Engineering, Vol. BME-33, No. 2* 1986
7. Akin, T. and Najafi, K.; A Micromachined Silicon Sieve Electrode for Nerve Regeneration Applications. In *IEEE* 1991
8. Cocco, M. and Dario, P. and Toro, M. and Pastacaldi, P. and Sacchetti, R.; An Implantable Neural Connector Incorporating Microfabricated Components. In *Micro Mechanics Europe, Neuchâtel* 1993
9. Meyer,J.-U. and Valderrama,E. and Dario,P.; Perforated Silicon Dice with Integrated Nerve Guidance Channel for Interfacing Peripheral Nerves. In *Proceedings of Eight IEEE International Workshop on Micro Electromedical Systems (MEMS'95), Amsterdam* 1995
10. Meier, J.; Selectivity and Design of Neuro-Electronics Interfaces. In *Doctoral Dissertation University of Twente* 1992
11. Jutten, C. and Hérault, J.; Blind separation of sources, Part I: An adaptive algorithm based on neuromimetic architecture. In *Signal Processing, Elsevier, Vol. 24* 1991
12. Speckman, H. and Raddatz, G. and Rosenstiel, W.; Improvement of learning results of the selforganizing map by calculating fractal dimensions. In *ESANN'94* 1994
13. Bogdan,M. and Rosenstiel,W.; Artificial Neural Nets for Peripheral Nervous System - remoted Limb Prostheses. In *Neural Networks & their Applications, Nanterre* 1994

Information Processing and Cellular Systems

LONG-RANGE METABOLIC SIGNALLING IN HETEROGENEOUS CATALYTIC SYSTEMS: COMPUTER SIMULATION STUDY

P. MENDES
Institute of Biological Sciences, University of Wales, Aberystwyth, Dyfed SY23 3DA, United Kingdom.

P. MARMILLOT
Department of Chemistry, Bowling Green State University, Bowling Green, Ohio 43403, U.S.A.

J.-F. HERVAGAULT
Unité de Recherche Associée, Centre National de la Recherche Scientifique 1442, Université de Compiègne, BP 649, 60206 Compiègne, France.

G. R. WELCH
Department of Biological Sciences, University of New Orleans, New Orleans, LA 70148, U.S.A.

1 Introduction

Information processing in living cells is largely a matter of the facilitation of adaptational metabolic responses to changing physiological situations. Cell metabolism, in itself, consists primarily of just two elementary events: enzyme-catalyzed (electro)chemical transformation of matter and mass-transport (which, in the simplest case, involves Fickian diffusion). The regulation of metabolism, particularly in multicellular differentiated organisms, often involves a hierarchy of over-riding control mechanisms. At the cellular level, the most common *modus operandi* entails activation/inhibition of enzyme activity by diffusible ligands. Analysis of the uniformity in the size of prokaryotic cells (~1μm), of eukaryotic cells (~10μm), and of the components (~0.1μm) contained within eukaryotic cells shows that the scale of most present-day cells and their substructures is maximized for the coupling of metabolic operation to diffusion.[1] From what we know today about the kinetic properties of enzymes and the transport characteristics of cytoplasmic media, metabolism could not work efficaciously if the sizes were greater.

A dilemma is posed by the existence of certain macroscale cellular systems, such as those found in striated muscle and nerve, whose size is often 0.1-1.0 m. Simple diffusion is not fast enough for the coordination of metabolic enterprises in these cell-types. We must seek alternate mechanisms for signal-transmission within the cytoplasm of these systems. In so doing, we must abandon the classical biochemical view, whereby intermediary metabolic processes are regarded as homogeneously "dissolved" in the aqueous cytosol. It is now known that larger eukaryotic cells contain an intricate cytomatrix, consisting of numerous membrane surfaces and a dense, variform array of proteinaceous fibrous elements.[2-6] There is abundant evidence that much of intermediary metabolism takes place in association with such particulate structures.[5,7-12] Adsorption of enzymes to subcellular structures is accompanied, generally, by changes in kinetic characteristics, arising from a variety of microenvironmental influences.[13-16] Interest in this organizational motif has heightened with the observation, in a number of cases, that the reversible association of enzyme activity with the cytomatrix is regulated by metabolic effectors.[12,17,18] Enzymes that partition between the bound and soluble states have been termed "ambiquitous".[19,20]

Among the systems catalogued for their "ambiquitous" character, the glycolytic pathway has been studied most extensively. Histochemical and biochemical investigations in muscle fibers have shown that the majority of the glycolytic enzymes are localized along the actin-filaments, as well as near the *I-band* which delineates the sarcomeric units of the muscle ultrastructure.[21-23] Also, glycolytic enzymes have been found to associate reversibly with filamentous elements in neuronal cells.[12,24]

There are theoretical indications that glycolysis generates "limit-cycle" oscillations in cell metabolism; and a number of potential physiological roles have been attributed thereto.[25-27] Glycolytic oscillations have, indeed, been observed experimentally in a number of cellular systems (see for example refs 25 and 28). Most of the modelling on this phenomenon has focused on the nonlinear kinetics of the allosteric enzyme, phosphofructokinase (PFK, EC 2.7.1.11), as the primary "oscillophor". Theoretical studies have shown this system capable of generating, not only limit-cycle oscillations, but also multiple steady-state (hysteresis) behavior and abrupt-transition action.[28,29] All such investigations, though, have been based on homogeneous, bulk-phase enzyme/metabolite populations.

Our attention concerns the theoretical PFK model developed by Goldbeter and Lefever,[30,31] involving the spatiotemporal behavior of the *regulatory adenylate* (ATP/ADP) *system* in a limit-cycle regime. Taking diffusion into account, these workers observed spatiotemporal order in the form of homogeneous oscillations, standing or propagating chemical waves, and stationary spatial structures. We have employed the theoretical background of this model, in order to study reaction-diffusion phenomena in a glycolytic system operating under "ambiquitous" intracellular conditions, *viz.*, with part of the PFK enzyme-population free in the cytosolic aqueous bulk-phase and part bound to cytomatrix structure. Couching the results in the "heterogeneous-catalysis" approach of Ross and co-workers,[32] we find a potential role of the limit-cycle mode in regulatory signal-transmission in the larger eukaryotic cells (see also ref 33).

2 Theory

PFK catalyzes the following reaction: fructose 6-phosphate + ATP $\rightarrow$ fructose 1,6-bisphosphate + ADP. (In this work we consider only the ATP/ADP couple in our simulations, which is equivalent to forcing the concentration of the other two metabolites constant.) The basic kinetic model for PFK, free in the bulk solution, is described in detail elsewhere.[30,31] We recount here the essential elements thereof. The allosteric, dimeric enzyme is assumed to exist in two conformational states: "relaxed" (R) and "tense" (T); transition between the two is fully concerted.[34] The substrate (ATP) binds to the two forms with different affinity, and only the R form of the enzyme substrate complex decomposes to yield product (ADP). The product-ligand acts as a positive effector; it binds exclusively to the R form of the enzyme. The substrate is supplied to the system at a constant rate, while the product is removed by a first-order reaction. In the present case, we take the existence of two populations of PFK: PFK_f, the enzyme free in the bulk solution, and PFK_b, the enzyme bound to a subcellular structure.

The form of the reaction rate, catalyzed by PFK_f in the bulk-phase solution is taken from previous studies,[30,31,33] as follows:

$$v_f = \frac{\left[2 \cdot E_0 \cdot \varepsilon/(\varepsilon+1)\right] \cdot \alpha \cdot \left[1+\alpha/(\varepsilon+1)\right](1+\beta)^2}{L \cdot (1+\alpha c)^2 + (1+\beta)^2 \cdot \left[1+\alpha/\ (\varepsilon+1)^2\right]} \tag{1}$$

where α and β denote, respectively, the concentrations of substrate (ATP) and product (ADP), divided by the respective equilibrium dissociation constants, $K_{S(R)}$ and $K_{P(R)}$, for the enzyme-substrate complexes (assuming $K_{S(R)} = K_{P(R)} = d/a$, where d and a denote the unitary kinetic constants, respectively, for the dissociation and binding of substrate and product to the enzyme); E_0 is the total enzyme concentration; L is the allosteric constant; c is the nonexclusive binding coefficient for the substrate; and ε is the (irreversible) normalized catalytic turnover of the active (R form) enzyme substrate complex.[30] When bound to subcellular structures, PFK exhibits altered kinetic properties - in some cases losing its allosteric (sigmoidal) behavior, in favor of a quasi-Michaelian (hyperbolic) rate form.[17,35] Here, we take v_b, the reaction rate for the "bound" enzyme PFK_b as equal to Eq. 1, with $L = 0$.

The evolution of the ATP/ADP concentrations in space and time is given by the following system of partial differential equations:

$$\begin{aligned} \frac{d\alpha}{dt} &= a \cdot \left[\sigma_1 - f(\alpha,\beta)\right] + D_\alpha \cdot \frac{\partial^2\alpha}{\partial r^2} \\ \frac{d\beta}{dt} &= a \cdot \left[f(\alpha,\beta) - \sigma_2 \cdot \beta\right] + D_\beta \cdot \frac{\partial^2\beta}{\partial r^2} \end{aligned} \tag{2}$$

where D_α and D_β are the diffusion coefficients of ATP and ADP, respectively, σ_1 is the injection rate of the substrate and σ_2 is the first order kinetic constant for the outflow of product.[31,33] The function $f(\alpha,\beta)$ corresponds to v_b, in the spatial domain where the enzyme is bound, whereas in the bulk phase, $f(\alpha,\beta)$ conforms to v_f. We consider diffusion in two dimensions as in our previous work.[33]

We make some specific, simplifying assumptions concerning the "heterogeneity" of the system. The bound enzyme (PFK_b) is localized in finite-dimensional "strips" adjacent to bulk phase areas. The concentration of PFK is assumed to be the same in the two regions. A wide range of ratios of kinetic constants for bound vs. free forms of PFK (and other glycolytic enzymes) has been reported in the literature, depending on the cell type and on the experimental conditions. Here, we suppose (in addition to the aforementioned assumption that $L = 0$) that the normalized catalytic turnover constant (ε) of PFK is lowered by a factor of 50 in the bound state (previous simulations showed that results are qualitatively the same over a broad range of such values[33]). Also, the diffusion coefficients of substrate and product in the bound phase are lowered by a factor of 5, compared to the values in the bulk phase.

3 Material and Methods

All the simulations were carried out on a Dan Technology computer based on the Intel 80486 DX 33 MHz running the Microsoft Windows NT 3.5 operating system.

Since all the generic metabolic simulators (see for example ref 36) known to us can only deal with models in which the solvent is well stirred, *i.e.* where diffusion is negligible, we wrote one specifically to deal with the model described above. It was written in the C programming language, using a 5-point central finite differences method for the spatial component and a fourth order Runge-Kutta method for the temporal component. The program was compiled with the Microsoft Visual C/C++ compiler version 2.1. Three-dimensional plots were produced using the public domain program *gnuplot*.

4 Results

In all the simulations shown here the spatiotemporal behavior of two-phase (free and bound) PFK system is studied in two-dimensions. The area in question is a 0.3 x 0.3 cm square (0.09 cm^2). For the numerical integration of Eqs. 2 the square is represented as a 60 x 60 grid of equidistant points. Each of these points represents the behaviour of a small square (0.005 cm side) centred around the point.

We have used the set of kinetic parameters described previously by Goldbeter [31] and us.[33] This set of parameters leads to an unstable steady state if the diffusional terms of Eqs. 2 are negligible (for example by stirring). Since this steady state is

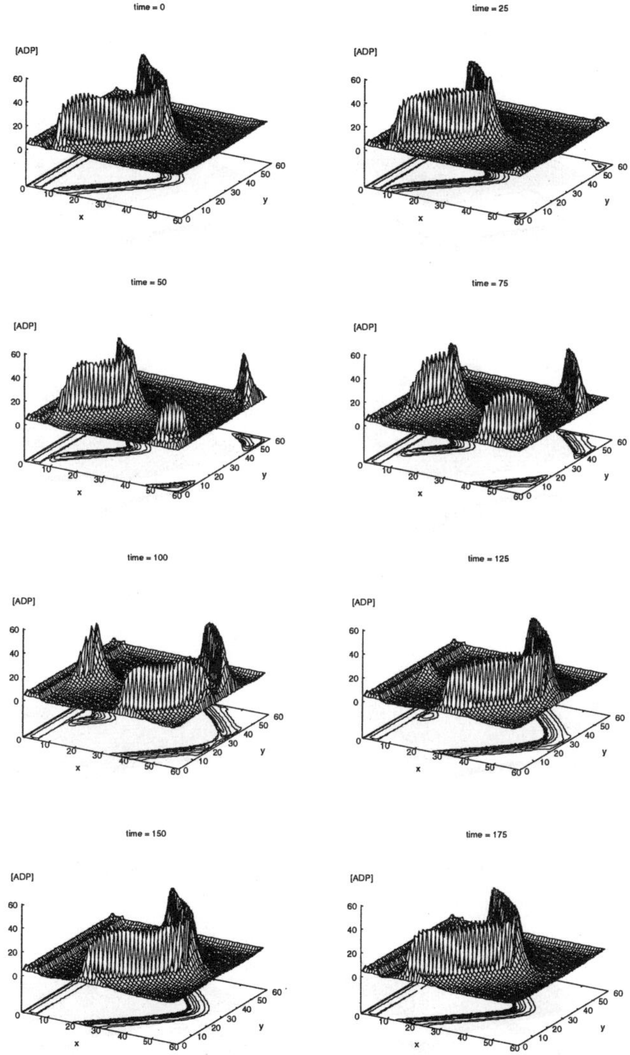

Figure 1. Spatial distribution of ADP concentration at successive times, for a 200-second periodicity. The case of one strip of bound PFK. Each three dimensional plot represents the reduced concentration of ADP (z-axis) as a function of position in the square (x- and y-axis). Plots are separated by 25 seconds each. The space is partitioned in a strip of bound PFK (x between 0 and 6, along the y-axis) and an area of PFK in aqueous phase. Details of the simulation in the Results section.

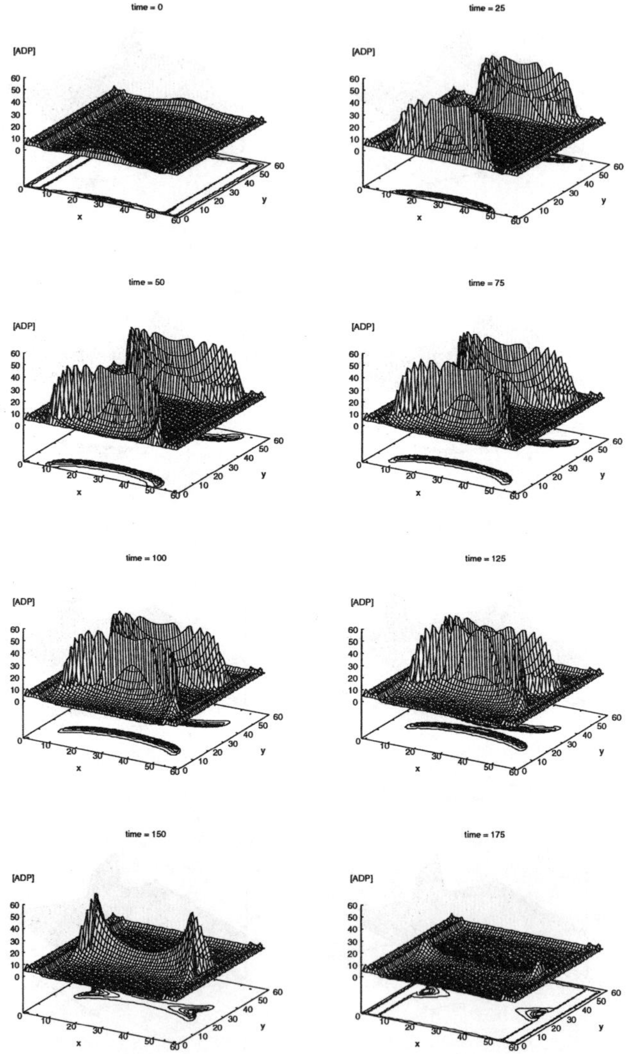

Figure 2. Spatial distribution of ADP concentration at successive times, for a 200-second periodicity. The case of two strips of bound PFK. Details as in Fig 1, except for the existence of an extra strip of bound PFK (x between 54 and 59, along the y-axis).

unstable the system goes into a limit cycle around the steady state (sustained oscillations). We consider only the case in which the diffusional terms are not negligible, introducing for this 4 diffusion coefficients for the diffusible species (ADP_f, ATP_f, ADP_b and ATP_b). The parameters are as follows: $D_{\alpha f} = D_{\beta f} = 10^{-6}$ cm^2s^{-1}, $D_{\alpha b} = D_{\beta b} = 2 \times 10^{-7}$ cm^2s^{-1}, $E_0 = 5 \times 10^{-6}$ M, $L = 5 \times 10^6$, $c = 0.01$, a = 10^8 $M^{-1}s^{-1}$, $\varepsilon_f = 0.1$, $\varepsilon_b = 0.1$, $\sigma_1 = 5 \times 10^{-9}$ Ms^{-1}, $\sigma_2 = 10^{-9}$ s^{-1}. At the edges of the square (*i.e.* in the segments of co-ordinates $(x,0)$, $(x,59)$, $(0,y)$, $(59,y)$) α and β are made constant and take values of $\alpha = 40.1953$ and $\beta = 5.0008$ (which give rise to an unstable steady state in the case of no diffusion effects,[33]).

Following Marmillot *et al.*[33] we partition the area of reaction in one or more strips of bound phase, here always 0.03 cm wide, and strips of bulk-phase of variable width.

4.1 1 strip of bound PFK

One strip of bound PFK is placed along one edge of the square (y axis), from points $x = 0$ to $x = 5$. The integration started from a homogeneous state where in every point in the square $\alpha = 40.1953$ and $\beta = 5.0008$. A first integration is carried out for 10000 seconds to allow the system to settle down to a stable state. At this point we investigated the behavior of the system and found that it had stabilized in a state in which waves of ADP (β) concentration travel from the edge opposite the strip of bound PFK towards it. This behavior is depicted in Fig. 1 where 8 frames sampled at 25 seconds are displayed. The waves are formed as two "hills" of high [ADP] at the corners opposite to the strip, which then join forming a smooth V shaped wave. The latter travels towards the strip and disappears when in contact with its edge. At the same time another wave is already being formed at the opposite edge. These waves occur with a temporal periodicity of around 200 seconds and have a spatial wavelength of around 0.027 cm.

4.2 2 strips of bound PFK

The arrangement described in the previous section is extended by adding one extra strip of bound PFK on the edge of the square opposite to the first strip. After 10000 seconds, the system has stabilized and displays travelling waves of ADP concentration. In this case two waves are formed, one in each edge of the square that is in contact with the aqueous phase. These waves travel along the two strips of bound PFK towards the center of the square. Once they reach the center, they interfere with each other and cancel out. At this time, two more waves are already being formed at the edges and the process repeats itself. In Fig. 2 we show 8 frames sampled at 25 seconds. The temporal period of this phenomenon is, like in the previous case, around 200 seconds and the spatial wavelength is approximately 0.015

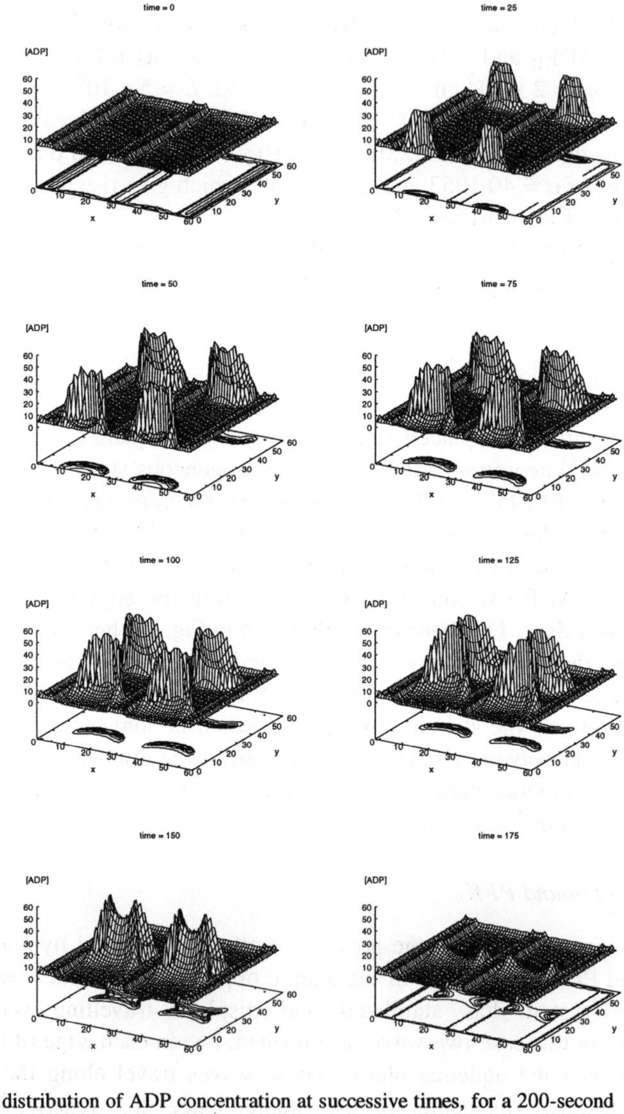

Figure 3. Spatial distribution of ADP concentration at successive times, for a 200-second periodicity. The case of three strips of bound PFK. Details as in Fig 2, except for the existence of an extra strip of bound PFK (*x* between 27 and 32, along the *y*-axis).

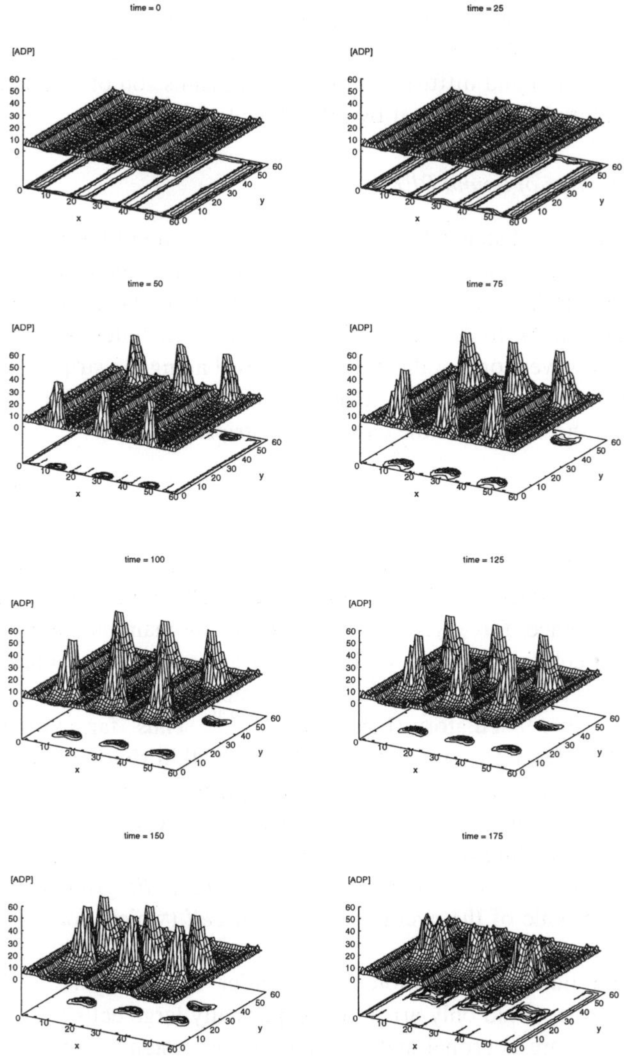

Figure 4 Spatial distribution of ADP concentration at successive times, for a 200-second periodicity. The case of four strips of bound PFK. Details as in Fig. 2, except for the existence of two extra strips of bound PFK (x between 18 and 23 and x between 36 and 41, along the y-axis).

cm. The front of each wave travels through 0.015 cm during approximately 100 seconds (the other 100 seconds are spent forming the waves at the edge and in the interference process in the center). Using the Einstein relation we estimate that the average time for the ADP molecules to travel this distance is 11250 seconds, so the coupling of reaction and diffusion allows the transmission of a signal at a speed two orders of magnitude faster than if by diffusion alone.

4.3 3 and 4 strips of bound PFK

We now shorten the distance between the strips of bound PFK by adding extra strips between the two extreme ones. Figs. 3 and 4 show the behavior of the systems with three and four strips, respectively. We observe that these arrangements produce behaviors very similar to the case of two strips. In each "inter-strip" space of aqueous phase a pair of waves form at the edges and travel along the strips towards the center, where they cancel out. The temporal period and spatial wavelengths are approximately the same as in the case of two strips.

5 Discussion

Glycolysis is known to exhibit oscillatory behavior in a variety of systems. The best example is yeast, where oscillations have been well-studied in cells and in cell-free extracts.[28] Evidence for glycolytic oscillations in mammalian systems has been obtained from cell-free extracts of beef heart and rat muscle, intact mouse *L*-cells, ascites tumor cells and rat fat cells.[37-40] Various physiological functions have been proposed for this oscillatory behavior.[25-27,29,38] Thus far, virtually all of the experimental evidence for glycolytic oscillations within *single cells* attests only to the *temporal* existence thereof. Sustained *spatial* order has been observed experimentally only in supracellular systems, *viz.*, in cell-free extracts and in large-scale, cell-cell signalling (*e.g.* morphogenetic) schemes. Consistent with this finding, theoretical analysis [33,41] has shown that *spatially-periodic* glycolytic patterns are not possible on the scale of the average, *individual cell* (which, for eukaryotes, is ~10^{-3} cm (~10 μm).

Thus, the limit-cycle spatiotemporal order in glycolysis, if it is to play a physiological role, could only arise in exceptionally large cells (or cellular syncytia). Likely examples are neuronal and skeletal-muscle systems, both of which are often macroscopic in length and both of which depend heavily on glycolysis for energy production. The geometry of the cellular system in our computer simulations is modelled broadly after the morphology of striated muscle, with the glycolytic enzymes associated with the parallel filaments, as well as free in solution. In such a design, rapid wave-like processes of the kind treated herein may serve as a faster-than-diffusion, bulk-phase *communication* (or *signal-transmission*) *device*, for example, when the muscle goes from a state of rest to one of sustained contraction.

Based on a PFK limit-cycle model, Goldbeter [42] suggested that the sustained spatiotemporal oscillations in ATP and ADP may serve to govern the overall operation of metabolic pathways.[43] Considering the structural complexity and enzyme "ambiquity" of large eukaryotic (e.g., muscle, neuron) cells, simple diffusion is not apt to be a sufficient mechanism for effective communication between (and co-ordination of) the many, heterogeneous, energy-dependent metabolic processes. Our results, viewed in the light of the Goldbeter model, signify a means by which *local* perturbations (*e.g.,* sources, sinks) in the adenylate-ratio can be communicated throughout the cell (see also ref. 44). In the absence of external (*viz.*, neuronal-electrical, hormonal) stimuli, the adenylate-ratio appears to be the key kinetic (allosteric) modulator of energy metabolism in the muscle sarcoplasm.[45] Coincidentally (?), a sizeable proportion of the "ambiquitous" glycolytic enzymes (e.g., PFK) are found to be free in the "bulk-phase" in the nonstimulated state; the external stimuli elicit a greater partitioning of the enzymes with the filamentous structures in muscle.[17] Also, we note that, morphologically speaking, the "transverse tubules" which conduct the nervous-electrical stimulus into the sarcomere are located at the I-band region which delimits the boundary thereof.

Each subunit of PFK contains an adenylate regulatory site, by which the enzyme is inhibited when ATP is bound; ADP, then, serves to activate PFK by displacing the bound ATP. Kinetic constraints *in vivo* buffer the concentration of ATP very high relative to that of ADP and AMP. Interestingly, when muscle goes from rest to sustained contraction, the ATP level is seen to drop by less than 10% - far less of a decrease than can account for the dramatic activation of glycolytic flux (and PFK activity).[45] Our model suggests a possible explanation for this puzzling situation. Pursuant thereto, we note that the physical association of PFK (and other glycolytic enzymes) with cytomatrix structures often entails cooperative effects amongst the various protein components.[12] For large cooperative networks of interacting proteins subject to an oscillating regulatory ligand, a *molecular field approximation* shows that protein clusters can be maintained in a stationary nonequilibrium allosteric configuration - even though the mean value of the ligand in the bulk is subthreshold.[44]

The model results reported herein are very approximate and lacking in a number of biological concerns. We shall continue to explore the computer-simulation study of glycolysis under more exact physiological conditions of geometry, kinetic constraints, diffusivity effects, etc.

Acknowledgements

We thank Douglas Kell for his comments on this work and for critical reading of this manuscript. P. Mendes thanks the Biotechnology Directorate of the BBSRC for financial support.

References

1. G.R. Welch, *Prog. Biophys. Molec. Biol.* **57**, 71 (1992).
2. K.R. Porter, *J Cell Biol.* **99**, s3 (1984).
3. K.R. Porter and K.L. Anderson, *Eur. J. Cell Biol.* **29**, 83 (1982).
4. K.R. Porter and J.B. Tucker, *Sci. Am.* **244**, 56 (1981).
5. J.S. Clegg, *Am. J. Physiol.* **246**, R133 (1984).
6. K. Luby-Phelps *et al.*, *Ann. Rev. Biophys. Biophys. Chem.* **17**, 369 (1988).
7. G.R. Welch ed., *Organized multienzyme systems. Catalytic properties,* (Academic Press, New York, 1985).
8. G.R. Welch and J.S. Clegg eds. *Organization of cell metabolism.* (Plenum Press, New York, 1987).
9. P.A. Srere, *Ann. Rev. Biochem.* **56**, 21 (1987).
10. G.N. Somero and S.C. Hand, *Physiol. Zool.* **63**, 443 (1990).
11. J.S. Clegg, *Curr. Top. Cell. Reg.* **33**, 3 (1992).
12. H.R. Knull and J.L. Walsh, *Curr. Top. Cell. Reg.* **33**, 15 (1992).
13. A.P. Minton, *Biopolymers* **20**, 2093 (1981).
14. N. Siegbahn *et al.* in *Organized multienzyme systems,* ed. G.R. Welch (Academic Press, New York, 1985).
15. J. Ricard, *J. Theoret. Biol.* **128**, 253 (1987).
16. H.V. Westerhoff and G.R. Welch, *Curr. Top. Cell. Reg.* **33**, 361 (1992).
17. C.J. Masters, *CRC Crit. Rev. Biochem.* **11**, 105 (1981).
18. B.I. Kurganov in *Organized multienzyme systems. Catalytic properties,* ed. G.R. Welch (Academic Press, New York, 1985).
19. J.E. Wilson, *Trends Biochem. Sci.* **3**, 124 (1978).
20. J.E. Wilson, *Curr. Top. Cell. Reg.* **16**, 2 (1980).
21. H. Arnold and D. Pette, *Eur. J. Biochem.* **6**, 163 (1968).
22. P. Sigel and D. Pette, *J. Histochem. Cytochem.* **17**, 225 (1969).
23. S.M. Dagher and H.O. Hultin, *Eur. J. Biochem.* **55**, 185 (1975).
24. H.R. Knull, *Biochim. Biophys. Acta* **522**, 1 (1978).
25. A. Goldbeter and S.R. Caplan, *Ann. Rev. Biophys. Bioeng.* **5**, 449 (1976).
26. P.H. Richter and J. Ross, *Science* 211, 715 (1981).
27. J.F. Hervagault *et al.*, *Proc. Nat. Acad. Sci. USA* **80**, 5455 (1983).
28. A. Boiteux *et al.*, *Curr. Top. Cell. Reg.* **17**, 171 (1980).

29. J.G. Reich and E.E. Sel'kov, *Energy metabolism of the cell.* (Academic Press, New York, 1981).
30. A. Goldbeter and R. Lefever, *Biophys. J.* **12**, 1302 (1972).
31. A. Goldbeter, *Proc. Natl. Acad. Sci. USA* **70**, 3255 (1973).
32. P. Ortoleva and J. Ross, *Adv. Chem. Phys.* **29**, 49 (1975).
33. P. Marmillot *et al.*, *Proc. Natl. Acad. Sci. USA* **89**, 12103 (1992).
34. J. Monod *et al.*, *J. Mol. Biol.* **12**, 88 (1965).
35. N.S. Karadsheh and K. Uyeda, *J. Biol. Chem.* **252**, 7418 (1977).
36. P. Mendes, *Comput. Applic. Biosci.* **9**, 563 (1993).
37. R. Frenkel, *Biochem. Biophys. Res. Comm.* **21**, 497 (1965).
38. K. Thornheim and J.M. Lowenstein, *J. Biol. Chem.* **250**, 6304 (1975).
39. K.H. Ibsen and K.W. Schiller, *Arch. Biochim. Biophys.* **143**, 187 (1971).
40. E.W. Lipkin *et al.*, *Biochemistry* **22**, 792 (1983).
41. G. Nicolis and I. Prigogine, *Self-organization in non-equilibrium systems.* (Wiley, New York, 1977).
42. A. Goldbeter, *FEBS Lett.* **43**, 327 (1974).
43. D.E. Atkinson, *Cellular Energy Metabolism and Its Regulation.* (Academic Press, New York, 1977).
44. G.R. Welch, *Prog. Biophys. Molec. Biol.* **32**, 103 (1977).
45. E.A. Newsholme and C. Start, *Regulation in metabolism.* (Wiley, London, 1973).

ELEMENTARY MODES OF FUNCTIONING IN BIOCHEMICAL NETWORKS

S. SCHUSTER
Department of Biology, Humboldt University Berlin, Invalidenstr. 42, D-10115 Berlin, Germany

C. HILGETAG
Department of Psychology, University of Newcastle, Newcastle upon Tyne NE1 7RU, U.K.

J.H. WOODS and D.A. FELL
School of Biological & Molecular Sciences, Oxford Brookes University, Gipsy Lane, Headington, Oxford OX3 0BP, U.K.

1 Introduction

The kinetic parameters of enzymic reaction systems are often imperfectly known and subject to frequent changes. In contrast, the structure of these systems (that is, the topology of how substances are linked by reactions) remains virtually constant, unless evolutionary time scales are studied. Therefore, the modeling of any biochemical system should begin with the analysis of its structural invariants. These include conservation relations among concentrations, detailed balanced subnetworks, and restrictions on fluxes imposed by balance equations [4, 5, 14, 16]. Moreover, thermodynamic aspects may be included in this analysis. Irreversibility of a reaction, for example, is equivalent to a constraint on the sign of the flux.

The question of how to delimit metabolic pathways has repeatedly been posed. The difficulty results from the essential interdependency of all reactions in the cell. One possible way of approaching this problem is by looking for simple patterns of fluxes fulfilling a steady-state condition for metabolic intermediates, which verifies that these do not accumulate nor deplete in the course of time. Steady-state flux patterns may be represented by specific vectors in the so-called nullspace of the stoichiometry matrix [8, 14].

Structural (topological) analysis in many fields is often done by using graph theory. As for biochemical networks, problems arise when reactions other than monomolecular ones are studied, because they cannot simply be represented by arcs. Indeed, several attempts have been made to adapt graph theory to biochemical networks by introducing auxiliary vertices [3, 9]. In our view, however, structural analysis of metabolic systems can be tackled more elegantly by using a matrix formalism than by graph theory.

Leiser and Blum [11] proposed identifying cyclic and non-cyclic fundamental modes of systems containing substrate cycles, by invoking that any steady-state flux pattern could be decomposed into a superposition of these modes and by requiring that these modes comply with sign constraints for fluxes (if any). We generalized these ideas and related concepts of Fell [6, 7], and recently introduced a concept of 'elementary flux modes' [15]. In the present contribution, we elaborate on this approach. We give a detailed presentation of an algorithm for detecting all elementary modes for any given reaction network. We will discuss how elementary modes can be interpreted, with special reference to information processing by enzymes.

A number of biochemical systems will be analysed by the help of our method. These systems include simple schemes accounting for calcium oscillations and for activation of the adenylate cyclase by the G-protein, which are important examples of intracellular signalling.

2 Elementary modes

Some biochemical reactions may proceed in either direction under physiological conditions, such as the reactions shared by glycolysis and gluconeogenesis. Where some reactions are reversible and some are irreversible, we may decompose the flux vector, $v = (v_1,\ldots,v_r)^{\mathrm{T}}$, into the subvectors v^{irr} and v^{rev}. By appropriate definition of the orientation of reactions, the irreversibility constraint can be written as

$$v^{\mathrm{irr}} \geq \mathbf{0}\,. \tag{1}$$

Consider, for example, the reaction system shown in Scheme 1 and assume that reactions 1 and 3 through 6 are irreversible in the direction indicated, while the net flux of reaction 2 can go in any direction.

At steady state, the reaction rates must fulfil the equation

$$\frac{\mathrm{d}S_i}{\mathrm{d}t} = \sum_{j=1}^{r} n_{ij} v_j = 0\ ,\ \ i = 1,\ldots,n\,, \tag{2}$$

with S_i and n_{ij} denoting the metabolite concentrations and stoichiometric coefficients, respectively. Eq. (2) can be written in matrix notation as

$$\mathbf{N}\boldsymbol{\nu} = \mathbf{0} \ . \tag{3}$$

The stoichiometry matrix of Scheme 1 reads

$$\mathbf{N} = \begin{pmatrix} 1 & -1 & 0 & -1 & 0 & 0 \\ 0 & 1 & -1 & 0 & 0 & 1 \\ 0 & 0 & 0 & 1 & -1 & 0 \\ 0 & 0 & 0 & 0 & 1 & -1 \end{pmatrix} . \tag{4}$$

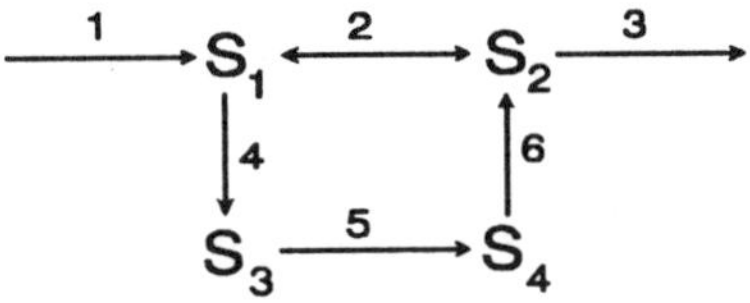

Scheme 1

It is intuitively clear that the net flux in the system shown in Scheme 1 can go via two different routes and that, in addition, a cyclic flux can occur. These routes and the cycle can be represented by the vectors

$$\boldsymbol{\nu}_1 = \begin{pmatrix} 1 & 1 & 1 & 0 & 0 & 0 \end{pmatrix}^{\mathrm{T}}, \qquad \boldsymbol{\nu}_2 = \begin{pmatrix} 1 & 0 & 1 & 1 & 1 & 1 \end{pmatrix}^{\mathrm{T}}, \tag{5a,b}$$

$$\boldsymbol{\nu}_3 = \begin{pmatrix} 0 & -1 & 0 & 1 & 1 & 1 \end{pmatrix}^{\mathrm{T}} \tag{5c}$$

(taken in arbitrary units). These vectors allow to express, by non-negative linear combination, any flux pattern that can occur in the scheme. The flux vector $\begin{pmatrix} 3 & 1 & 3 & 2 & 2 & 2 \end{pmatrix}^{\mathrm{T}}$, for example, can be decomposed into $\boldsymbol{\nu}_1$ plus two times $\boldsymbol{\nu}_2$.

We now formalize the above reasoning by the following definitions.

Definition 1. A *flux mode*, $\boldsymbol{M}$, is defined as the set

$$\boldsymbol{M} = \{\boldsymbol{\nu} \in \mathrm{R}^r \mid \boldsymbol{\nu} = \lambda \boldsymbol{\nu}^*, \ \lambda > 0\}, \tag{6}$$

where $\boldsymbol{\nu}^*$ is an r-vector (unequal to the null vector) fulfilling the following two conditions:

(C1) Steady-state condition. The flux vector $\boldsymbol{\nu}^*$ satisfies eq. (3).

(C2) Sign restriction. If the system involves irreversible reactions, then the corresponding subvector, v^{irr}, of v^* fulfils inequality (1).

According to this definition, a flux mode is sufficiently characterized by one representative of ***M***.

Definition 2. A flux mode ***M*** with a representative v^* is called an *elementary flux mode* if, and only if, v^* fulfils the condition
(C3) Simplicity (non-decomposability). v^* cannot be represented as

$$v^* \neq \lambda_1 v' + \lambda_2 v'' ; \quad \lambda_1, \lambda_2 > 0 , \tag{7}$$

with two flux vectors v' and v'' (unequal to the null vector) having the properties:
(i) v', v'' obey restrictions (C1) and (C2),
(ii) both v' and v'' contain zero elements wherever v^* does so, and they include at least one additional zero component each.

Condition (C3, ii) is a formalization of the concept of 'genetic independence' introduced in [17]. This condition states that a decomposition of routes into two other modes must not involve additional enzymes. The flux mode running through reactions 1, 4 through 6, and 3 in Scheme 1 is to be considered elementary although it can be decomposed into the mode using reactions 1, 2, and 3 and the cyclic mode using reactions 4 through 6 and 2 (in reverse direction). This follows from condition (C3, ii) and can also be understood by the reasoning that the number of reactions depends on the level of description. Reaction 2 might in fact be composed of several elementary steps. Thus, condition (C3) enforces the reasonable condition that elementary modes should be invariant to splitting up a reaction into several steps.

Definition 3. A flux mode ***M*** is called a *reversible flux mode* if, and only if, $M' = \{-v \mid v \in M\}$ is a flux mode as well. Otherwise, ***M*** is called an *irreversible flux mode*. The same distinction can then be made for elementary flux modes.

It is sensible to consider oppositely directed reversible flux patterns as separate modes since opposite fluxes correspond to different biological functions. For example, the two directions of operation of the H^+-ATPase are related to ATP production and generation of a proton gradient. Scheme 1 is an example of a system that has irreversible modes only, although one reaction is reversible.

3 The Algorithm

We developed an algorithm for detecting all elementary modes in reaction networks of any complexity. It is related to the Gauss-Jordan elimination method. In an earlier paper [15], we have outlined a sketch of this algorithm and have given some geometrical ideas to substantiate it. A complete justification of the algorithm will be given elsewhere.

A sequence of matrices (tableaux) is consecutively computed, starting from the stoichiometry matrix augmented with the n x n identity matrix. The algorithm comprises the following steps:

1. Start from a tableau containing the transposed stoichiometry matrix, $\mathbf{N}^{\mathrm{T}}$, and the identity matrix,

$$\mathbf{T}^{(0)} = \begin{pmatrix} \mathbf{T}^{(0)}_{\mathrm{irr}} \\ \mathbf{T}^{(0)}_{\mathrm{rev}} \end{pmatrix} = \left(\begin{array}{c|cc} \mathbf{N}^{\mathrm{T}}_{\mathrm{irr}} & \mathbf{I} & \mathbf{0} \\ \mathbf{N}^{\mathrm{T}}_{\mathrm{rev}} & \mathbf{0} & \mathbf{I} \end{array}\right), \tag{8}$$

where the decomposition of $\mathbf{N}$ into $\mathbf{N}_{\mathrm{rev}}$ and $\mathbf{N}_{\mathrm{irr}}$ is done according to the decomposition of $\boldsymbol{v}$ into $\boldsymbol{v}^{\mathrm{rev}}$ and $\boldsymbol{v}^{\mathrm{irr}}$.

2. From the tableau $\mathbf{T}^{(j)}$ ($j = 0, 1, 2,..., n-1$), the elements of which are denoted by $t^{(j)}_{i,h}$, calculate the next tableau, $\mathbf{T}^{(j+1)} = \left(\mathbf{N}^{(j+1)^{\mathrm{T}}} \quad \mathbf{M}^{(j+1)}\right)$ in the following way:

a) For each row, $\boldsymbol{m}^{(j)}_{i\cdot}$, of $\mathbf{M}^{(j)}$, determine the set $\mathrm{S}(\boldsymbol{m}^{(j)}_{i\cdot})$ defined by

$$\mathrm{S}(\boldsymbol{m}^{(j)}_{i\cdot}) = \left\{k:\ m^{(j)}_{ik} = 0\right\}, \tag{9}$$

b) First, all rows of $\mathbf{T}^{(j)}_{\mathrm{rev}}$ with a zero in the $(j+1)$th column go into $\mathbf{T}^{(j+1)}$. Then, new rows formed by allowed linear combinations of pairs of rows of $\mathbf{T}^{(j)}_{\mathrm{rev}}$ with non-zero elements in this column consecutively go into $\mathbf{T}^{(j+1)}$ if they fulfil the conditions

$$t^{(j)}_{i,j+1} \cdot t^{(j)}_{m,j+1} \neq 0, \tag{10}$$

$$\mathrm{S}(\boldsymbol{m}^{(j)}_{i\cdot}) \cap \mathrm{S}(\boldsymbol{m}^{(j)}_{m\cdot}) \not\subseteq \mathrm{S}(\boldsymbol{m}^{(j+1)}_{l\cdot}) \tag{11}$$

for all row indices l belonging to the tableau $\mathbf{T}_{\text{rev}}^{(j+1)}$ as it has been compiled until that moment. If this tableau does not yet contain any row, condition (11) is taken to be fulfilled. The combination of rows should be done so as to give a zero element in the $(j+1)$th column,

$$\boldsymbol{\theta} = t_{i,j+1}^{(j)} \cdot \boldsymbol{t}_{m\cdot}^{(j)} - t_{m,j+1}^{(j)} \cdot \boldsymbol{t}_{i\cdot}^{(j)} . \tag{12}$$

Relation (11) implies that combination of the rows will lead to a new row containing a set of zeroes not yet generated. $\mathbf{T}_{\text{rev}}^{(j+1)}$ finally encompasses all rows of $\mathbf{T}_{\text{rev}}^{(j)}$ with $t_{i,j+1}^{(j)} = 0$ as well as all the vectors $\boldsymbol{\theta}$.

c) The first part of $\mathbf{T}_{\text{irr}}^{(j+1)}$ is constructed by using all rows of $\mathbf{T}_{\text{irr}}^{(j)}$ with $t_{i,j+1}^{(j)} = 0$. Now, transfer into $\mathbf{T}_{\text{irr}}^{(j+1)}$ appropriate linear combinations of all pairs of rows, $\boldsymbol{t}_{i\cdot}^{(j)}$ and $\boldsymbol{t}_{m\cdot}^{(j)}$, of $\mathbf{T}_{\text{irr}}^{(j)}$ containing non-zero elements of opposite sign in the $(j+1)$th column, that is,

$$t_{i,j+1}^{(j)} \cdot t_{m,j+1}^{(j)} < 0 \tag{13}$$

and also satisfying relation (11) for all row indices l belonging to $\mathbf{T}_{\text{irr}}^{(j+1)}$ as it has been compiled until that moment. Appropriate combination means addition after scaling to ensure the $(j+1)$th column of the result will be zero, that is,

$$\boldsymbol{\phi} = \left| t_{i,j+1}^{(j)} \right| \cdot \boldsymbol{t}_{m\cdot}^{(j)} + \left| t_{m,j+1}^{(j)} \right| \cdot \boldsymbol{t}_{i\cdot}^{(j)} . \tag{14}$$

d) Similarly, for all pairs of rows, $\boldsymbol{t}_{i\cdot}^{(j)}$ and $\boldsymbol{t}_{m\cdot}^{(j)}$, with $\boldsymbol{t}_{i\cdot}^{(j)}$ belonging to $\mathbf{T}_{\text{rev}}^{(j)}$ and $\boldsymbol{t}_{m\cdot}^{(j)}$ belonging to $\mathbf{T}_{\text{irr}}^{(j)}$ and fulfilling relations (10) and (11) for all row indices l belonging to $\mathbf{T}_{\text{irr}}^{(j+1)}$ as it has grown until that moment, calculate the vectors

$$\tau = \operatorname{sgn}\left(t_{i,j+1}^{(j)}\right)\left(t_{i,j+1}^{(j)} \cdot \mathbf{t}_{m\cdot}^{(j)} - t_{m,j+1}^{(j)} \cdot \mathbf{t}_{i\cdot}^{(j)}\right) \tag{15}$$

consecutively and append them to $\mathbf{T}_{\text{irr}}^{(j+1)}$. $\mathbf{T}^{(j+1)}$ is obtained by

$$\mathbf{T}^{(j+1)} = \begin{pmatrix} \mathbf{T}_{\text{irr}}^{(j+1)} \\ \mathbf{T}_{\text{rev}}^{(j+1)} \end{pmatrix}. \tag{16}$$

3. The previous phase of the algorithm ends with $\mathbf{T}^{(n)}$. The reversible elementary modes are represented by the rows of $\mathbf{M}_{\text{rev}}^{(n)}$ and $-\mathbf{M}_{\text{rev}}^{(n)}$, and the irreversible elementary modes are represented by the rows of $\mathbf{M}_{\text{irr}}^{(n)}$.

To understand this algorithm, it is helpful to realize that at the beginning, the hypothetical situation is considered that all metabolites are external. In this case, every reaction represents an elementary mode on its own. This is reflected by the submatrices $(\mathbf{I} \quad \mathbf{0})$ (reversible modes) and $(\mathbf{0} \quad \mathbf{I})$ (irreversible modes) in the initial tableau, $\mathbf{T}^{(0)}$. In each step of the algorithm, one metabolite, S_i, is "internalized", that is, it is no longer considered external. Its concentration must then satisfy the steady-state condition (2). This is just achieved by the linear combinations of preliminary elementary modes as given in eqs. (12), (14) and (15). One can include the external metabolites in the stoichiometry matrix without zeroing the corresponding columns in the tableau. At the end of the algorithm, this gives the stoichiometric coefficients of the externals in the elementary modes.

For illustration, we first consider a branched system in which a substance S_1 is produced in a reversible reaction and utilized in two competing, irreversible reactions. The initial tableau reads

$$\mathbf{T}^{(0)} = \left(\begin{array}{c|cc} \mathbf{N}_{\text{irr}}^{\text{T}} & \mathbf{I} & \mathbf{0} \\ \mathbf{N}_{\text{rev}}^{\text{T}} & \mathbf{0} & \mathbf{I} \end{array}\right) = \left(\begin{array}{c|ccc} 1 & 1 & 0 & 0 \\ \hline -1 & 0 & 1 & 0 \\ -1 & 0 & 0 & 1 \end{array}\right). \tag{17}$$

To construct the tableau $\mathbf{T}^{(1)}$, we can here combine all pairs of rows possible. Note that if reactions 2 and 3 were irreversible, the second and third rows could not be combined, because $(-1)\cdot(-1) > 0$, so that condition (13) would not be fulfilled. We obtain

$$\mathbf{T}^{(1)} = \left(\begin{array}{c|ccc} 0 & 1 & 1 & 0 \\ 0 & 1 & 0 & 1 \\ \hline 0 & 0 & 1 & -1 \end{array}\right). \tag{18}$$

Since $n = 1$ for the considered system, we can proceed to step 3 of the algorithm, and obtain the irreversible elementary modes $(1 \;\; 1 \;\; 0)^{\mathrm{T}}$ and $(1 \;\; 0 \;\; 1)^{\mathrm{T}}$, and the reversible elementary modes $(0 \;\; 1 \;\; -1)^{\mathrm{T}}$ and $(0 \;\; -1 \;\; 1)^{\mathrm{T}}$.

To demonstrate the relevance of condition (11), we analyze the reaction system shown in Scheme 2, which is composed of irreversible reactions only. The tableau $\mathbf{T}^{(2)}$ reads

$$\mathbf{T}^{(2)} = \left(\begin{array}{ccc|ccccc} 0 & 0 & 0 & 1 & 1 & 0 & 0 & 0 \\ 0 & 0 & -1 & 0 & 1 & 0 & 0 & 1 \\ 0 & 0 & 1 & 1 & 0 & 1 & 1 & 0 \\ 0 & 0 & 0 & 0 & 0 & 1 & 1 & 1 \end{array}\right). \tag{19}$$

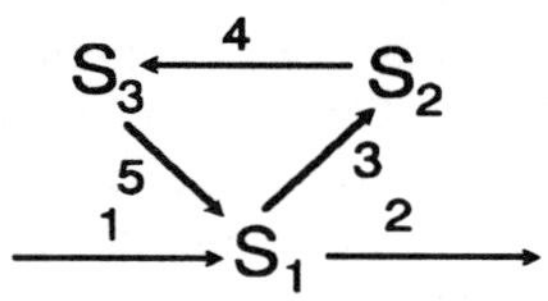

Scheme 2

In the next step, one might think of combining the second and third rows, because the elements 1 and −1 have opposite signs. However, these rows do not satisfy condition (11), because the intersection of the sets indicating the positions of their zeroes is empty and is, hence, a subset of the respective sets of rows 1 and 4, which go first into $\mathbf{T}^{(3)}$. The final tableau therefore contains two rows only,

$$\mathbf{T}^{(3)} = \mathbf{T}^{(4)} = \left(\begin{array}{ccc|ccccc} 0 & 0 & 0 & 1 & 1 & 0 & 0 & 0 \\ 0 & 0 & 0 & 0 & 0 & 1 & 1 & 1 \end{array}\right). \tag{20}$$

If we had combined the second and third rows in $\mathbf{T}^{(2)}$, we would have obtained the mode $(1 \quad 1 \quad 1 \quad 1 \quad 1)$, which is not elementary.

Now consider a reaction system with the stoichiometry matrix

$$\mathbf{N} = \begin{pmatrix} 1 & 1 & -1 \\ 2 & 1 & 0 \end{pmatrix} \tag{21}$$

with all reactions being reversible. The tableau $\mathbf{T}^{(1)}$ reads

$$\mathbf{T}^{(1)} = \left(\begin{array}{cc|ccc} 0 & 1 & 1 & -1 & 0 \\ 0 & 2 & 1 & 0 & 1 \\ 0 & 1 & 0 & 1 & 1 \end{array}\right). \tag{22}$$

At the beginning of the next step, the tableau $\mathbf{T}^{(2)}$ is empty. Therefore, the first and second rows can be combined to give rise to the elementary mode $(1 \quad -2 \quad -1)$. It is reversible, that is, $(-1 \quad 2 \quad 1)$ is an elementary mode also. Combinations of the first and third rows as well as of the second and third rows are now excluded, because they fail condition (11) against the row already generated in $\mathbf{T}^{(2)}$. Here condition (11) ensures that no duplicate modes are calculated.

We automated the given algorithm as computer programs in Turbo-Pascal for PC's and Smalltalk for UNIX workstations. The stoichiometry matrix and the information about which reactions are irreversible are the only input required. The output of elementary modes is then given in the form of vectors. It has turned out that the run time strongly depends on the order in which the columns of the tableau are zeroed. It appears that it is favourable to consider those columns that are most sparse first. Moreover, building the solution along linear pathways (that is, arranging the stoichiometry matrix so as to encompass band-diagonal structures) often shortens run time. When analysing relatively big systems (e.g., the model of fat synthesis in adipose tissue proposed in [5] comprising 42 reactions), we met with the difficulty that, in the intermediate tableaux, very huge integer numbers occured so that the Turbo-Pascal program produced overflow errors. The Smalltalk program, however, is able to handle such numbers.

A similar algorithm for computing metabolic routes leading from some given substrate(s) to some given product(s) was given by Mavrovouniotis *et al.* [12], improving an earlier approach of Seressiotis and Bailey [17] (for a detailed presentation see [13]). Their approach was motivated by the important question: what enzymes need to be expressed in a biotechnologically manipulated microorganism to obtain a desired product. A distinction is made between required substrate(s), required product(s), allowed substrates, allowed products and allowed intermediates. In determining the elementary modes, we only distinguish between internal and external metabolites. Importantly, cyclic modes in the sense of futile cycles cannot be obtained by the algorithm of Mavrovouniotis *et al.* [12]. Moreover, that method does not encompass the very efficient test criterion (11). Rather, it eliminates redundant and superfluous modes only after they have been calculated at each step.

4 Examples

Consider the scheme of the hexose monophosphate pathway, as shown in Fig. 1. Reactions 1 and 3 can be regarded as irreversible in the direction indicated, because they are accompanied by hydrolysis of ATP (which here is treated as a metabolite with constant concentration). Condition (1) reads

$$v_1,\ v_3 \geq 0\ . \tag{23}$$

The nullspace is spanned, for example, by the two vectors

$$\boldsymbol{k}_1 = \begin{pmatrix}1 & 1 & 1 & 1 & 1 & 2 & 0 & 0 & 0 & 0 & 0 & 0 & 0\end{pmatrix}, \tag{24a}$$

$$\boldsymbol{k}_2 = \begin{pmatrix}1 & -5 & -1 & -1 & -1 & 0 & 6 & 6 & 2 & 4 & 2 & 2 & 2\end{pmatrix}, \tag{24b}$$

which we calculated using the NullSpace command of the MATHEMATICA software package. The problem arises that neither $\boldsymbol{k}_2$ nor $-\boldsymbol{k}_2$ are possible thermodynamically, since they both violate condition (23). An appropriate means to compute flux vectors which do satisfy condition (23) is to apply the algorithm set out above. It gives three elementary modes represented by the vectors $\boldsymbol{m}_1 = \boldsymbol{k}_1$,

$$\boldsymbol{m}_2 = \frac{\boldsymbol{k}_1 + \boldsymbol{k}_2}{2} = \begin{pmatrix}1 & -2 & 0 & 0 & 0 & 1 & 3 & 3 & 1 & 2 & 1 & 1 & 1\end{pmatrix}, \tag{25a}$$

$$\boldsymbol{m}_3 = \frac{5\boldsymbol{k}_1 + \boldsymbol{k}_2}{2} = \begin{pmatrix}3 & 0 & 2 & 2 & 2 & 5 & 3 & 3 & 1 & 2 & 1 & 1 & 1\end{pmatrix}. \tag{25b}$$

$\boldsymbol{m}_2$ and $\boldsymbol{m}_3$ are relatively complicated and probably cannot be deduced just by inspection of the scheme. The first elementary mode can be interpreted as the upper part of glycolysis. The two other modes correspond to the hexose monophosphate shunt. The fact that this pathway is represented by two modes, rather than by one, might be surprising. It is due to the interrelation with glycolysis. The balance of G6P and F6P can be achieved either by using the phosphoglucoisomerase reaction in the reverse direction or by using the phosphofructokinase and aldolase reactions. This is a good example for a case where elementary modes overlap by including the same subset of reactions.

We also analysed a more complex model of the red blood cell as given by Joshi and Palsson [10]. This system contains 42 reactions for 33 metabolites and is clearly too complex to allow conclusions about its structural and functional organisation to be drawn by unguided intuition. The reactions of the model comprise glycolysis and the pentose pathway and also contain pathways for nucleotide metabolism, and for interactions of the cell with the external milieu.

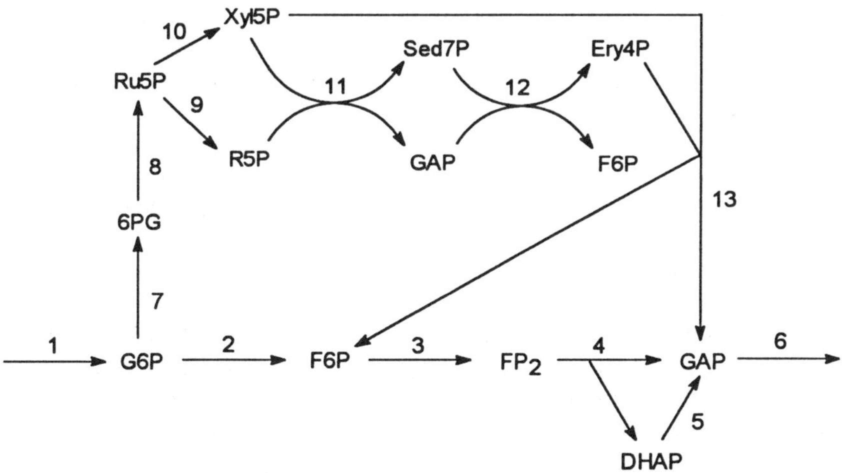

Figure 1: Scheme of the hexose monophosphate pathway together with the upper part of glycolysis.

Using the algorithm outlined above, 25 elementary modes could be found for the system. These modes show a large variety in the combinations of reactions used. They all, however, contain those six glycolytic reactions that are catalysed by the hexokinase, glyceraldehyde phosphate dehydrogenase, phosphoglyceromutase, enolase, pyruvate kinase, and lactate dehydrogenase.

All elementary modes also include an export reaction for lactate. These reactions of the energy metabolism therefore appear to be essential for the cell. Beside the glycolytic reactions, the elementary modes for this model also represent the other metabolic subsystems of the erythrocyte, and interactions between them. There exist combinations of the glycolytic reactions with reactions from the pentose pathway, as above, of glycolysis and parts of the nucleotide metabolism, and also combined fluxes using reactions from all three subsystems. Some elementary modes also represent the activity of the Na/K-pump combined with fluxes in the glycolytic and pentose pathways. As for the other examples above, all 25 elementary modes represent irreversible fluxes, even though some of the reactions they use are principally reversible.

These examples show that elementary modes can often be identified with pathways that have been intuitively recognized, like glycolysis. Other elementary modes, however, represent routes that are not obvious as main pathways of a metabolic system. They are nevertheless valid descriptions for elementary functional states that could work at stationary reaction rates after a sufficient number of the reactions not included in the elementary modes had been inactivated. Moreover, such modes demonstrate the potential for metabolic exchange and signalling between different reaction compartments of a cellular system.

A scheme including glycolysis and gluconeogenesis has been studied earlier [15]. We derived nine elementary modes, with one them corresponding to a futile cycle composed of the metabolites pyruvate, phosphoenolpyruvate and oxaloacetate.

We now address the question whether the concept of elementary modes may help elucidate principles of intra- and intercellular signalling. We first consider cyclic AMP, which is known to be an important intracellular messenger. It is formed from ATP by adenylate cyclase. This enzyme gets activated by the G-protein in the presence of specific hormones. This mechanism can be written as follows (cf. [1]):

$$\mathrm{RH} + \alpha\beta\gamma\text{-GDP} \rightarrow \mathrm{RH}\text{-}\alpha\beta\gamma\text{-GDP}\,,$$

$$\mathrm{RH}\text{-}\alpha\beta\gamma\text{-GDP} + \mathrm{GTP} \rightarrow \mathrm{RH}\text{-}\alpha\beta\gamma\text{-GTP} + \mathrm{GDP}$$

$$\mathrm{RH}\text{-}\alpha\beta\gamma\text{-GTP} \rightarrow \beta\gamma + \alpha\mathrm{GTP} + \mathrm{RH}\,, \qquad \alpha\text{-GTP} + \mathrm{AC} \rightarrow \alpha\text{-GTP-AC}$$

$$\alpha\text{-GTP-AC} \rightarrow \alpha\text{-GDP-AC}\,, \qquad \alpha\text{-GDP-AC} \rightarrow \alpha\text{-GDP} + \mathrm{AC}$$

$$\alpha\text{-GDP} + \beta\gamma \rightarrow \alpha\beta\gamma\text{-GDP}\,, \tag{26}$$

where α, β and γ stand for the three subunits of the G-protein, and R, H and AC denote the receptor, a hormone, and adenylate cyclase, respectively. A more detailed reaction scheme was recently analysed in [18]. Taking GTP and GDP as external metabolites here, one can assume that while the hormone is present, the system reaches a stationary state when. We obtain one elementary mode, which can be represented by a vector composed of ones. That is, at steady state all rates in this system are equal. This mode represents a futile cycle, driven by the hydrolysis of GTP. During operation of the cycle, the active form of adenylate cyclase is kept at a higher level than in the absence of the hormone. It catalyses a reaction which can also be considered as part of a futile cycle. The latter serves to maintain cAMP at a higher level then normally.

Cellular signalling is often linked with intracellular calcium oscillations. At first sight, it may be questionable whether the concept of elementary modes, which is based on the steady-state assumption, can be applied to oscillations. However, biological oscillations have a remarkable stationarity property in that the mean values of the state variables are stationary over long time spans. Reaction rates then fulfil balance equations as well, since for each intermediate, averaged inputs have to balance averaged outputs.

Consider a scheme of the following four reactions: (1) calcium uptake into the cell, (2) calcium efflux out of the cell, (3) ATP-consuming calcium uptake into intracellular stores (e.g. sarcoplasmatic reticulum), and (4) autocatalytic calcium release from these stores. The calcium concentrations in the cytoplasm and inside the stores are to be considered as variable concentrations. This scheme gives rise to the two elementary modes $\begin{pmatrix}1 & 1 & 0 & 0\end{pmatrix}$ and $\begin{pmatrix}0 & 0 & 1 & 1\end{pmatrix}$. The latter mode corresponds to the futile cycle which is responsible for the oscillations by the calcium-induced calcium release mechanism. The method of signalling is again based on a cyclic elementary mode.

5 Interpretation

Steady-state fluxes of metabolic systems can be decomposed into elementary modes. This approach is similar to the decomposition of sounds into pure harmonics. The number of elementary modes may be an important index characterizing biochemical systems. It indicates the 'metabolic richness' of the given system, by showing the variety of its physically realizable functions. Which of these functions are operative or in what proportions they operate simultaneously is determined by the extent of inhibition and activation of enzymes, that is, by the actual values of kinetic parameters. From a biotechnological point of view, elementary modes can be interpreted as synthetic routes for metabolic products.

Biochemists usually discern different pathways in cellular metabolism, although it is by no means clear how such pathways should be delimited properly. For this question, the distinction between internal and external metabolites (that is between substances with variable and constant concentrations, respectively) certainly plays an important role. Often, some concentrations are virtually constant, which results from special thermodynamic and kinetic properties, such as irreversibility of reactions, saturation of enzymes, and separation of time constants. Elementary modes as simple routes that connect external metabolites through internal ones may give a clue to a formal approach to the concept of pathway. This approach would have the advantage that elementary modes are uniquely defined (unlike basis vectors of the nullspace to the stoichiometry matrix), but would also involve the implication that an overlap of simple pathways had to be allowed, as can be seen in the hexose monophosphate pathway example.

Our methodology illustrates another interesting characteristic of biochemical reaction systems since futile cycles, for example the well-studied pyruvate/phosphoenol-pyruvate/oxaloacetate cycle or the fructose-1,6-bisphosphate cycle, can be detected as (cyclic) elementary modes. These have to be distinguished from cycles with a net throughput such as the tricarboxylic acid cycle. Cyclic modes in the sense of futile cycles are not found by the algorithms presented earlier by other authors [12, 13, 17].

It may occur that all elementary modes to a given network contain a zero at some position identical for all modes. This implies that the respective flux is always zero when the system subsists at steady state. This result makes practical sense in some cases, for example for the adenylate kinase reaction.

Although there are many biochemical reactions that can proceed in both directions, reversible flux modes rarely occur in living cells. Indeed, many biochemical transformations can proceed in opposite direction, but not on exactly the inverse routes. It seems that in biochemical systems, irreversible reactions are located in sufficient number and at appropriate positions to exclude occurrence of reversible modes.

It is an intriguing question whether the concept of elementary modes might be useful for elucidating paths of intra- and intercellular signalling. The examples of cAMP and calcium signalling considered in the previous section show that the operation of energy-consuming cycles (which then are not literally 'futile') is an important mechanism in cellular information processing. The catalytic action of many enzymes serves to process information by transforming a signal encoded in the concentration of substrate or effector to another signal encoded in the concentration of product or the reaction rate [2]. Of particular interest are cascades of enzymes modifying each other. These cycles are often interconnected in that a substance involved in one circuit catalyses a reaction belonging to another circuit. Analysis in terms of elementary modes then gives separate modes for each cycle. We believe that structural analysis can help to study the interdependence of these modes in the process of signalling.

References

1. B. Alberts, D.Bray, J.Lewis, M. Raff, K. Roberts and J.D. Watson, *Molecular Biology of the Cell.* Garland Publishing, New York 1989.
2. D. Bray, *Nature* **376** (1995) 307-312.
3. B.L. Clarke, *J. Chem. Phys.* **60** (1974) 1481-1492
4. B.L. Clarke, in: *Advances in Chemical Physics*, Vol. 43, eds. I. Prigogine and S.A. Rice (Wiley, New York, 1980), pp. 1-216.
5. D.A. Fell and J.R. Small, *Biochem. J.* **238** (1986) 781-786.
6. D.A. Fell, *Comments on theor. Biol.* **6** (1990) 1-14.
7. D.A. Fell, in: *Modern Trends in Biothermokinetics*, eds. S. Schuster *et al.* (Plenum Press, New York and London, 1993), pp. 97-101.
8. F. Horn and R. Jackson, *Arch. Rational Mech. Anal.* **47** (1972) 81-116.
9. C. Hyver, *J. theor. Biol.* **42** (1973) 397-409.
10. A. Joshi and B.O. Palsson, *J. theor. Biol.* **141** (1989) 515-528.
11. J. Leiser and J.J. Blum, *Cell Biophys.* **11** (1987) 123-138.
12. M.L. Mavrovouniotis, G. Stephanopoulos and G. Stephanopoulos, *Biotechn. Bioengng.* **36** (1990) 1119-1132.
13. M.L. Mavrovouniotis, *Ind. Eng. Chem. Res.* **31** (1992) 1637-1653.
14. C. Reder, *J. theor. Biol.* **135** (1988) 175-201.
15. S. Schuster and C. Hilgetag, *J. Biol. Syst.* **2** (1994) 165-182.
16. S. Schuster and R. Schuster, *J. Math. Chem.* **6** (1991) 17-40.
17. A. Seressiotis and J.E. Bailey, *Biotechn. Bioengng.* **31** (1988) 587-602.
18. Y. Tang and H.G. Othmer, *Phil. Trans. R. Soc. Lond. B* **349** (1995) 179-195.

COMPUTATIONAL MODELS OF CELLULAR PROCESSING

A. BELL and M. HOLCOMBE,
Department of Computer Science,
University of Sheffield.
Sheffield, S1 4DP, UK.

1. Introduction

Machines and metabolism. The use of finite state machines to model specific metabolic systems originated with Krohn, Rhodes and Langer, [1], whc developed a method for representing a metabolic system as a finite state machine by regarding substrates as states, coenzymes and ATP as inputs and hydrolysed ADP and augmented coenzymes for outputs. Enzymes were not modelled and the assumption was that the appropriate enzymes were available wherever needed. Reaction rates were also ignored. The model was a translation of the metabolic data flow system into a simple machine model. It also did not reflect the parallelism inherent in such a metabolic environment.

The benefit of using this approach was that the theory of finite state machines was quite well developed, in particular there was a decomposition theory which the authors wished to apply. Essentially the theory of finite state machines, for some purposes, can be regarded as part of the theory of semigroups. The Krohn-Rhodes decomposition theory allows us to take an arbitrary finite state machine and to replace it by a collection of machines constructed from simple, finite groups together with some very simple aperiodioc or reset machines assembled together in simple and natural ways - essentially by using parallel and series interconnections. The hope was that by decomposing the machine defined by a metabolic system we would discover some relationship between the simple groups that appeared in this decomposition and the biochemistry.

The theory was generalised in Eilenberg and then by Holcombe [1]. In Holcombe [2] a simple model of the Krebs cycle was analysed using this decomposition method and the groups associated with the Krebs cycle computed. No clear connection with the biochemistry was established, however.

Other machine models of metabolic systems and cellular organisations have been studied. Many make gross assumptions about the way the systems work, none of which seem justified. For example, the role of enzymes as a control mechanism is often omitted, the specific active sites are also neglected. The assumption that each metabolic system acted independently of all others, rather than as part of a complex concurrent system with mutual control processes, is another major weakness.

2. The mathematical basis

We consider a very general type of machine that can model the most general types of computation so far considered and is thus useful in the mathematical theory of metabolic organization and computation. This general setting is the X-machine introduced by Eilenberg [5]. This machine, in its simplest form comprises a fundamental data type, X, which is the basic environment in which computations are carried out together with a set of functions or relations which are the mechanisms which "do" the computations are defined as operating on X. We identify a set, Φ, of such relations on X. A set of possible internal states of the system, Q, are postulated and a structure within the state space is defined by a relation F : Q x Φ --> Q. If q, $q_1 \in Q$ and

q_1 = F(q,) then there will be an arc from q to q_1 labelled by an element $\phi \in \Phi$. Thus the function or relation F will define a labelled, directed graph with nodes from Q and labels from Φ.

Essentially an X-machine looks like a finite state machine with one important exception, the labels of the arcs are functions that operate on a fundamental data type, X, (actually the labels could be relations in theory, in practice we usually use functions). The way in which a computation is carried out is defined as follows:

select a start state and an initial value x_0 for the data type X;

construct a path from the start state to a terminal state (in many examples every state is considered a terminal state and so all we are doing there is defining a path through the state space);

the path labels form a sequence of functions that operate on X;

we apply them in turn to the initial value, x_0 of X and end with a final value, say $x_1$of X.

The transformation from the initial value to the final value represents the computation carried out by the machine.

One benefit of the model is the clear distinction that it makes between the control state of the system and the state of the data being processed at any particular moment, this data will reflect a number of important parameters within and around the cell, principally the availability and concentration of important chemical species together with environmental factors such as temperature. All of these factors will be regarded as data and the cell products also. This is reflected in the way in which computations are defined by sequences of functions that are applicable in certain states only and to data satisfying specific conditions.

These machines generalise most of those previously studied and do provide a general framework in which to discuss computational issues. The complete theory of X-machines attempts at providing a framework within which we try to describe the computational theory of any structure. The results from this theory will be useful in defining a computational model of the function and organisation of many levels of metabolic activity.

3. The computational organisation of the cell

In [7], one author discussed the issue of modelling different levels of computation within the cell and proposed the X-machine model as a framework for a hierarchical approach using higher level metabolic machines operating concurrently. Some analogies with the modelling of VLSI in computer hardware were made.

The fundamental property of the X-machine is that it is completely general and thus we have potential for a substantially unified theory for the subject. (See [3], [4].) However the benefits of X-machines do not end there. As we mentioned before, we wish to relate processing with processors, that is function with structure. Following [7] we demonstrate how X-machines can achieve this. The basic problem with many models in this area is their treatment of the cell as a ‘bag of chemicals’ with no account taken of their internal structure and the specific active sites that are involved. Our next model tries to address this failing.

3.1 A generalised cellular architecture

Let us consider a simple, idealised cellular architecture illustrated in Fig. 1. The general information flows are indicated using the thick arrows and are meant to be indicative of the sort of functional relationships that might apply in different processing sites.

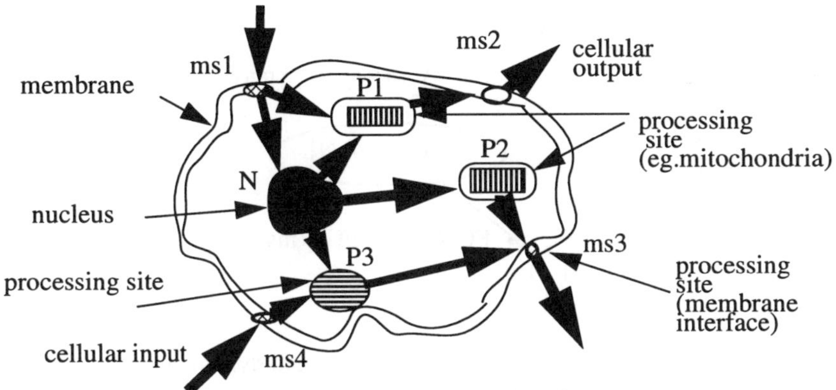

Fig. 1. A hypothetical cell indicating active sites.

The analogy with the VLSI model is strengthened when we consider some of the sort of processing that might be involved, for example we have simple metabolic processes that might correspond to the behaviour of a register, that is information (molecules) might be stored temporarily at a location prior to being further transformed by another processing site, we might have sites performing fairly simple processing akin to that of an adder which operates when its input places contain appropriate values and then there are structures like multipliers that operate continuously. The overall model of the system could then be determined with reference to the data types involving the various active sites and the functions operating on the system described at this level.

The fundamental data type for this simple generic example will be of the form:

X= INms1 x INms4 x [MS1] x [MS2] x [MS3] x[MS4] x [N] x [P1] x [P2] x [P3] x OUTms2 x OUTms3

where each element defines the type of values that the site can deal with and would be defined in a formal manner following our previous work ([4]), INms1 and INms4 are the input datatypes at ms1 and ms4 respectively and OUTms2 and OUTms3 are the output data types at ms2 and ms3 respectively.

The next stage is to identify when the various sites are active and from this we construct a top level model of the system using the specific processing functions that determine what happens to the data. This is drawn from [7].

In the diagram below we note the existence of functions indicated by the arrows, thus we can list the functions and the appropriate data types. As a convention let us indicate input datatypes and output datatypes to sites as follows:

for site K we define IN(K) and OUT(K) to be the appropriate types and

COMP(K) : IN(K) -+-> OUT(K) to be the specific processing function (which might be a partial function allowing for processing to be undefined under certain data conditions - the processor then stops until the situation is resolved. Thus, for example, considering an example of an input system to a cellular system we could con-

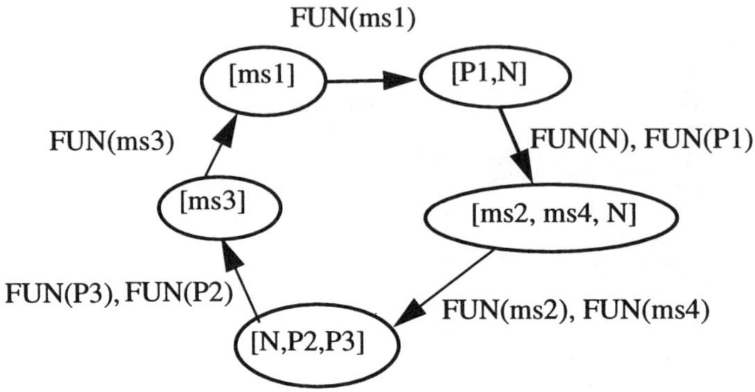

Fig. 2. A possible state space describing the functional dynamics of the cell.

sider the neurotransmission process. If we were to regard an active input site as a collection of acetylcholine receptors at a postsynaptic dendrite in a neuron we would wish to model the binding of acetylcholine at α subunits of the site and the transmission of Na^+ ions across the membrane. Further analysis of the origin of the neurotransmitters as synaptic vesicles in an originating synaptic knob would be achieved by virtue of a model of the internal processing of the preceding cellular system.

It should be remarked that, although discrete models were the inspiration for this approach, we are making no assumptions about the type of the functions and types being considered here. What we do need is some concept of the start and the end of processing activity by each component site. This may raise issues, later, of time and its role in these models. It is possible to augment the X-machine model with time information but we will not deal with this added complication here. An example dealing with this is to be found in [8].

Returning to our hypothetical cell, we note that in some cases there are relationships between types, for example., the environmental inputs and outputs are: INms1, INms4, OUTms2, OUTms3.

where:
IN(ms3) = [P3] x [P2],
IN(P3) = [ms4] x [N],
IN(P1) = [N],
etc.

The component functions are described thus:
COMP(ms1) : INms1 -+-> [ms1],
COMP(ms2) : [P1] -+-> OUTms2,
COMP(ms3) : [P2] x [P3] -+-> OUTms3,
COMP(ms4) : INms4 -+-> [ms4],
COMP(N) : [ms1] -+-> [N],
COMP(P1) : [N] -+-> [P1],
COMP(P2) : [N] -+-> [P2],
COMP(P3) : [ms4] x [N] -+-> [P3],

The system functions are all defined on X and operate on the elements of X as follows:

FUNms1 : (i1, - , a, - , - , - , - , - , - , - , - , -) -->(i1, - , i1, - , - , - , - , - , - , - , - , -);
FUNms4 : (- , i2 , - , - , - , a , - , - , - , - , - , -) --> (- , i2 , - , - , - , i2 , - , - , - , - , - , -);
FUNN : (- , - , a , - , - , - , - , - , - , - , - , -) --> (- , - , - , - , - , - , b , - , - , - , - , -)
where COMP(N)(a) = b [N];
FUNP1 : (- , - , - , - , - , - , a , - , - , - , - , -) --> (- , - , - , b , - , - , - , - , - , - , - , -)
where COMP(P1)(a) = b[P1];
FUNP2 : (- , - , - , - , - , - , a , - , - , - , - , -) --> (- , - , - , - , - , - , - , - , b , - , - , -)
where COMP(P2)(a) = b[P2];
FUNP3 : (- , - , - , a , - , - , b , - , - , - , - , -) --> (- , - , - , - , - , - , - , - , c , - , - , -)
where COMP(P3)(a, b) = c[P3];
FUNms2 : (- , - , - , - , - , - , - , a , - , - , - , -) --> (- , - , - , - , - , - , - , - , - , - , b , -)
where COMP(ms2)(a) = b OUTms2;
FUNms3 : (- , - , - , - , - , - , - , - , a , b , - , -) --> (- , - , - , - , - , - , - , - , - , - , - , c)
where COMP(ms3)(a , b) = cOUTms3;

Note that we have, at ms3, an example of "fan-in", whereby the data paths diverge at N and come back together at ms3. This is a rather undesirable feature in VLSI design, it is not clear if it occurs naturally in cellular metabolism and, if so, what consequences it has.

The processing is determined by what sites are active and what functions are processing, we can postulate an abstract scenario for such a system in Fig. 2. However, we are not claiming this to be a realistic example, it is just an indication of the possibilities of the method. It would be possible, once the specific details of the various functions were known, to simulate the situation such a system operating and to analyse various aspects of its behaviour.

3.2 State refinement

A model such as we have developed is unsatisfactory in a number of ways. Firstly it has "bundled up", in an inconvenient form, much of the concurrent processing that is a feature of such systems. We need to "unpack" this activity and we do this by a process of refinement of the processing functions. This means introducing intermediate states and breaking the functions into parallel components using a simple interleaving semantics.

Let [P1, N, ms3] mean that each of the sites P1, N, ms3 are active and processing. From this we can detect a certain amount of simple parallelism, for example there is simultaneous processing by N and P2 and this can lead to a refinement of the state space by introducing new states and interleaving the functions as described in the VLSI example. A refinement of the model in Fig. 2. is shown in Fig. 3. This new model expands the concurrency and demonstrates how we can replace the parallel processing by sequential processing at the expense of requiring a larger collection of paths through the state. This might assist in modelling some aspects of the system. other aspects, such as time factors, will need other types of model and will be an interesting area for further study.

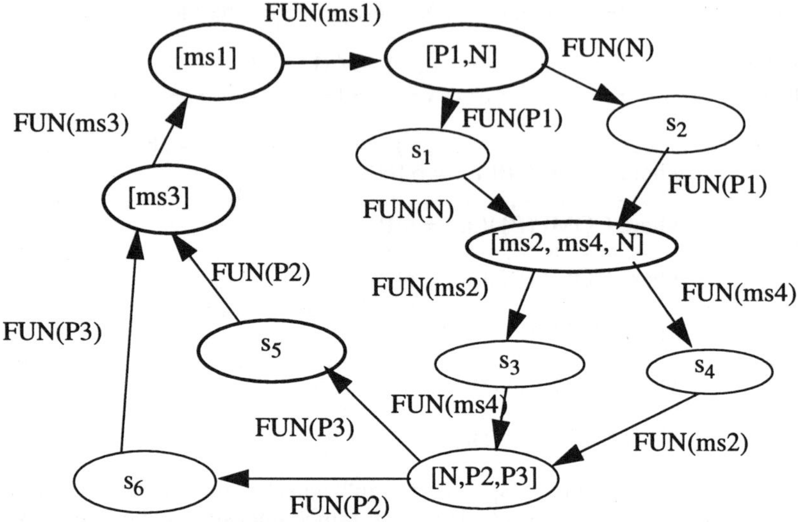

Fig. 4. A possible refined state space.

[s_1, s_2, ..., s_6 are new states introduced during the refinement.]

Naturally, the state space for a real cell will be an order of magnitude more complex. Having produced a general framework for the overall cellular processing we need to consider in some detail what is happening at specific active sites. This is the subject of the next section.

4. Specific, resource dominated metabolic processing

We now consider how we might model the functions of the form FUN(P1) etc.

4.1 Mathematical models of biochemical systems

Our intention is to create a completely general formal method for the specification of parallel processing models of biochemical systems in vivo. This method is intended to be easily adapted, expanded and understood, and aims to give quantitative data about the system being modelled. It is also intended as a basis for further, more abstract analysis using mathematical, logical and machine theories. By keeping the treatment general we can allow for further features and assumptions to be integrated at a later date.

Our model works at the inter-molecular scale, and has two levels of abstraction, the first is intended to be intuitively understandable to a biologist and is used as an intermediate step in the construction of the second, which is more explicit and accurate. In biochemistry distinctions are drawn between enzymes, cofactors, substrates, energy molecules etc. These distinctions are artificial since a molecule of a particular type in one part of a system may act as another type in another part of the system. It is the interactions of the molecules which are important to this model. It may be argued that certain types of molecule should be treated differently due to their capacity for computation, however, we claim that all molecules have this capacity, and the large macromolecules are just an interestingly powerful special case. Whatever the truth of this, the model is constructed in such a way as to allow for refinement at any scale should it prove necessary. For these reasons the model treats all chemicals identically as labelled registers recording the quantity of that chemical in the region under consideration. This quantity is called the *abundance* of the chemical. A separate register is kept for each different conformation of each chemical, and the transformation from one conformation to another is treated as any other interaction.

The interactions of the chemicals are also, in principle, of the same type. That is, they are all quantum mechanical interactions between electron states. It is necessary in order to improve the comprehensibility of our models, however, to distinguish between interactions which cause changes in register values (substrate or product behaviour) and those which do not (enzyme type behaviour). It is also necessary to treat the actions of hormones, globally important factors and membrane conformations in different ways.

Local interactions (those occurring within the confines of a particular compartment of an organelle) in the abstract version are treated identically as algebraic equations which add or subtract numbers to or from the abundance of participating chemicals. These equations will be functions of substrate concentrations and global factors such as temperature, pH and water content. (We contend that the use of differential rate equations cannot be relied upon; (Rosen [9]). These global factors are handled as external registers linked to all functions throughout the system. Both versions of the model consider the interactions of chemicals and reaction sites in the various compartments of an organelle as a single independent system, acting in parallel with other compartments of the organelle, and the rest of the cell, and connected to them by certain transport mechanisms modelled by functions which are similar in form to those for the "local" interactions[1]. This is along the lines of earlier work by Holcombe, [4].

The membranes separating compartments have not been considered, except in their plasticity, which is dealt with in a simplistic way. All reaction sites are considered to be amalgamated into a single "average site" for each compartment. Distances between sites are average distances recorded in a two dimensional array. It is possible that these might not be physical distances, but time relations for the dispersal of reactants, their form is as yet unknown. No consideration has yet been given to the correct spatial arrangement of products of interactions, or to the construction of cellular structures. Time is not explicitly represented, but is a consideration in the form of the equations representing the rates of reactions between chemicals. It may be necessary to incorporate an explicit time component to account for the gradual dispersal of reactants and delays in certain external mechanisms.

4.2. A hierarchy of models.

In this first level of abstraction chemicals are represented by labelled circles or vertices and interactions by various edges from vertex to vertex, vertex to edge or

1.It would be more realistic to consider the whole thing as a single continuous equation, but this would limit the scope for later refinement and seriously damage comprehensibility.

edge to vertex. The diagram representing the reactions within a cer-

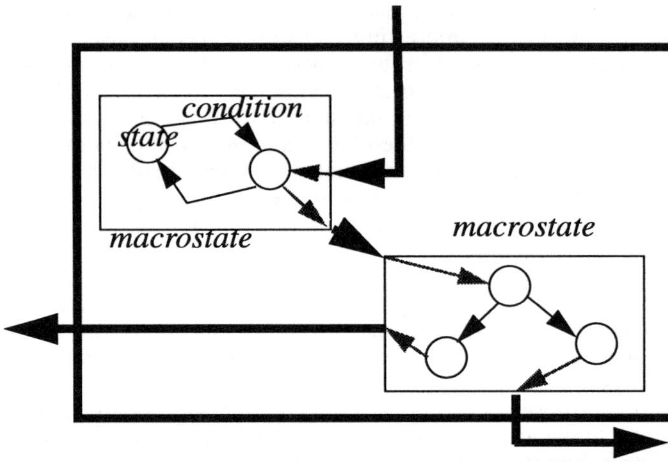

Fig. 4(a). Macrostates composed of states subject to various processing conditions that determine changes of state

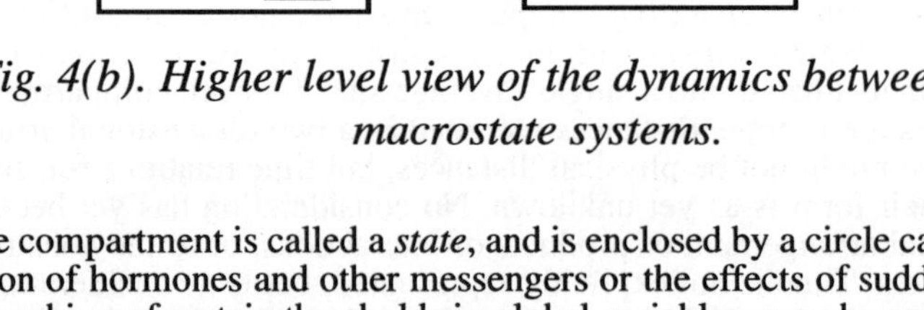

Fig. 4(b). Higher level view of the dynamics between macrostate systems.

tain organelle compartment is called a *state*, and is enclosed by a circle called its *border*. The action of hormones and other messengers or the effects of sudden changes, such as the reaching of certain thresholds in global variables, may be represented by an instantaneous change in the reaction equations. This is represented by a new state connected to the first by an arrow from border to border, labelled with the conditions necessary for the change[1] [8].

The states representing an organelle compartment are bounded by a rectangle or *boundary*, and is called a *macro state*. Communication between macro states (the action of transport mechanisms) is represented by a special type of reaction involving a vertex drawn on the boundaries of the communicating states, connected to each

1.A full biochemical model of the action of the messenger should remove this simplification.

other by a heavy arrow, and to the appropriate vertices of the macro states by elongated edges. The plasticity of membranes alters distance relations and can result in the redistribution of reaction sites between compartments. This is represented by placing the affected macro states in another rectangle connected, in the same way as states in a hormonal change, by an arrow labelled with the necessary conditions, to a rectangle enclosing the new macro states. See Fig. 4.

4.3 Diagrammatic "Flow Chart".

Fig. 5. represents a flow chart representation of part of glycolysis. This process takes place in the cytoplasm and feeds into mitochondria. We can convert such a flow chart to a machine formalism involving register machines according to a number of rules which describe how these flow charts are constructed from component reactions.

4.4. Resource-based machine model.

This has the same notation as the "flow chart" model except at the level of the single state. At this level vertices are replaced by rectangular boxes, representing registers. The interactions are represented by arrows between the registers and a rounded box labelled with a function.

An arrow to a function box represents the input of the *availability* of the chemical (A product of its abundance, its affinity to the reaction (normally 1, for non-enzymes) and a distance relation based on either the separation of chemical source and reaction site or on concentration. The function uses the smallest input availability to determine the rate of the reaction, that is how much of each substrate is used, and product created in a given time based on stochastic considerations. Arrows to a function box also denote that the result of the function is subtracted from the register. Arrows from a function box denote that the result of the function is added to the register.

We now consider some examples of representations of various types of interaction. Each type of interaction will be represented by a basic component machine with a simple combinator mechanism for constructing a complete metabolic model. The reactions are introduced in the standard flow chart representation and register machines described that implement them. The principle idea is that the registers act as an internal store of the available resource - the availability of the appropriate substrate. Registers are annotated with the reaction equation controlling the reaction and the substrate flows are described by letters on the connecting arcs.

We treat enzymes in a simplified way at this stage, making an assumption that the appropriate species are available as required. However, the incorporation into the model of a more realistic approach is not difficult, but would detract from the clarity of the exposition at this stage.

4.4.1. Spontaneous reactions.

The direction of the arrow shows the direction of the reaction.

4.4.2. Register or resource-based reaction.

Here *a* is the availability of substrate S_1 and *b* is the quantity converted. If S_2 is not produced on a one for one basis from S1 then b is multiplied by an appropriate constant.

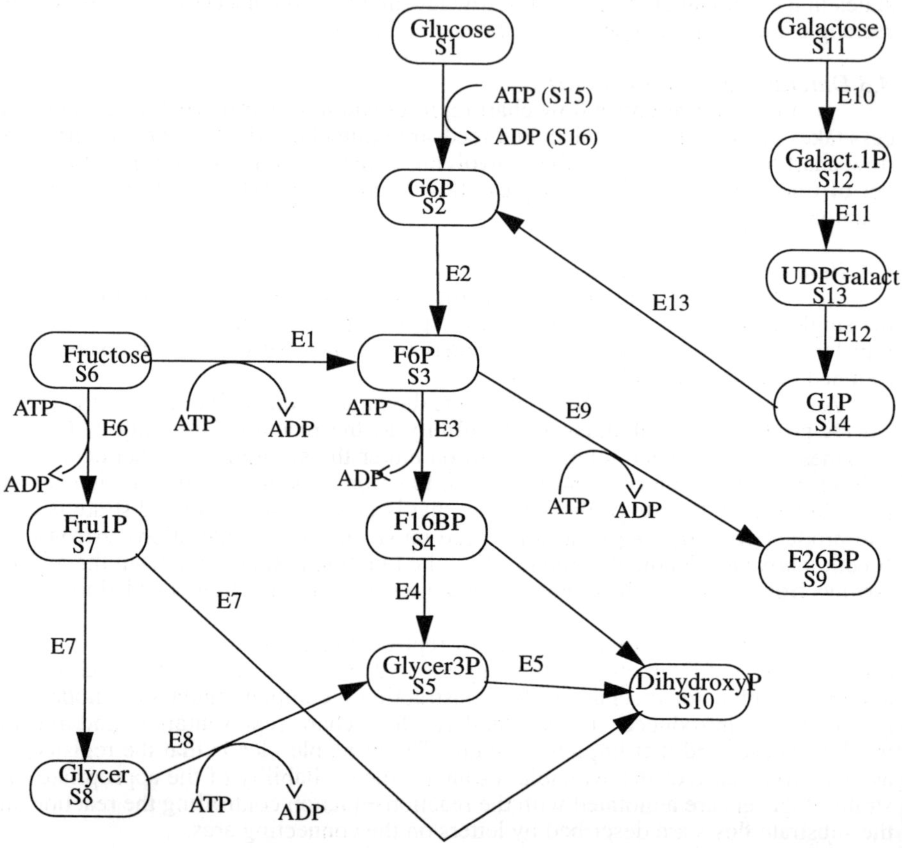

Fig. 5. Glycolysis (The specific substrates and enzymes are listed in the appendix).

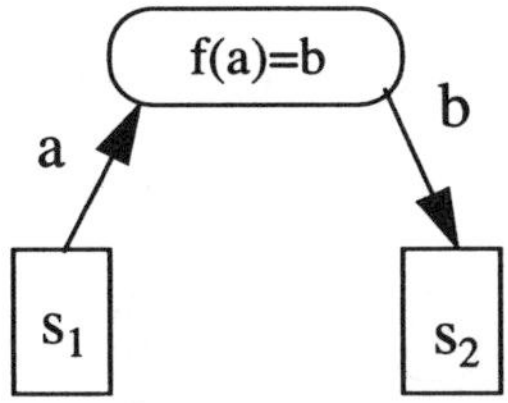

NB. Since this reaction is reversible there should be a second function box with arrows arranged in reverse to represent the reverse equation.

4.4.3. A fuller treatment allowing reverse reaction.

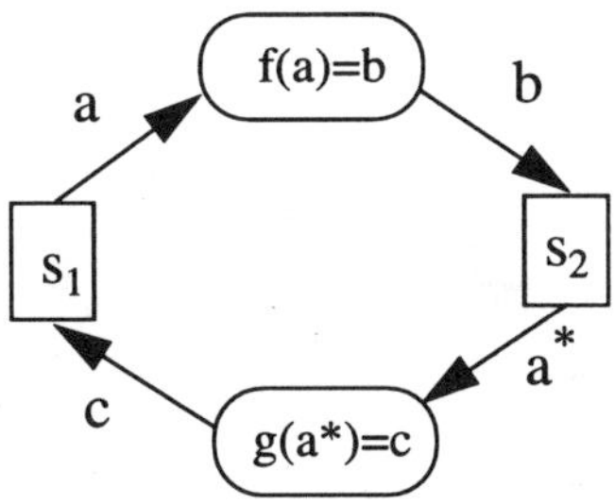

Henceforth we shall omit such reverse reactions for ease of comprehension, but we must not forget them as they may be significant in reaching equilibrium situations or in regimes where some product removal mechanism breaks down.

4.4.4. Catalysed reactions.

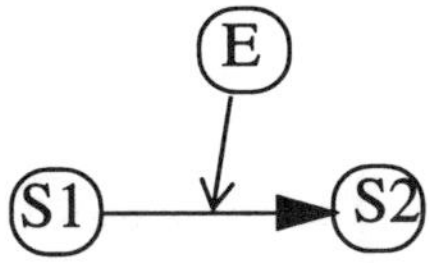

Register machine model

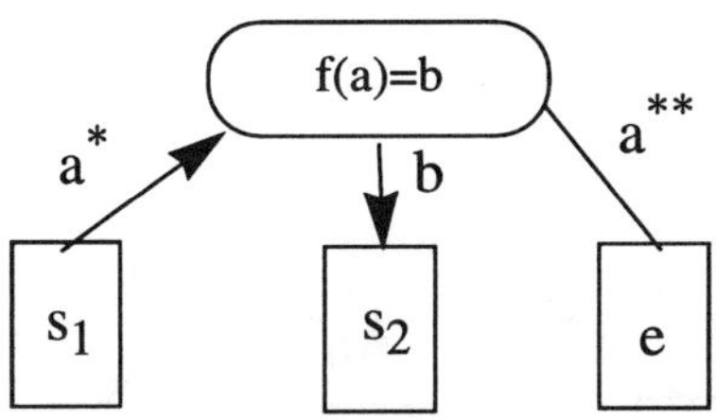

Here a* is the availability of substrate S1, a** is the availability of enzyme (or other catalyst) E and a is the minimum of {a*, a**}. Note that the edge attached to E has no arrow, this means that the abundance of E is not changed.

4.4.5. Branching.
A single substrate decomposing into two or more products.

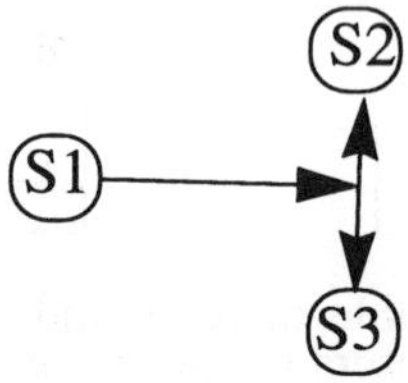

The drawing of edge to edge prevents confusion with pairs of competitive reactions. Note that the arrow heads are in the "forward" direction. It should be noted that edges do not have to intersect at right angles, this has been done purely for clarity.

Register machine:

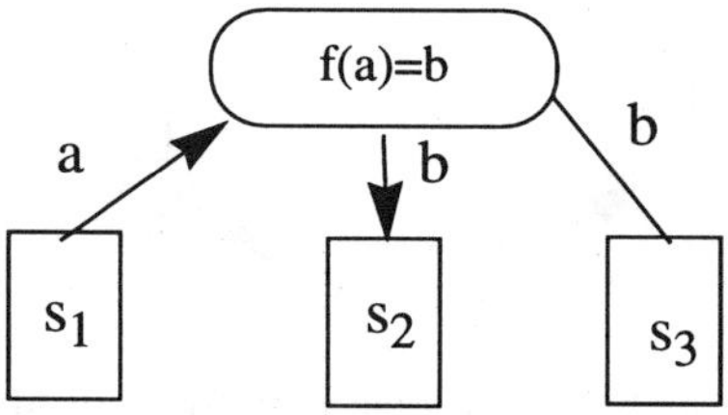

It is clear that this is not really different to the simple spontaneous reaction above.

4.4.6. Additions.

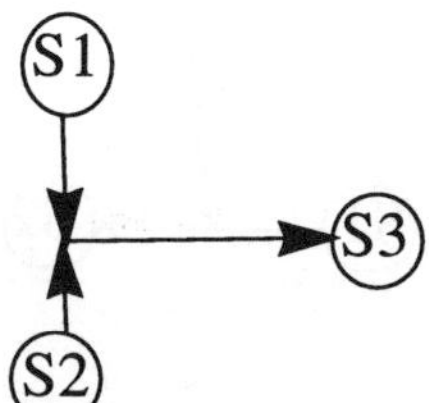

Register machine:

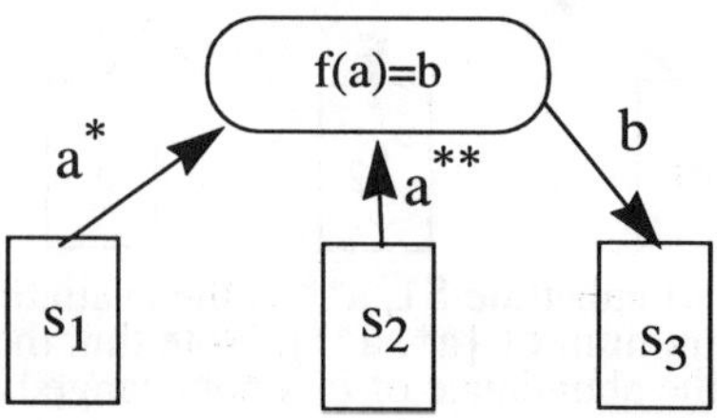

Here a* is the availability of S1, a** is the availability of S2 and a is the minimum of

{a*, a**}.
In these diagrams, where two lines cross, if there is no arrow head, there is no connection.

4.5. The register machine system for glycolysis

Putting all these constructs together using the flow chart in Fig. 5. gives us the system illustrated in Fig. 6. The model, so far, does not describe those aspects of the processing system that are bidirectional. Some reactions proceed in both directions at the same time. This phenomenon can easily be derived, as mentioned in 4.3.3. by doubling up the appropriate register machine.

Returning briefly to the question of how to model the availability of enzymes at these locations we can adapt the model in a simple way, by adding further registers and defining the production and availability of enzymes, possibly stipulating thresholds for different levels of enzyme.

5. Conclusions and further work.

Logical models tend to be very powerful in their qualitative results, but do not give the quantitative data required for many applications. Mathematical approaches tend to be either over simplified, or unworkable in practice due to the complex recursive nature of the systems. Both approaches tend to be unable to deal with the parallel nature of the processes, and generally ignore the effects of physical structure, competition for resources, temperature, etc. The approach that we have outlined here promises to provide a number of useful features that will help us to model both the organisational aspects of these systems as well as the resource and dynamic aspects. The issue of what happens in more detail at each active site can be explored by using the same technique, albeit at a lower level. Thus an individual mitochondrion can be modelled more explicitly, identifying specific sites of active processing and building a machine model whose computational function equals the higher level function (such as COMP(P1)) identified in the model described in section 3.

The more generalised model of the hybrid machine which allows for the explicit representation of the register or resource level of a substrate together with the specific equations governing its activity and the formal Hybrid Projection Temporal Logic, see [8], which provides the semantics for systems of concurrent hybrid machines, will provide us with a real formal basis for investigating these complex systems. Our ultimate goal is to codify the rules that control these systems in the logic and to use this mechanism to both simulate the systems and to formally prove results about their ultimate states.

References.

[1]. Krohn. K., Langer. R. and Rhodes. J.L."Algebraic principles for the analysis of a biochemical system." *J. Theor. Biol.* 116. 1967, 399-426.
[2]. Ji. S. "Biocybernetics: a machine theory of biology." in "*Molecular theories of cell life and death.*" Ed. S. Ji, 1991, Rutgers Univ. Press. USA.
[3]. Holcombe. M. "Mathematical models of cell biochemistry." in "*Molecular theories of cell life and death.*" Ed. S. Ji, 1991, Rutgers Univ. Press. USA.
[4]. Holcombe. M. "Towards a formal description of intracelluar biochemical organization." *Comp. Math. Applic.* 20. 1990, 107-115.
[5]. Eilenberg. S. "*Automata, languages and machines.*" Vol. A.1974, Academic Press.

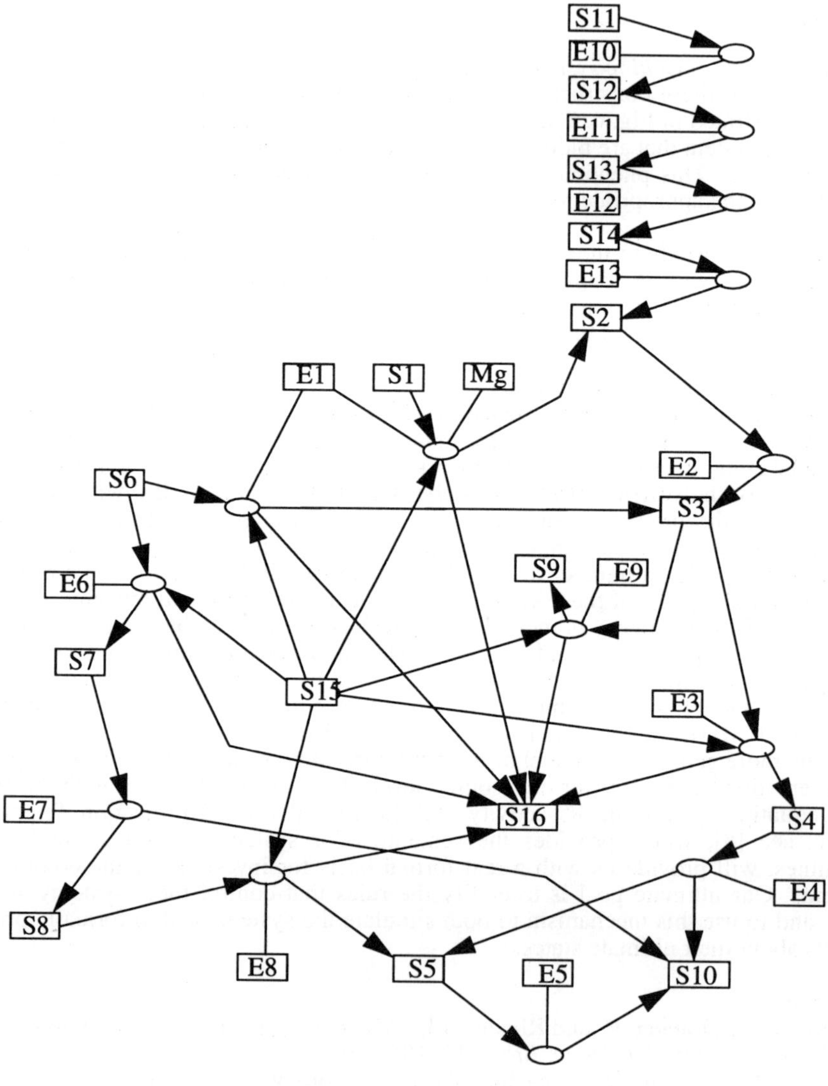

Fig. 6. The simplified machine diagram for glycolysis.

[6]. Welch, G.R. & Kell, D.B. "Not just catalysts - molecular machines in bioenergetics". In "*The fluctuating enzyme.*" Ed. G.R. Welsh. Wiley, 1986, 451- 492.
[7]. Holcombe. M. "From VLSI through machine models to cellular metabolism." in "*Computing with biological metaphors.*" Ed. R. Paton, Chapman & Hall, 1994. 11-25.
[8]. Holcombe. M. & Duan. Z. A hybrid projection temporal logic for hybrid systems.
University of Sheffield internal report Nov. 1995 CS-95-01
[9]. Rosen. R. "Some comments on activation and inhibition." *Bull. Math. Biol.* 41, 1979, pp427-455.
[10]. Snoussi. E.H. & Thomas. R. "Logical identification of all steady states." *Bull. Math. Biol.* 55 1993, p973.
[11]. Thomas. R. & Thieffry. D. "Developing a logical tool to analyse biological regulatory networks." in *Computing with biological metaphors.* 1994, Ed. R.Paton. Chapman & Hall.
[12]. Welch. G.R. "The computational machinery of the living cell." in *Computing with biological metaphors.* Ed. R.Paton. 1994, Chapman & Hall.

Appendix. Key to Abbreviations in Simplified Glycolysis. (Figs. 5 & 6.)

E1 Hexokinase	S1 Glucose
E2 Phosphoglucoseisomerase	S2 Glucose 6P
E3 Phosphofructokinase	S3 Fructose 6P
E4 Aldolase	S4 Fructose 1,6BP
E5 Triose P isomerase	S5 Glyceraldehyde 3P
E6 Fructokinase	S6 Fructose
E7 Fructose 1P aldolase	S7 Fructose 1P
E8 Triosekinase	S8 Glyceraldehyde
E9 Phosphofructokinase2	S9 Fructose 2,6BP
E10 Galactokinase	S10 Dihydoxyacetone
E11 Galactose 1P uradyl 1 transferase	S11 Galactose
E12 UDP Galactose 4 epimerase	S12 Galactose 1P
E13 Phosphoglucomutase	S13 UDP Galactose
Mg Magnesium	S14 Glucose 1P
	S15 ATP
	S16 ADP

THE CELL AS A PROBLEM-SOLVING "ENGINE"

PEDRO C. MARIJUAN
Departamento de Ingeniería Eléctrica y Electrónica, C.P.S. Universidad de Zaragoza, Zaragoza 50015, Spain (email: marijuan@mcps.unizar.es)

1. An organizational view of the cell

Eighteenth century biologists had already acknowledged cells as the fundamental building blocks of living beings. The "cellular theory", developed by Oken, Bichat and Schwann, and later completed by Virchow, was crucial to approach biology to its modern evolutionary track.[1] During recent decades, however, the NeoDarwinian emphasis on the gene as the actual subject of evolution and the bias of molecular biologists towards the structural facets of their work have tended to reduce cells to almost faceless components of organisms --tantamount to "bags of genes". Curiously, physiologists and embryologists have long ago insisted that multicellular organization is an almost untractable subject, particularly regarding the functional aspects.

Here, we are going to characterize cells as genuine *problem-solving systems*. We will do that by putting together two strands of thought: one close to molecular biology, particularly dealing with the recent integrative achievements on signaling systems and the cell-cycle (including two "negative" topics previously forgotten: apoptosis and protein degradation); and another strand corresponding to the second thought-wave that computers have provoked within the biological field,[2] as materialized in new interdisciplinary fields such as molecular computing, bioelectronics, biocomputing, bioinformatics, etc. In the long run, both strands of thought should converge and produce a coherent interdisciplinary point of view we could term as *bioinformation.*[3]

Our analysis will begin by considering the cell's active elements, enzyme-proteins (including the phenomenon of their "endogenous absence" or degradation), and will continue towards the overall functioning of the system: metabolism, DNA-world, signaling system, cell cycle, multicellularity. As we will discuss, a *bona fide* problem-solving dynamics can be sketched, based upon the distinction of three generic properties: self-organization, self-modification, and self-reconfiguration. Through the overlapping of these properties, the cell may distribute the global problems of its functioning (or what we will call "functional voids") down to the microscopic dimension of the individual work-cycles of its constituent enzymes and proteins.

We will argue that a viable cell, endowed with a metabolic network, a coding DNA-world, a proteolytic system, and a signaling system, represents the evolutionary emergence of a quasi-universal *problem-solving "engine"*. An informational engine which is capable of mesoscopic/microscopic scale-changes, based upon a permanent internal "bath" of dynamic information-gradients.

Accompanying the evolution towards multicellularity, an important improvement in the cellular "engine" occurs. Problems at the new macroscopic dimension of multicellular organisms may be broken down to the mesoscopic scale of individual cells by means of new specialized cellular devices in charge: signaling systems. Thereafter, the full-blown evolutionary process of invention of new organismic solutions and informational problem-solving organizations (nervous systems, animal societies...) may take place.

2. Enzyme function: The active elements of the cell

Enzymes and proteins are the active elements of the cell, its processors or *transistors* (anecdotally, enzymes and transistors share similar types of functional curves, this was eagerly noted by Monod, Jacob, and other early workers in the field.[4]) Enzymes achieve their peculiar active function, and are produced themselves, in three basic steps: transcription from the DNA repository, translation at the ribosome factory, and folding within the chaperone "cradle". All of these formative processes are tightly regulated according to the whole cellular functioning. The end of the enzyme's active life, which preferentially occurs at the proteasome "coffin", is carefully regulated too.[5]

Protein folding can be taken as the guiding phenomenon to analyze the function of the enzyme. It is a singular process which allows a unidimensional distribution of structural information to become a tridimensional one endowed with processing capabilities (specific recognition, plus "intelligent" catalytic work). Weak forces (hydrophobic, van der Waals, hydrogen bonding forces) are the main organizers of this yet not-well-understood process.[6,7] The catalytic peculiarities of the native enzyme can be understood as instances of symmetry breaking and symmetry restoration in the achieved folding: the electronic transformation of substrates at the active site, the impact of effectors at binding sites, the occurrence of coherent vibrations without heat dissipation, the capacity to briefly take energy loans from the environment, the occurrence of quantum-tunneling effects, etc. In quantum terms, this overall dynamics is a non-picturable one.[8,9]

Because of these singular properties, enzymes have often been dubbed *Maxwell demons*.[10] Indeed they are "ecological machines" capable of negotiating complex energy exchanges with their internal and external environment so that they substantially lower the activation energy necessary to transform their substrates.

They abide to what has been called the principle of "minimal price of action".[11] Without this minimizing principle, the host of chemical processes (around 600 metabolic reactions take place within a single prokaryotic cell, and more than 400.000 enzyme-proteins of 2.000 different types are simultaneously at work) would literally burst any permanent structure. This principle helps to explain why enzymes are so big in relation with their substrates (the active site only represents around 5% of the enzyme structure) and why metabolic reactions are broken down in multiple steps and intermediaries (more than 20 in the case of glucose).

An interesting formalization of the enzyme operations corresponds to the *molecular automata hypothesis*.[12-14] By thermodynamically distinguishing different states in the work-cycle of the enzyme, and by drawing the adequate transition table with the corresponding state variables and probabilities of occurrence --in a stochastic fashion, not deterministically-- then the kinetics of the enzyme is faithfully reproduced just with a few binary variables within a Monte-Carlo simulation.[3]

This automata characterization is well in accordance with the criticisms that physiologists have addressed to the traditional deterministic approach of theoretical biochemistry (e.g., control theory). It is molecular impacts, the number of encounters between reactants and catalysts (enzymes) per unit of time, and not abstract macroscopic concentrations, what counts at the intracellular and intercellular milieu.[15]

3. Self-organization and enzyme networks

In the automata perspective, every enzyme work-cycle can be understood as a microscopic problem-solving action. The individual enzyme molecule "feels" the composition of the medium surrounding it and reacts with a succession of transitions, cycles, outputs, etc. They represent the whole answer that the enzyme gives to the initial conditions felt --the solution to the biochemical and thermodynamical problem introduced by the initial variables.[3] The modification of the probabilities of transition as a function of the transitions performed by the automaton itself, following the changes within the microenvironment surrounding the molecule, is one of the aspects that gives to this stochastic model, an in general to the action of the enzyme, its flavor of "intelligent machine."

The path towards chemical equilibrium that every enzyme transition is aimed at, becomes extraordinarily complex when different enzymes mutually exchange their substrates, products, and effectors (activators and inhibitors) within an overall networking of processes. The relative presence or absence of the different substances "informs" each enzyme, each molecular automaton, about the results of the operations performed by the other automata. A complex metabolic network is generated, the persistence of which is granted, in general terms, by the inflow of primary nutrients, the accumulation of strategic intermediaries, and an adequate network of effector feedbacks balancing the different productions. If these conditions are met, a *self-organizing* dynamics (both in the sense of non-equilibrium thermodynamics and computer science --a "network of synergetic automata") dominates upon the whole metabolic system.

The trunk, the spine of the metabolic network is, in almost every living cell, the anaerobic degradation of the glucose; together with the Krebs cycle and the pentose cycle, they comprise more than half of all the dedicated metabolic enzymes. Studies about the connectivity of the whole metabolic network have found that about 80% of the appearing metabolic intermediates have just one use in the cell,[16] and the whole system, far away from being a randomly connected structure, actually resembles the assembly lines of a factory: a graph theory analysis of a real enzyme network of 12 components yields a complexity factor only 5.6 % of a randomly connected system, indicating that metabolic pathways have evolved into efficient metabolic systems.[17] Elegant cases of evolutionary optimization in metabolic processes (e.g., the pentose phosphate cycle and the glycogen molecular design) comprising symmetries and trade-off between different parameters have been described.[18,19]

A necessary condition for the advance of further complexity in any metabolic network is to free its enzymes from the particular energy barriers of their specific reactions. Metabolic kinases and phosphatases, respectively the donors and acceptors of phosphate groups to other molecules, were the enzymic devices evolved by the cell in order to fulfill this necessity. Curiously, they were primordial enzymes occupying a central role in the degradation of glucose; later, by coupling the ATP produced in this reaction to other metabolic regions, they made possible the expansion of the whole network. At the same time, they have become the most extensively regulated enzymes: 7 effectors have been authenticated in animal phosphofructokinase (PFK), and a fantastic number of 23 different substances could be impinging on its functioning.[20] Glycogen phosphorylase is another fundamental energy-transducer enzymic device endowed with a very elegant and precise system of allosteric control.[21]

The self-organization implicit in the whole reaction pathways and control feedbacks of the metabolic network makes the system capable of a steady accumulation of nutrients, components, etc. By providing the energy and raw materials needed to solve the alien problems confronted by the system within its environment, this *self-organization* property is what actually supports the emergence of any further organizational complexity.[3] Let us emphasize that every cellular solution, even the most sophisticated communicational event, has to be finally supported by enzymic "work horses" running in metabolic pathways.

4. Protein degradation: A necessary adaptive condition

Returning to the function of the enzyme molecule, the loss of correct folding beyond motions relatively allowed by the conformational boundaries, such as enzyme "breathing" and segmental flexibility,[22] usually implies the end of function. Irremediable losses of folding occur due to chemical and thermal aggressions, and also with the inevitable "wear and tear" caused by the normal functioning. The unfolded or partially unfolded enzymes and proteins become immediate targets of the proteolytic enzymes associated with the protein complex known as proteasome (the "coffin"), which liberates the constitutive aminoacids of the protein, to be recycled by the cell. The proteasome, already present in the most ancient bacteria (archebacteria), is among the most conserved cellular substructures --like chaperons themselves.[5,23]

Therefore, protein degradation becomes a sort of inescapable fate: every protein and every enzyme have a specific life-span. When this particular time arrives, they "pass away" in their effective functioning and their aminoacids are recycled. This apparent cause of havoc in the "society of enzymes" is at the same time a necessary adaptive condition. It can be discussed whether a cell made up of everlasting proteins could adapt to a changing environment and could develop many of its sophisticated processing functions (e.g., the very cell cycle).

The point is that the cell has developed a number of adaptive mechanisms exploiting this phenomenon of "endogenous absence" in its general order of processes: the families of specialized proteases, the already mentioned proteasome, the ubiquitin system, the N-Rule, the "destruction box", the controlled expression of genes, etc.[23-26] The adaptation is so extreme that protein degradation has become a necessary condition for a viable cell, particularly in multicellular organisms. The list of multicellular processes depending on total (or partial) protein degradation by proteases is almost endless: intracellular and extracellular protein transport, cytoskeleton and extracellular matrix maintenance, cell to cell communication, cell cycle control, apoptosis, multicellular developmental programs, reproductive functions, digestive functions, immune system, synaptic plasticity at the nervous system, complement system, circulatory system, defense and predatory systems (venoms), etc. Significantly, and in spite of its intrinsic importance, protein degradation was passing almost unnoticed to theoreticians and experimentalists, until very recently.[27]

5. Self-modification and the DNA-world

Through the structural information of DNA (and RNA) the cell can modify its own population of enzymes and proteins on a vast scale, subsequently incorporating new sets of agents and compensating for their degradation --this is the basis of the organizational property of *self-modification.*

The DNA-world does not only constitute a reservoir mechanism for heredity and self-construction of the system, but a physiological problem-solving device continuously used by every individual cell: the variable expression of genes, coupled with the protein degradation phenomenon, constitutes the main source of adaptation to the demands posed by the internal and external environment --the "fundamental core" that supports most of the adaptive processing of the cell.[3]

The coding of the non-picturable function of every enzyme within a succession of nucleotides allows for the evolutionary process of optimization of this function to occur by means of very simple manipulations upon the succession of nucleotides. The simple symbols of DNA become entitled to encode real laws of nature. Therefore, because of the reproductive capability --the culmination of the cell cycle-- together with the process of natural selection, the DNA-world becomes a formidable tool for evolutionary problem-solving and organizational exploration.

Nucleic acid sequences also contain additional information about protein transcription, translation, transport, modification and ultimately degradation. As the *chromosome field theory*[28] has put it, the functional domains of DNA imply a very large variability of structures, expanding from the single base up to the whole chromosome (in the millions): base, codon, restriction site, operator, promoter, exon, intron, enhancer, gene, split gene, transposon, operon, repetitive DNA, gene territory, centromere & telomere, chromosome arm, chromosome. This complex structure of the DNA-world greatly enhances the evolvability properties of the system (e.g., allowing the flexible integration of signals implying concurrent transcriptional needs, making possible the creation of different iso-proteins, enzymes and channels by alternative splicing and post-translational modification, etc.).

The flexible combination of protein synthesis and protein degradation represents a functional property, *self-modification*, only found in the living cell. The capacity to self-modify the population of active elements according to the needs of the system within a life-cycle trajectory carefully regulated, is what distinguishes the living organization from any other self-organizing process found in nature.

The "problems" of the living system become solved, at the height of this new property, by the synthesis of adequate active elements (coupled with the necessary destruction of other elements) filling-in the occurring *functional voids* within the life-cycle of the entity as detected and measured by the active elements themselves. It is the gradients of dynamic information generated at the multiple symmetry-breakings within the overall "network of automata" what triggers and orientates the use of the structural information accumulated in the DNA world. Without this dynamic information produced by the network of active elements, the DNA becomes a useless molecule, an inert chemical structure.

6. The signaling system: The self-reconfiguration property

The signaling system has been explored and charted only partially as yet. It comprises hundreds of protein-kinases, protein-phosphatases, and other classes of *converter* enzymes which act both upon the channels and receptors, the gene expression, and upon the enzymes of the metabolic network. (Fig.1) It comprises the so called "second messengers" too (cAMP, cGMP, Ca, $InsP_3$, diacylglicerol, ceramide, arachidonic acid, NO).

Metabolic kinases and phosphatases had already been the enzymes in charge of coupling ATP to the energetically unfavored reactions --the real "work horses" of metabolic expansion. It is curious that, again, most of the processing expansion of the cell was built around the evolving capacity of kinases and phosphatases. By externalizing their "phosphorilating lip", previously within an internal pocket or cavity, and by developing a sensing surface around the "lip" to recognize the new macromolecular targets,[29,30] these old *metabolic* enzymes originated a very broad family of *protein* kinases and phosphatases (around 1% of the eukaryotic genome, in correlation with the number of cell types); and with them, a genuine signaling system was established.

The signaling system represents the informational apparatus that has made possible the evolution of multicellular organisms. It massively conveys the logical property of *self-reconfiguration* for the populations of enzymes,[3] preparing them for the manipulation of information (amplification and damping) and for the solution of the new problems implicit in the incoming informations from a complex multicellular environment. In the differentiated cell, the response to growth signals, the culmination of the cell cycle, the development of its specialized function, even the programmed death are all of them harmoniously integrated with respect to the metabolic and genetic subsystems by, basically, the action of these converter enzymes upon other enzymes, aided by their cohort of specialized messengers.[31]

It is remarkable that during its ontogenetic process, every specialized cell has to tailor its own signaling system in order to appropriately "understand" the incoming informations, and to respond to them. One is lead to speculate, following recent developments on parallel processing and signaling pathways,[32] that the cross-talk between signaling pathways, the sharing of second messengers, and the successive chaining of converter kinases and phosphatases configures a topologically distributed net of very basic "computing" operations, the logic of which has not been solved yet in its intrinsic elegance.[33]

Particularly interesting is the complex integration of the temporal code of signals, and the careful balance --at different levels-- between opposed signaling pathways. By means of this generalized overlapping between different sampling operations, we can say, the cell obtains a ponderate unified perspective of its environment --and correspondingly acts.

Through the signaling system, cells can exchange signals loaded with information content about their own internal states. The signal implies a "call to action", the mobilization of specialized enzymic pools of the target cell in order to contribute to solve the problems or functional voids of the cell who has send the signal. In this regard, the limited rate of protein synthesis and the limited available intracellular space and water (solvatation rate) have been powerful motive forces for cells to differentiate and specialize their individual enzymic pools, increasing their conjoint metabolic efficiency. (And the related necessity to bridge the gap between the microscopic dimension of the received signal and the mesoscopic or macroscopic dimension of the target metabolic network is the reason for the existence of long chains of amplificating protein-kinases in the signaling system).[34]

In the wake of the specialization strategy followed by multicellular colonies, we find that biological information has truly emerged. Cells communicate and exchange meaningful information. As the *metabolic code hypothesis*[35] has put it: there is an intercellular *signal* (symbol) which conveys a *meaning* (content), but only to those endowed with the correct *decoding* apparatus (signaling system).

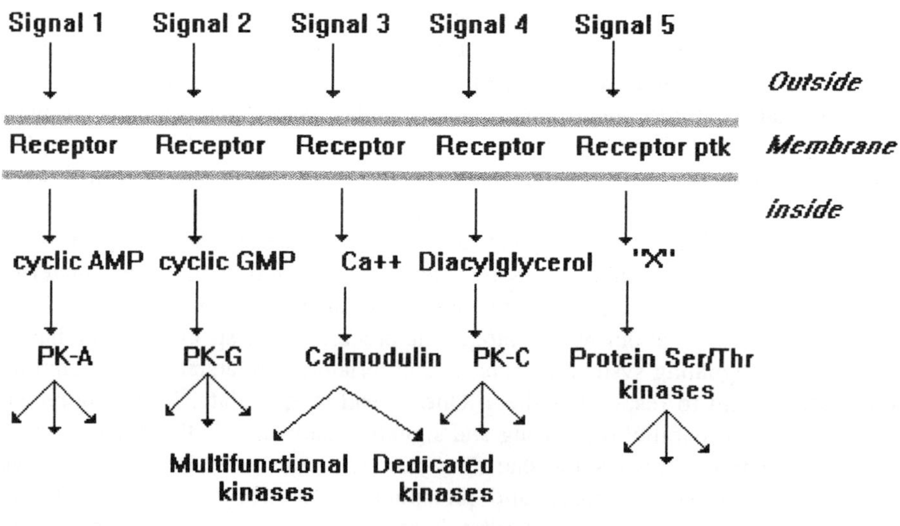

Figure 1. Five of the principal signaling pathways that operate in eukaryotic cells. The representation does not include the intermediate effectors between receptors and second messengers, the networking of paths, and the phosphatases. The signaling paths in the left are often associated with the development of the cell function; the ones in the right with the differentiation process and cell fate (the stages of the cell cycle, including apoptosis); calcium may participate in both. Steroid hormones (non represented) bind to intracellular receptors and take a direct path to the nucleus; they are also associated with cell fate and differentiation. *(modified, from P. Cohen, TIBS 17, 408, 1992)*

7. Organismic problem solving

The evolutionary process has pushed the cellular processing mechanisms towards higher and higher performances: quasi-universality in the global metabolism, in the DNA-world, in the proteolytic system, and particularly in the signaling system. These improvements support the evolutionary reliance on information processing as a dominant tool for intercellular problem-solving, overcoming the restricted possibilities of local diffusion and brute mass-transportation.

The control of the cell-cycle "engine" was a necessary multicellular achievement. Taming the molecular processes involved in the reproduction of cells made them amenable to the development of specialized functions for the whole organism. It implied a coherent integration among the above subsystems, the reproductive machinery, and a new, sophisticated cell-death machinery (apoptosis). (Fig.2) One of the most interesting integrative notions, *cellular checkpoints*, launched not far ago in relation with the control structure of the cell cycle[36,37] seems well in accordance with the problem-solving interpretation we are advocating here.

For every cell a "problem" is any deviation (functional void) from the normal train of events which constitute its life-cycle trajectory.[33] Adequate solutions are found by its own processing resources or by cooperating with other cells. The term *functional void* is a label for the biochemical work to be performed by the system after the arrival of signals or after the occurrence of external or internal events, in order to restore its life-cycle trajectory. It naturally encompasses the "meaning" of the incoming signals. When the cell has to send a signal conveying an unsolved functional void, the recipient needs an appropriate apparatus (signaling system) in order to decode and build the "meaning", and solve it. It is to say, after a signal has been sent through an appropriate channel indicating the class and dimension of the occurring functional void, once the signal has arrived to the target cell, it has to be decoded and amplified in order to recreate the original functional void (or its abstract equivalent within the computational space of the signaling system), then the problem is solved by specialized enzyme pools, and the solution exported.

It is a process which occurs in both directions: mesoscopic needs or problems (functional voids) of individual cells are integrated into global organismic signals, and macroscopic organismic needs (e.g., in the interaction with its environment) are decomposed into mesoscopic signals generating problems or functional voids in individual cells. From the point of view of cells, organisms become extensive markets for functional voids exchange: adequately endowed with evolutionary inventions to detect, measure, amplify, solve, export, etc., the occurring functional voids --needs-- of the cellular individuals in the development of their life cycles. These inventions maximize the local problem-solving capabilities and minimize the related bulk processes of transportation, energy expenditure, waste, entropy production, etc. As a result, a coherent organismic life is made possible. The global problems of the organism may be broken down among populations of specialized cells working at differentiated subsystems: digestive, circulatory, respiratory, muscle-skeletal, hormonal, neuro-endocrine, immune system, nervous system, etc.

Multicellular life implies, by its own, the appearance of new instances of "absence", similar to protein degradation in the case of cellular life. Apoptosis (controlled cell-death) becomes a fundamental problem-solving tool for organisms. It has been recently recognized as a normal developmental event that is critical for organogenesis in a wide variety of contexts. The inhibition of apoptosis is lethal for even the simplest multicellulars (e.g., *C. elegans*); its partial malfunctioning causes severe illnesses.[38] Unsurprisingly, like in the protein degradation case, very little research was devoted by molecular biologists to this "negative" topic until very recently. Nilly-willy, there has been a flurry of convergent achievements during recent years connecting all these previously isolated topics: signaling pathways, cancer, cell-cycle "engine", apoptosis, protein degradation. This situation has prompted unexpected discussions on "biological universals" among experimentalists; their *motto* as someone has elegantly put it: *e pluribus unum.*[27]

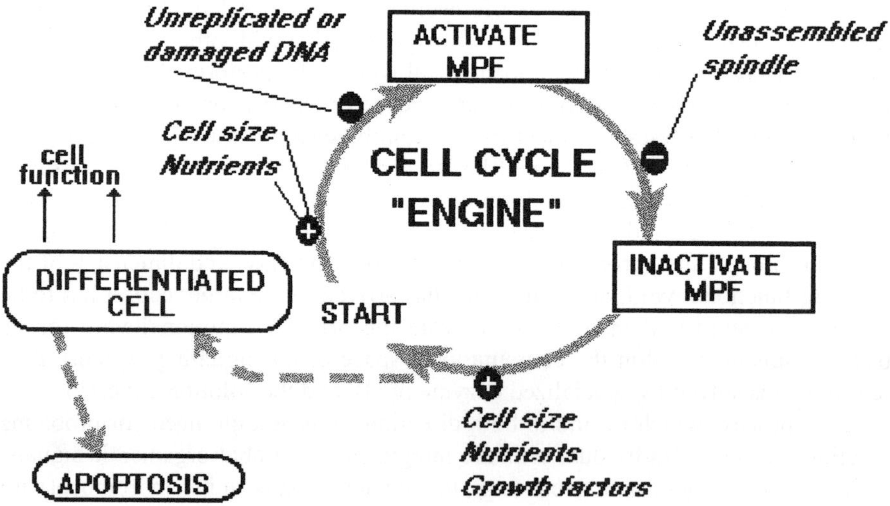

Figure 2. The cell-cycle engine and its regulation. The MPF (maturing promoting factor) is a combination of cyclins and associated kinases. Cyclins, fundamental control elements, are degraded throughout the cycle in an ubiquitin-dependent process. Other proteins such as p21, p16, p27, and p53 participate in the control structures known as checkpoints. Every checkpoint instantiates a signal transduction system in its own. They receive a plurality of signals, integrate them, and transmit the results to other regulatory components. The transit towards a differentiated cell type implies a permanent interruption of the cycle in one checkpoint--the cell becomes committed to develop its specialized function. Apoptosis will occur due to senescence or to the arrival of specific signals.

8. New avenues of thought --bioinformation

The recently discovered cellular mechanisms, so unexpected for hardcore molecular biologists, clearly pave the way for non-traditional interpretations. Information is the key term. The cell, we may say, appears as configuring a new type of quasi-universal information processing "engine". The overarching goal of this type of informational engine is its own permanence and reproduction, achieved by means of the above panoply of cellular responses in front of the infinite multiplicity of internal and external changes. It is a problem-solving dynamics distinctively organized around filling-in the occurring "functional voids" or absences by the synthesis of adequate enzymes and proteins, and actually overlapping the three mentioned general properties of self-organization, self-modification, and self-reconfiguration.

Cells and multicellulars have evolved their quasi-universal information processing mechanisms in order to counteract any problems they might find within their life cycles. Or in more formal terms, interpretations based on symmetry breaking[39] and symmetry restoration within a cycle-trajectory may be sketched too. Functional voids would represent instances of symmetry breaking within gauge symmetries.[40] And cells and organisms would "struggle" to counteract them by means of their quasi-universal information processing (and productive) mechanisms. From this point of view, cells, organisms, nervous systems, and societies share a number of information processing strategies --and countless metaphors and parallels have been drawn between ATP, second messengers, neuronal entropies, currencies, etc.[2,41] As we have already said in the case of the cell, information in these "social" entities is what triggers and orientates the productive counteracting machineries.

The interpretation advocated here, as noted at the Introduction, seems to imply not only the use of new formal tools and the emphasis on new experimental fields, but also the long-term development of a new discipline at a pair with biochemistry and biophysics: *bioinformation*.[14] As previously with the application of the chemical and physical perspectives --biochemistry and biophysics respectively-- it is now the turn of the recently framed information sciences to provide a new perspective to illuminate the study of life.

REFERENCES

1. F. Jacob, *The logic of life: A history of heredity* (Pantheon books, New York, 1973).
2. R.C. Paton, in *Computing with Biological Metaphors,* ed. RC Paton (Chapman & Hall, London, 1994).
3. P.C. Marijuán, in *Computing with Biological Metaphors,* ed. RC Paton (Chapman & Hall, London, 1994).

4. P.C. Marijuán, *BioSystems* **25**, 259 (1991).
5. J.S. Weissman *et al.*, *Science* **268**, 523 (1995).
6. H. Frauenfelder and P.G. Wolynes, *Physics Today*, feb. 94, 58 (1994).
7. J. Frydman *et al*, *Nature* **370**, 111 (1994).
8. M. Conrad, *Computer* **25** (11), 11 (1992).
9. A.U. Igamberdiev, *BioSystems* **31**, 65 (1993).
10. A. Elitzur, *J. theor. Biol.* **168**, 429 (1993).
11. E.A. Liberman, *Biofizika* **19**, 148 (1974).
12. M. Conrad, *Communic. ACM* **28**, 5, 464 (1985).
13. M. Sugita, *J. theor. Biol.* **1**, 415, (1961).
14. PC. Marijuán and J. Westley, *BioSystems* **27**, 97 (1992).
15. D.N. Wheatley and J.S. Clegg, *BioSystems* **32**, 83 (1994).
16. P.A. Srere, *Ann. Rev. Biochem.* **56**, 89 (1988).
17. M. Kohn and S. Bedrosian, in *Information Processing in Biological Systems,* ed. SL Mintz and A Perlmutter (PLenum Press, New York, 1985).
18. E. Meléndez-Hevia, *Biomedica Biochimica Acta* **49**, 903 (1990).
19. E. Meléndez-Hevia *et al.*, *Biochem. J.* **295**, 1 (1993).
20. A. Sols, *Curr. Topics Cellul. Regul.* **19**, 77 (1981).
21. M.F. Browner and R.J. Fletterick, *TIBS* **17**, 66 (1992).
22. A. Fersht, *Enzyme structure and mechanisms*, (Freeman & Co., New York, 1985).
23. J. Driscoll, *Histology and Histopathology* **9**, 1, 197 (1994).
24. A. Hershko, *TIBS* **16**, 265 (1991).
25. A. Bachmair *et al.*, *Science* **234**, 186 (1986).
26. H. Neurath, *TIBS* **14**, 268 (1989).
27. M. Kirschner, *TIBS* **17**, 281 (1992).
28. A. Lima-de-Faria, *Evolution without selection,* (Elsevier, Amsterdam, 1988).
29. F. Zhang *et al.*, *Nature* **367**, 704-710
30. A. Barford *et al.*, *Science* **263**, 1397 (1994).
31. S.E. Egan and R.A. Weinberg, *Nature* **365**, 781 (1993).
32. D. Bray, *J. theor. Biol.* **143**, 215 (1990).
33. P.C. Marijuán, *Optical memory and neural networks* **4**, 2, 63 (1995).
34. M. Conrad in *Advances in Computing*, ed M Yovits (Academic Press, NY, 1990).
35. G.M. Tomkins, *Science* **189**, 760 (1975).
36. T.A. Weinertand L.H. Hartwell, *Science* **241**, 317 (36).
37. A.V. Murray, *Nature* **359**, 599 (1992).
38. G.C. Thompson, *Science* **267**, 1456 (1995).
39. K. Matsuno, *BioSystems* **17**, 179 (1985).
40. D. Alves and P.C. Marijuán, *Symmetry: Culture and Science* **6**, 1, 38 (1995).
41. G.R. Welch, *TREE* **2**, 10, 305 (1987).

MODELS OF CYTOPLASMIC ARCHITECTURE AND FUNCTION

M.A. AON, S. CORTASSA
Instituto Superior de Investigaciones Biológicas (INSIBIO, CONICET-UNT), Departamento de Bioquímica de la Nutrición and Instituto de Química Biológica "Dr. Bernabé Bloj", Facultad de Bioquímica, Química y Farmacia, Universidad Nacional de Tucumán, Chacabuco 461, 4000 - San Miguel de Tucumán, Argentina

A. CÁCERES
Instituto de Investigación Médica "Mercedes y Martín Ferreyra", 5000 Córdoba, Argentina

1 Introduction

In the last two decades, the ideas concerning the organization and functional properties of the cytoplasm of cells have grown around two main concepts. The first one is that the cytoplasm is a crowded[1] though organized protein crystal[2,3]; whereas the second one deals with the fact that metabolism and the gene expression machinery are not indifferent but profoundly influenced by that organization[4-6].

Regarding cellular proteins, at least one third of them take part in the macromolecular networks and laminae of the cytoplasm and nucleus[4]. The replication and transcription of DNA and the processing and translation of RNA[4,7-9]; endocytosis and secretion, occur deeply associated with reorganization of nuclear and cytoplasmic macromolecular lattices[2,4,10].

Intimate association of enzymes and ultrastructure as well as their dynamics have been considered of utmost importance in metabolic regulation. Several workers before were interested in the elucidation of enzyme binding properties with actin, its effect on enzyme kinetic properties, the interaction between enzymes in supramolecular aggregates and its influence upon transition times in consecutive catalytic steps[3,11-14]. Similarly, studies have been reported for interaction of glycolytic enzymes with tubulin or microtubules[15,16] and their effect on enzyme kinetics[6,17,18].

Overall, experimental evidence increasingly favors the idea that metabolism occurs strongly associated to the cellular scaffolds[3,11,19-22] and that these insoluble matrices and their assembly strongly affect the dynamics of chemical reactions[3,5-6,20,21,24-25].

2. Modelling functional global properties of cells: mechanisms of coherence and transduction

At present, within the framework of the "structured view" of the cytoplasm[2-3,25], there are at least two distinguishable models concerning cell function. These are the "tensegrity"[21] and the "fractal percolation"[23] models. We will compare two aspects of these models: *the mechanism of coherence* and *how is this coherence transduced* into functional properties.

By *mechanism of coherence* we mean how is effected the passage from microscopic to macroscopic order in cell function, i.e. how the activity of millions of molecules or architecture of supramolecular structures, become synchronized in space and time in an apparently "purposeful", functional, way. *Transduction* deals with the mechanisms through which cells transduce environmental stimuli (e.g. hormones, neurotransmitters) into differential gene expression, metabolism and cellular energetics. Both aspects represent tightly linked concepts since the mechanisms of coherence play a central role in the cellular machinery of transduction. On this basis let us briefly compare the two models.

2.1 The tensegrity model.

For Ingber's model[21,24], coherence is given by changes in the "tensional integration" of components of the cytoskeleton and its transduction is essentially conformational[21,27]. In essence, the tensegrity model proposes that the entire cytoskeleton responds to stress as a single, *tens*ionally inte*grated* (tensegrity) structure. Experimental evidence in support of this model came from experiments of stress-strain relation measured with magnetic microbeads attached to the surface of living endothelial cells[27]. The mechanical stress applied to the surface of cells with a twisting device showed that the cytoskeleton response to the applied stress was a property of the integrated system. The partial disruption of the cytoskeleton with depolymerizing agents such as cytochalasin D or nocodazole did not completely suppress the cytoskeleton stiffening, i.e. a large resistance to mechanical deformation was evident at high levels of applied stress[27]. Using a tensegrity model to interpret the data, the stiffness-stress (force) linear response of the cytoskeleton of living cells could be simulated and it was further shown that, when increasing the force, the mechanically interdependent structural elements rearranged without topological disruption or loss of tensional continuity. Thus, a molecular continuum of mechanically interdependent struts and tensile elements structurally stabilize by global rearrangement rather than local deformation in response to stress[21,24]. In this light, the cytoskeleton response to applied stress appeared as a property of the integrated system and not a characteristic of any individual part[27].

According to the tensegrity model, coherence would be given by the use of tensegrity architecture by cells. In this way, a mechanism is provided to distribute mechanical stresses to regulatory molecules and enzymes that are immobilized on or interact with cytoskeletal and nuclear scaffolds and thereby to integrate cell structure and function. Mechanical stress in turn, is transduced into biochemical function through conformational changes, i.e. a mechano-chemical transduction[21].

2.2 The fractal percolation model.

The conception of the cellular cytoplasm as a fractal proposes that the cytoskeleton is organized according to the principles of fractal geometry. A random fractal - percolation clusters - has been hypothesized as a plausible model on the basis of quantification of the fractal dimension in micrographs[23,28]. Iteration of an invariant pattern that spans several length scales might describe the complex forms shown by cell's architecture[23]. From a dynamic perspective, a main feature of percolation clusters is given by the existence of a *percolation threshold* for which a percolation process undergoes a transition from *local* to *global connectedness* [23,29].

In the fractal model, coherence arises when the dynamics of cellular processes become unstable at sites in the cytoplasmic lattice belonging to the percolation cluster, i.e. further from the "percolation threshold". Transduction occurs because instability is triggered by environmental stimuli acting through multiple pathways[22-23,30]. Enhanced catalytic performance may be one of the outcomes of transduction. In heterogeneous fractal media such enhanced catalytic performance may be achieved through two mechanisms: *(i)* segregation of reactants in disjointed fractal topologies; *(ii)* increase of the encounter probability between substrate or effector, and enzyme or target, by their coexistence in the same (2D) topological dimension[23].

2.3 (Macro)Molecular Crowding

Another view of cytoplasmic structure emphasizes the "macromolecular crowding" properties of the ground plan of living cells. Cells contain a large variety of macromolecules in their cytoplasm, largely proteins, which taken together will occupy a substantial fraction of the total cytoplasmic volume, called *volume occupied* or *crowded* [1,31-32]. Following changes in cell volume during osmotic regulation, *molecular crowding* and *confinement* directly affect the form and function of intracellular macromolecules. From data obtained in red blood cells, it has been suggested that cells may use macromolecular crowding as a means of sensing their volume; what appears to be cell volume regulation is really a regulation of cytoplasmic macromolecular crowding[33-34]. Apparently, protein concentration is what the cells seek to maintain to approach an optimal degree of cytoplasmic nonideality[33].

3. Heterogeneous distribution of cytoskeletal proteins as a mechanism of metabolic fluxes modulation

The view of the cytoplasm as a crowded object is not excluded by either tensegrity or fractal percolation models. For instance, *crowding* may alter the average degree of self-association of an enzyme, in turn changing its average catalytic activity[1,6,31]. This effect was shown to be specifically induced by microtubular protein or actin in the pyruvate kinase/lactate dehydrogenase enzymatic couple[22,35]. Furthermore, the supramolecular organization of microtubular protein could entrain the dynamics of enzymatic reactions through mechanisms of self-association-dissociation of enzymes toward higher or lower oligomeric forms with higher or lower levels of activity, respectively[6].

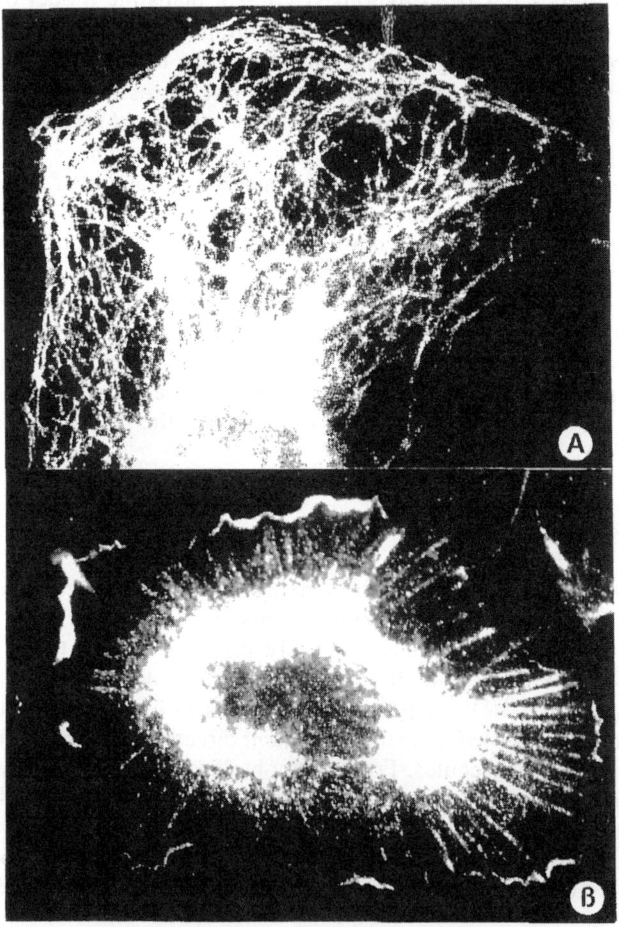

Figure 1. Heterogeneous distribution of anti-tubulin and anti-actin immunofluorescent staining at the cellular level.
Micrographs showing the distribution of microtubules (A) or microfilaments (B) in astrocytes. Microtubules and microfilaments extends throughout the cell cytoplasm. Microtubules were labeled with an antibody against tyrosinated α-tubulin, and microfilaments were labeled with fluorescein-phalloidin. Note the straight actin bundles radiating from the nuclear periphery alternating with dots, and the regionalized staining adjacent to the plasma membrane. The tubulin lattice appears as an intricated meshwork of fibers highly dense around the nuclear periphery.

We have also presented experimental evidence that suggests that the *same* enzyme may be *differentially* modulated at different locations in the cytoplasm depending on the cytoskeletal protein (and their associated proteins) present at that location[35]. According to this proposal, a precise biochemical coupling can be postulated to exist between the highly sophisticated mechanisms by which cells rearrange their (supra)molecular architecture (Fig. 1) and the dynamics of biochemical reactions taking place concomitantly.

This is a relevant matter since the information presently available strongly suggests that the cytoskeleton is one of the endogenous determinants of cell form, and that its (supra)molecular asymmetries are involved in the development of, e.g. neuronal polarity[36-38]. Furthermore, recently it has become apparent the existence of distinct cytoplasmic domains in neurons intimately associated with the ability of these cells to create microtubule subsets that differ in molecular composition, and spatio-temporal distribution[37-38]. Thus, a clear link is established between cell polarity, form and the underlying dynamics of cellular biochemistry, by the proposal attributing a modulatory role to the heterogeneous cytoplasmic distribution of cytoskeletal proteins[35] (Fig. 1). Considering the entrainment of enzymatic reactions by the dynamics of the cellular architecture, a whole set of observations are distinctly enlightened. Actin and tubulin as major structural proteins in all eukaryote cells, change their distribution along the cell cycle. In the budding yeast *Saccharomyces*, the cell cycle-dependent patterns of microtubules and microfilaments localization are well studied[39-40]. Reported experimental data show that cytoskeleton components continuously assemble/disassemble along the length of the axon of cultured neurons or rearrange its spatial distribution in response to growth factors[10,41-43].

A prediction of the fractal model of cytoplasm is that a cell by regulating its geometry or architecture may in turn regulate the level of its percolation threshold, p_c. In percolation lattices, transitions between local to global connectedness happen at the threshold level of occupancy of the lattice[29]. Different p_c would determine when *a local instability in the dynamics of an enzymatic reaction may extend to the whole cellular field if it spreads in a domain belonging to the spanning cluster.*

4. A rheological view of tensegrity and fractal models

Tensegrity and fractal models do not exclude each other although they belong to different preconceptions[44]. A structural preconception based on geometry and self-assembly is evident in the tensegrity model whilst the fractal model follows kinetic principles of systems organization away from equilibrium for generating pattern and form.

There exists compelling rheological evidence which shows that cytoplasmic polymers behave as a *reversible non-covalent gel network* which is constantly being assembled and disassembled[25,45-46].

The tensegrity model describes cytoplasm as a prestressed molecular continuum of independent struts and tensile elements (see section above: The tensegrity model). Cytoskeleton stiffness and apparent viscosity change in parallel following application of mechanical stress because of the postulated molecular continuum. Shifts in cytoplasmic rheology, i.e. viscosity, result from filament alignment or interfilamental friction. Altering the global architecture of the cell, changes the force balance of the cytoskeleton that in turn elicits changes of its stiffness and apparent viscosity[47].

The supramolecular network of the cytoplasm effects rheological shifts involving sol-gel transitions. The connectivity of such a network would certainly depend on the concentration of assembling filaments, the strength of their interaction, or the number of cross-linkers, e.g. MAPs[46]. Besides, geometrical arrangements certainly influence the connectivity of the gel network. As already stressed (see: The fractal percolation model), a common ingredient in these systems is that at a given value of a characteristic parameter, a macroscopic connected cluster forms in the system, e.g. a cluster of connected polymer molecules results in gel formation. When the latter happens, the physical properties of the system can change drastically. In the case of sol-gel transitions, the polymer solution has liquid properties before the gel forms, that is, a well-defined value of the viscosity, whereas in the gel phase, the viscosity is practically infinite and the system has well-defined elastic properties. The critical parameter at which these transitions happen is the percolation threshold, that for sol-gel transitions may be the concentration of polymers in solution.

Summarizing, a clear link exists between the behavior of a fluid in a percolation supramolecular network, and the rheological behavior of the molecular network itself. Otherwise stated, at the percolation threshold, p_c, not only the dynamics of the (bio)chemical reactions have the possibility to reach a global influence, but also the rheological character of the polymer network, i.e. may effect a sol-gel transition. This is utmost important since it integrates in a unified and coherent scheme, rheological and biochemical evidence with the self-organizing properties of non linear kinetic schemes. The latter is a potential source of self-emergent spatio-temporal behaviors, i.e. bistability, oscillations, thresholds, chaos or waves of chemicals. In the following we will show the possibility of the existence of such phenomena in the coupling of two dynamical subsystems such as the assembly-disassembly of microtubular protein and the kinetics of an enzymatic reaction.

5. Dynamic organization of biochemical reactions and cytoskeleton components

The more adequate models for cellular functional coherence will be those ones in which structure and dynamics are combined. In the way to shape those models we should not miss that cellular processes, scale in space and time, from the conformational change in the α-helix of a protein up to cell division. By the latter we mean that processes such as enzymatic reactions or cell division occur within characteristic spatial coordinates and when perturbed they will relax with a characteristic relaxation time[30]. We have proposed that microtubular protein which organizes as fractal percolation lattices[22,28] exhibits self-organization in the spatio-temporal "window" (micrometers and minutes) in which microtubules assemble-disassemble[48-49]. In percolation processes, at the threshold level of occupancy of the lattice, transitions between local to global connectedness happen. The percolation threshold of the fractal cytostructure may be related to the scaling exhibited by physiological processes at the 'transition point' in the spatio-temporal domain for which cytoskeleton lattices self-organize[6,30]. From these considerations, it has been proposed that the passage from microscopic to macroscopic functional coherence achieved through self-organization of cytoskeleton components, allows to postulate an analogy between the concepts of self-similarity in fractal cytostructures and that of spatio-temporal scaling exhibited by physiological processes at distinct levels of organization[30].

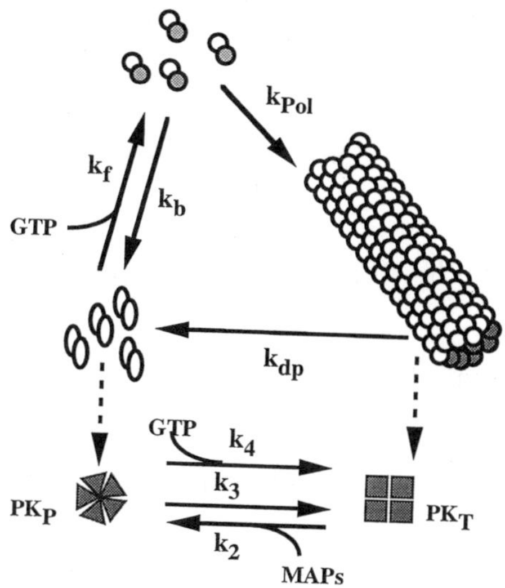

Figure 2. Scheme of the model which links microtubular protein polymerization-depolymerization dynamics and pyruvate kinase (PK) kinetics.
PK in the presence of MTP, may exist as a tetramer (PK_T) or pentamer (PK_P). The cycle of MTP polymerization-depolymerization assumes that tubulin may exist as polymerized (C_P) or non-polymerized, bound to GTP (C_T), or non-polymerized bound to GDP (C_D). The dashed arrows indicate that polymerized or depolymerized forms of MTP stabilize either the tetrameric or pentameric forms of PK (Reproduced from Aon et al., 1995).

In order to test if self-organization of microtubular protein (MTP) could entrain the dynamics of enzymatic reactions, we modelled the coupling between the assembly-disassembly of MTP and pyruvate kinase (PK) through a system of four ordinary differential equations (ODEs) (Eq. 1-4). Figure 2 shows a scheme of the model which links the dynamics of assembly-disassembly of MTP and PK kinetics. The equations which describe the activity of PK (Eq. 5-7) were obtained by fitting the kinetic behavior of the enzyme with respect to the concentration of its substrate phosphoenolpyruvate (PEP), in the presence of either polymerized or non polymerized MTP[6]. The enzyme, in the presence of MTP, may exist as a tetramer (T) or pentamer (P); the proportion of each form depending on the polymeric status of MTP, MAPs or GTP (through the 3rd term at the r.h.s. of Eq. 1). The MAPs bound to the microtubular lattice appears to promote the formation of the P form through the first term at the r.h.s. of Eq. 1; the VPK_{max} of the P form being larger than that of the T one[6].

The equations that describe the MTP polymerization-depolymerization assumes that tubulin may exist in one of three forms, i.e. either polymerized, or non-polymerized bound to GTP; or non-polymerized bound to GDP[50]. A conservation equation relates the three forms of MTP (Eq. 8). The polymerization term of the state variable C_P is autocatalytic with n_1 as the nuclei size (n_1=2). The coupling between the kinetic behavior of PK and the polymeric status of MTP is realized through the term of pentamer formation, PK_P, which is proportional to the amount of polymerized MTP, C_P, in Eq. 1 and 4. The mathematical formulation of the model is as follows:

$$\frac{d[P]}{dt} = 0.1\, k_2\, (PK_t - [P])\, C_P - k_3\, [P] - k_4\, [P][GTP] \tag{1}$$

$$\frac{d[PEP]}{dt} = -\, V_{PK} - k_1\, (PEP_o - [PEP]) \tag{2}$$

$$\frac{dC_T}{dt} = -\, k_P\, C_P^{n1}\, C_T + k_f\, C_D\, [GTP] - k_b\, C_T\, (GTPo - [GTP]) \tag{3}$$

$$\frac{dC_P}{dt} = k_P\, C_P^{n1}\, C_T - k_{dp}\, C_P \tag{4}$$

$$V_{PK} = \frac{V_{PK}^{max}\, [PEP]^n}{K_s^n + [PEP]^n} \tag{5}$$

$$V_{PK}^{max} = V_T^{max} + \left(V_P^{max} - V_T^{max} \right) \frac{[P]}{PK_t} \tag{6}$$

$$n = 4 + \frac{[P]}{PK_t} \tag{7}$$

$$C_o = C_D + C_T + C_P \tag{8}$$

The parameters GTP_o and PEP_o which represent the total intracellular pools of GTP and PEP, as well as the amount of intracellular tubulin, C_o, are main biological inputs of the model. Essential experimental results such as the kinetics of MTP polymerization and the increase in PK-catalyzed fluxes as a function of MTP concentration are simulated by the model[35].

The existence of coupled-irreversible transitions in the MTP polymeric status and in the activity of PK, is a main result obtained with the model (Fig. 3). The system is bistable, i.e. it shows two possible steady states of activity (polymerized or non-polymerized MTP, and a fully active or half-active enzyme) for the same value of the total MTP concentration, Co, and the rate of polymerization, K_{pol}, depending on the sense of variation of the latter. Generally speaking, bistable systems are reversible although under some conditions irreversible transitions may happen. In these cases the system once passed a limit point can not return to the previous branch of steady states. The irreversible jump occurs depending on the previous history of the system (e.g. the direction of change of K_{pol}). After the jump, the state variables, e.g. the non polymerized or polymerized MTP, remain indefinitely on the steady branch without possibility of coming back to the previous situation. These results may also provide a putative explanation to the well known phenomenon of dynamic instability shown by microtubules either *in vitro* or *in vivo* in the sense that shrinking microtubules seldom undergo a transition to the growing state[41,48,51-52].

Figure 3 emphasizes the multidimensional character of functional regulation in cells, i.e. the changes in steady state values of PEP versus the simultaneous variation of the total amount of MTP, C_o, and its rate of polymerization, K_{pol}. This results in a 'line of bifurcation points', i.e. joining the points that limit the stable and unstable branches of steady states in the individual bifurcation plots (Fig. 3). In fact, the variation of the parameter K_{pol}, may result from the 'lumping' of several physiological processes which, in turn, entrain different functional properties. For instance, actin or tubulin polymerization are sensitive to cytoplasmic Ca^{2+} and phosphoinositides levels[53] and pH changes[54]. These physiological variables will affect K_{pol}. Likewise, C_o may be modulated by changes in gene expression[55-56].

By inspecting Fig. 3, as the amount of protein increases either the irreversible transitions in PK activity (displayed by the level of the substrate PEP) or the amount of polymerized protein, C_p (not shown) occur at lower rates of polymerization, K_{pol}. The transitions in MTP polymeric status also entrain irreversible changes in the catalytic properties of PK as well as of its conformational state namely aggregation as a pentamer (Fig. 2).

In summary, we have been able to show that a self-organized phenomenon, i.e. bistability or multiple stationary states, may occur in the coupled dynamics of MTP and a biochemical reaction as a function of two relevant cellular parameters, C_o and K_{pol}. These two parameters in turn regulate the geometry of the lattice of MTP and through this the percolation threshold, p_c, and the fractal dimension, D[23,28,29]. This reinforces the idea that polymerization-depolymerization of cytoskeleton components may be involved in the passage from microscopic to macroscopic order at the cellular level. Besides, the existence of abrupt transitions at limit points in the dynamic behavior of MTP polymeric status together with PK activity, further suggests a link between p_C and self-organization of cytoskeleton components, and their involvement in the synchronization of cellular activities further apart in the cytoplasm of living cells.

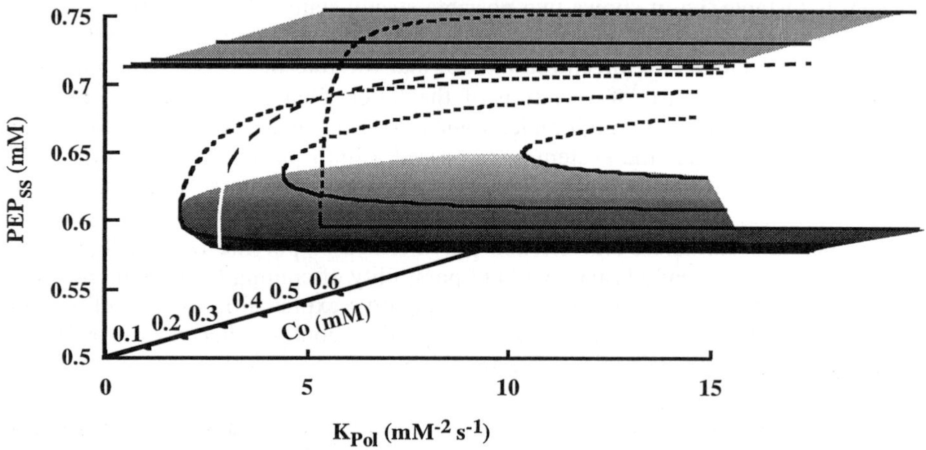

Figure 3 A 3-dimensional representation of the qualitative dynamic properties of two coupled systems.

The study of the qualitative dynamic behavior of a model based on experimental data[6] was performed with the program AUTO. The stability properties as well as the different type of steady states (ss) exhibited by the model are represented as a function of two bifurcation parameters of physiological relevance such as the rate of polymerization of MTP, K_{pol}, and the total amount of MTP species, Co, present in the system. The steady state values of phosphoenolpyruvate (PEP), the substrate of PK, were plotted.

The steady values of state variables represented by the system of Eq. 1-4 were computed with AUTO. The bifurcation behavior presented stable (———) and unstable (- - - -) branches of steady states with limit points. The diagrams were obtained with the following parameter values: $k_1 = 1.0\ s^{-1}$, $k_2 = 10\ mM^{-1}\ s^{-1}$, $k_3 = 0.05\ s^{-1}$, $k_4 = 0.02\ mM^{-1}\ s^{-1}$, $k_f = 3.0\ mM^{-1}\ s^{-1}$, $k_b = 2.5\ 10^{-3}\ mM^{-1}\ s^{-1}$, $k_{dp} = 2.5\ 10^{-3}\ s^{-1}$, $n_1 = 2$, $K_s = 0.6$ mM, $PEP_o = 10$ mM, $PK_t = 0.01$ mM, $GTP_o = 1.0$ mM, [GTP] = 0.9 mM, $V_P^{max} = 25\ s^{-1}$, $V_T^{max} = 14\ s^{-1}$. The shaded areas represent the domain (or 'surface') of stable behavior of the system separated by a 'gap' of unstable comportment. The parameter values of the subsystem which describes assembly-disassembly of MTP were taken from Carlier et al.,[50]. Those parameter values corresponding to PK kinetics were obtained from Cortassa et al.,[6].

6. Concluding remarks: models and metaphors of the cell

How does current models or views of the cytoplasm of cells, fit into current metaphors of the cells? What does this mean for modelling of functional properties of cells?

According to the tensegrity model, global changes in tissue structure alter the local cytoskeleton force balance hence eliciting an integrated cellular response. The remodelling molecular cascade would result from tension molding of the cytoskeleton, and internal oscillatory transformations between locally stable and unstable architectural states. The larger the magnitude of the applied stress, the larger the rearrangement of the cytoskeleton matrix. In this scheme, the mechanism of coherence results from the passage between global cytoskeleton reorganization to local (metabolic) behavior, rather than the inverse.

The view emerging from the fractal percolation model suggests that the assembly-disassembly of cytoskeleton components act as a macroscopic, coherent, inducer of changes in systemic properties such as metabolic fluxes. The supramolecular organization of the cytoskeleton as a percolation lattice adds a feasible physicochemical mechanism of coherence when the passage from local to global behavior in the cytoplasm is effected at the percolation threshold, p_c. Below p_c, the cytoplasm would behave as loosely coupled, independent dynamical subsystems. Structural microscopic features comprising molecular interactions subtly modulated by relevant physiological parameters such as pH, Ca^{2+} or second messenger levels, determine macroscopic geometrical characters of the lattice that change the p_c. The dynamics of chemical reactions can be either entrained by the architectural arrangements of cytoskeleton components, or at unstable parametric regions can give rise to self-organized behavior whose emergence induce new functional properties to cells, i.e. adaptive conduct ensuing changes of the intra- or extra-cellular environments.

Overall, we have the impression, as well as other researchers[57-58], that there is not a unique, 'royal', way to model systems whose functional properties are manifested at several simultaneous levels of organization such as cellular systems[30]. Even all metaphors, with the appropriate biological constraints, will have to be used such as the 'cell-as-machine', 'cell-as-society', 'cell-as-text' or 'cell-as-field' exhibiting parallel distributed processing[58]. Instead of a single model spanning all levels of organization to give a unified account of the whole complexity, one approach is to think at the coexistence of a 'chain of models' linking adjacent levels of organization which implies the coexistence of explanations at several levels.

Acknowledgements

We are indebted to Centro de Cálculo de la Universidad Nacional de Tucumán for computing facilities. This work was supported by grants from Fundación Antorchas and Consejo de Investigaciones de la Universidad Nacional de Tucumán. MAA, SC and AC are career investigators from CONICET, Argentina.

References

1. A.P. Minton, *Mol. Cell. Biochem.* **55**, 119 (1983).
2. A. Fulton, *Cell* **30**, 345 (1982).
3. J.S. Clegg, *Am. J. Physiol.* **246**, R133 (1984).
4. S. Penman *et al., Cold Spring Harbor Symp. Quant. Biol.* 1013 (1981).
5. K.J. Pienta and C.N. Hoover, *J. Cell. Biochem.* **55**, 16 (1994).
6. S. Cortassa *et al., J. Cell. Biochem.* **55**, 120 (1994).
7. R. van Driel *et al., J. Cell. Biochem.* **47**, 311 (1991).
8. K.A. Suprenant, *Cell Motil. Cytoskel.* **25**, 1 (1993).
9. A. Fulton, *J. Cell. Biochem.* **52**, 148 (1993).
10. N. Hirokawa in *The Neuronal Cytoskeleton*, (Wiley-Liss, New York, 1991).
11. C. Masters, *J. Cell Biol.* **99**, 222 (1984).
12. F. Clarke *et al.*, in *Regulation of carbohydrate metabolism*, ed. R Beitner (CRC Press, Boca Raton,1985).
13. M.A. Luther and J.C. Lee, *J. Biol. Chem.* **261**, 1753 (1986).
14. T. Keleti and J. Ovadi, *Curr. Top. Cell. Regul.* **29**, 1 (1988).
15. H.R. Knull and J. L. Walsh, *Curr. Top. Cell. Regul.* **33**, 15 (1992).
16. J. Ovádi and F. Orosz, *Curr. Top. Cell. Regul.* **33**, 105 (1992).
17. A. Lehotzky *et al., J. Biol. Chem.* **268**, 10888 (1993).
18. P. Marmillot *et al., Arch. Biochem. Biophys.* **315**, 467 (1994).
19. G.R. Welch, *Prog. Biophys. Molec. Biol.* **32**, 103 (1977).
20. J.S. Clegg, *Curr. Top. Cell. Regul.* **33**, 3 (1992).
21. D.E. Ingber, *Cell* **75**, 1249 (1993).
22. S. Cortassa and M.A. Aon, *Cell Biol. Internat.* **18**, 687 (1994).
23. M.A. Aon and S. Cortassa, *FEBS Lett.* **344**, 1 (1994).
24. D.E. Ingber *et al., Int. Rev. Cytol.* **150**, 173 (1994).
25. K. Luby-Phelps *et al., Ann. Rev. Biophys. Biophys. Chem.* **17**, 369 (1988).
27. N. Wang *et al., Science* **260**, 1124 (1993).
28. C. Rabouille *et al., J. Biomol. Struct. & Dynamics* **9**, 1013 (1992).
29. J. Feder, *Fractals*, (Plenum Press, New York, 1988).
30. M.A. Aon and S. Cortassa, *Mol. Cell. Biochem.* **120**, 1 (1993).
31. A.P. Minton, *Biophys. J.* **63**, 1090 (1992).
32. M.M. Garner and M.B. Burg, *Am. J. Physiol.* **266** (Cell Physiol. 35), C877 (1994).
33. J.C. Parker, *Am. J. Physiol.* **265**, C1191 (1993).
34. D. Haussinger *et al., Am. J. Physiol.* (Endocrinol. Metab.) **30**, E343 (1994).
35. M.A. Aon *et al., J. Cell. Biochem.* **59** (in the press) (1995).
36. F. Solomon, *J. Cell Biol.* **90**, 547 (1981).
37. A. Cáceres *et al., Dev. Brain Res.* **13**, 314 (1984).
38. A. Ferreira and A. Cáceres, *Dev. Brain Res.* **49**, 205 (1989) .
39. D.J. Lew and S.I. Reed, *J. Cell Biol.* **120**, 1305 (1993).
40. J. Chant, *Trends in Genet.* **10**, 328 (1994).
41. E.M. Tanaka and M.W. Kirschner, *J. Cell Biol.* **115**, 345 (1991).
42. G. Morfini *et al., J. Neurosc. Res.* **39**, 219 (1994).
43. M. DiTella *et al., Cell Motil. Cytoskel.* **29**, 117 (1994) .

44. L.G. Harrison, *J. theor. Biol.* **125**, 369 (1987).
45. K. Luby-Phelps *et al., J. Cell Biol.* **102**, 2015 (1986).
46. G. Forgacs and S.A. Newman, *Int. Rev. Cytol.* **150**, 139 (1994).
47. N. Wang and D.E. Ingber, *Biophys. J.* **66**, 2181 (1994).
48. M. Kirschner and T. Mitchinson, *Cell* **45**, 329 (1986).
49. H.P. Erickson and E. T. O'Brien, *Annu. Rev. Biophys. Biomol. Struct.* **21**, 145 (1992).
50. M.F. Carlier *et al., Proc. Natl. Acad. Sci. USA* **84**, 5257 (1987).
51. T. Mitchison and M. Kirschner, *Nature* **312**, 237 (1984).
52. T. Horio and H. Hotani, *Nature* **321**, 605 (1986).
53. P.A. Janmey, *Ann. Rev. Physiol.* **56**, 169 (1994).
54. K.A. Suprenant, *Cell Motil. Cytoskel.* **19**, 207 (1991).
55. A. Ben-Ze'ev, *Trends Biochem. Sci.* **475** (1986).
56. D.W. Cleveland, *Trends Biochem. Sci.* **13**, 339 (1988).
57. P.S. Churchland and T.J. Sejnowski, *Science* **242**, 741 (1988).
58. R.C. Paton, *Biosystems* **29**, 63 (1993).

MODELS OF INFORMATION PROCESSING WITHIN INHOMOGENEOUS CELLULAR STRUCTURES

P. HARTMANN
Technical University of Darmstadt, Alexanderstraße 10, D-64283 Darmstadt, Germany

1 Introduction

Besides *experiments* on biological systems, *simulations* are another method for achieving deeper insights into the principles of cell-cell interactions and about cellular information processing. Experiments are often restricted to snapshots of static situations observable at specific stages of development. And most experiments require that the observed organism is destroyed. Simulations, in contrast, allow us to analyse the spatio-temporal development of cellular structures. Furthermore, simulations give us the ability to verify complex theories which may be far beyond the scope of experiments based on a limited set of observable state values.

In order to perform a simulation, in a first step an adequate abstract model must be constructed. The model consists of two major components, the *initial configuration* and the *set of rules* which describe the spatio-temporal development. In the following it is assumed that cell-cell interactions shall be studied. In this case the cells are interpreted as the elementary building blocks of the model. The initial configuration describes the state of each individual cell (i.e. the cell's physical properties, concentrations of biochemical substances, switching state of genes, ...) as well as how the cells are composed to form a tissue. The set of rules can be interpreted as the *genome* of the cells and it describes the processing of information on the level of biochemical reactions, cell signalling mechanisms and switching of genes.

A highly simplified simulation can make use of a cellular automaton model (see [1]). A cellular automaton (CA) consists of an aggregate of a multitude of finite automata organized in a regular grid. In such a model each of the biological cells is assigned to a specific automaton cell, and the grid expresses the neighbourhood relationship between adjacent cells. The set of rules is called the *local transition function* of the automaton. For illustration, see the left-hand side of figure 1, which displays a single cell embedded into a rectangular grid.

In many cases the level of abstraction of a CA model is too high, because real cells are not arranged within a regular grid. More versatile methods based on *graphs* turn out to be better suitable for describing inhomogeneous cellu-

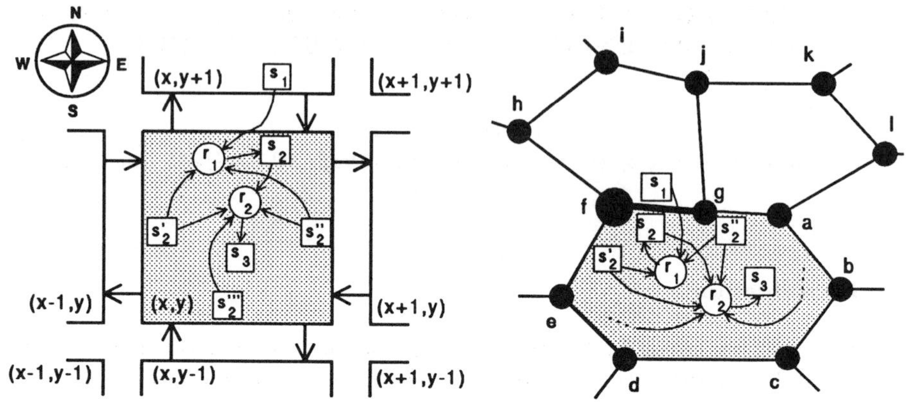

Figure 1: Comparison between the situation observed within cellular automata (left-hand side) and inhomogeneous cellular structures (right-hand side). The right-hand side displays the cellular hypergraph of the simple example used throughout this paper. The lowercase letters indicate the names of the graph's vertices.

lar structures. Furthermore, if questions related to morphogenesis shall be discussed, effects like *cell division*, *cell migration* and *cell death* [2] need to be formulated. Graphs allow us to simulate such temporal changes of the *structure* of a tissue, because *graph replacement systems* can be applied to them: The *productions* of a graph replacement system express how the state of some cell influences its shape, its cell division or how it aggregates with other cells.

With respect to the specific requirements of biological systems, the concept of *cellular hypergraphs* (CHG) and *replacement systems* (CHGRS) for these graphs has been developed [3]. Important steps towards an efficient implementation are documented in [4] and [5]. The novel CHG have specific properties which make them ideally suitable for building three-dimensional models of biological tissues. And the CHG concept allows us to describe inhomogeneous and dynamically changing cellular structures with an arbitrary number of dimensions. In this sense it generalizes the massively-parallel concept of cellular automata. The right-hand side of figure 1 illustrates how a cellular hypergraph can be used to describe a cellular structure.

In section 2 a simple example will be introduced which helps to illustrate principle questions and the solution offered by the presented approach. In combination with this example, the fundamentals about cellular hypergraphs and replacement systems are summarized. Based on the notation given in section 2, section 4 explains the terms *address* and *relative addressing*. In section 5 it will be discussed what the term *component type* means in conjunction with a cellular hypergraph. The usage of more complex expressions will be described in the subsequent section 6. Towards the end of this article it is demonstrated how these expressions can be used to formulate the local transitions function for the example presented at the beginning in section 2.

2 Example

This section introduces an example used throughout this article. In this example it is assumed that the (biological) cells perform signal processing on the level of biochemical reactions: Receptors of type r_1 are associated with the wall segments of each cell. These cell-surface receptors cross the cell walls. They are activated by a extracellular signalling molecules s_1, and they propagate this information towards the interior of the cell by mediating a chemical reaction. This reaction produces the intracellular substance s_2 *locally* at the same wall segment where the receptor r_1 is located. It shall be considered that the chemical reaction is inhibited by the same substance s_2 that is produced at the two neighbouring wall segments. This means that the difference between the levels of substance s_1 reaching adjacent wall segments needs to be computed. Furthermore the average over all entities s_2 shall be computed and assigned to the substance s_3. The substance s_3 is assumed to be located within the cytoplasm and therefore it can be seen to be *global* within each single cell.

Figure 1 visualizes two different methods for modelling this sample biological system, *cellular automata* and *cellular hypergraphs* [a]. On the left-hand side a *cellular automaton* (CA) model with its processing elements organized in regular grid is displayed. On the right-hand side a more adequate model is sketched out where a graph is used for symbolizing the cells and the relationships between the cells.

A cellular automaton consists of only one type of components, which are called the *cells* of the automaton. Each of the cells can carry an identical amount of state information. The underlying regular grid allows us to address the cells in relation to their position within the grid. The local transition function of the CA model can be formulated easily since it is possible to make use of the directions *north*, *east*, *south* and *west*. The constant s_2^{max} represents the maximum value of the variable s_2:

[a] It should be noted that the possible structural dynamics of the graph are *not* subject of this example, the related fundamentals have been presented in [3]. And cellular hypergraphs are not restricted to two dimensions, they allow us to construct three-dimensional models of a tissue as well.

r_1: $s_2(x,y).north := s_1(x,y{+}1).south \cdot (2 \cdot s_2^{max} - s_2(x,y).east - s_2(x,y).west)$

$s_2(x,y).east :=$ ... (similar to above)

$s_2(x,y).south := \ldots$

$s_2(x,y).west := \ldots$

r_2: $s_3(x,y) := (s_2(x,y).north + s_2(x,y).east + \ldots)/4$

Within inhomogeneous cellular structures the situation turns out to be totally different. There are no main axes (*north-to-south* and *east-to-west*) which allow us to address the cells relative to each other, and the addresses (x, y) of the cells cannot be denoted any longer. Different cells can even have a different number of neighbours. And furthermore different types of components need to be represented. For instance, for modelling *biological* cellular systems the component types *cell body* and *cell wall* are required.

Therefore the description how to address and process state values has a more complicated appearance. The following rules sketch out the equations that are equivalent to the CA rules given above:

r_1: $s_2.inner\ wall\ segment := s_1.outer\ wall\ segment \cdot$
$(2 \cdot s_2^{max} - s_2.inner\ preceding\ segment - s_2.inner\ succeeding\ segment)$

r_2: $s_3.actual\ cell := \dfrac{\sum_{all\ wall\ segments} s_2.actual\ inner\ wall\ segment}{total\ number\ of\ wall\ segments}$

These rules demonstrate that the new terms *component type* (i.e. *wall segment* for storing the values of s_1 and s_2, and *cell body* for storing the values of s_3) need to be introduced. And it must be possible to describe the *relative addressing* between these different component types, i.e. *succeeding/preceding wall segment* or *neighbouring cell*. In order to describe these relationships, the concept of cellular hypergraphs has been developed:

3 Cellular hypergraphs

Cellular hypergraphs are a special type of hypergraphs, which itself are a generalisation of undirected graphs. A hypergraph $h = (V, E)$ consists of vertices $V = \{a, b, c, \ldots\}$ and edges E, where each edge $e \in E$ is a subset of the set of vertices ($e \subset V$).

In contrast to hypergraphs, for within a *cellular hypergraph* (CHG) $g^D = \{e_1^{d_1}, e_2^{d_2}, \ldots\}$ each edge $e^d \in g^D$ is labeled with a number of *dimensions* d; $0 \le d \le D$ which is noted as an upper index. The integer number D denotes the dimensionality of the graph g^D. The dimensions of the edges reflect the number of dimensions of the related components within a cellular structure. Edges with 0 dimensions identify *points* (the corners between cell walls), 1-dimensional edges *line segments* (the cell walls) and 2-dimensional edges *area elements* (the cell bodies) [b].

The right-hand side of figure 1 displays the graphical interpretation of the sample CHG. The 0-dimensional edges are identical with the vertices of the graph. They are drawn as dots (labeled with letters 'a' to 'l'). The one-dimensional edges are drawn as straight line segments between a pair of vertices. The two-dimensional edges exist implicitly whenever one-dimensional edges enclose an area segment. In the notation of sets, the displayed section of the graph can be written down as follows (in this notation the dimensionality of the edges is noted as an upper index subsequently to the closing curly brackets):

$$\begin{aligned} g^2_{sample} = \{ & \{a\}^0, \{b\}^0, \{c\}^0, \ldots, \{l\}^0, \\ & \{a,b\}^1, \{b,c\}^1, \ldots, \{l,a\}^1, \\ & \{a,b,c,d,e,f,g\}^2, \{f,g,h,i,j\}^2, \{a,g,j,k,l\}^2 \}^2 \end{aligned}$$

The edges of a CHG with their different dimensions must comply with strong restrictions which ensure that some given CHG can only represent a valid cellular structure. A hypergraph g^D is called to be a *cellular* hypergraph if it conforms with the restrictions (G0)-(G4) [c]:

G0: Each 0-dimensional edge $e^0 \in g^D$ contains exactly 1 vertex and each 1-dimensional edge $e^1 \in g^D$ contains exactly 2 vertices.

G1: For each d-dimensional edge $e^d \in g^D; d < D$ there exists at least one $d{+}1$-dimensional edge $e^{d+1} \in g^D$ such that $e^d \subset e^{d+1}$.

G2: Each d-dimensional edge $e^d \in g^D; d > 0$ is totally covered by a connected set of d–1-dimensional edges $e_i^{d-1} \in g^D$.

G3: Between each pair of connected d-dimensional edges $e_1^d, e_2^d \in g^D$ there exists a connecting chain of edges with a specific structure.

G4: For every pair of edges $e^d, e^{d+2} \in g^D$ with $e^d \subset e^{d+2}$ there exist exactly 2 edges with $d{+}1$ dimensions $e_i^{d+1} \in g^D$ such that $e^d \subset e_i^{d+1} \subset e^{d+2}$.

In conjunction with relative addressing mechanisms the restriction (G4) is the most important one. It ensures that *elementary relative addressing steps*

[b]In a three-dimensional model, the cell bodies would be represented by three-dimensional edges, and the cell walls by two-dimensional edges, respectively.

[c]These restrictions are only sketched out in this paper, they can be looked up from [3]

can be performed in a specific way: If there are given three edges such that $e^{d-1} \subset e_1^d \subset e^{d+1}; 0 < d < D$, then (G4) expresses that there exists exactly one alternative edge e_2^d that can be used for replacing the edge e_1^d without invalidating condition (G4). The consequences of this restriction will become more clearly in section 4.

The properties of a CHG can be visualized if it is drawn in the form of a hierarchical graph. This has been done in figure 2 for a section of g^2_{sample}.

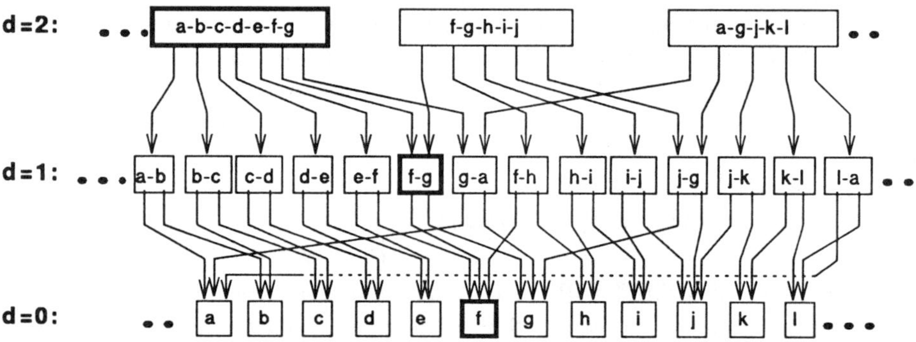

Figure 2: Part of the CHG g^2_{sample} drawn as a hierarchically structured graph. The arrows symbolize the relationship *'contains'* between the edges with different dimensions.

Those edges with the maximum number of two dimensions (representing the cells) have been placed in the top layer. The mid layer contains the edges with one dimension (the cell walls) and the bottom layer contains those with 0 dimensions (corners of the cells). The arrows pointing from edges with more dimensions to those with less dimensions symbolize the relationship 'contains' between these edges. Given some edge e^d in a graph g^D, this relationship is denoted by the sets $U(e^d, g^D) := \{e^{d+1} \in g^D | e^d \subset e^{d+1}\}$ of all *upper* edges and $L(e^d, g^D) := \{e^{d-1} \in g^D | e^{d-1} \subset e^d\}$ of all *lower* edges. For instance,

$U(\{f,g\}^1, g^2_{sample}) = \{\{a,b,d,e,f,g\}^2, \{f,g,h,i,j\}^2\}$ and

$L(\{f,g\}^1, g^2_{sample}) = \{\{f\}^0, \{g\}^0\}$.

4 Addresses and relative addressing

In conventional graphs some *address* is usually implemented as a pointer to some vertex of the graph. In contrast to this point of view, addresses of *cellular hypergraphs* have a more complex appearance. This becomes more clearly if the question *'How much information can we obtain about a specific location within a D-dimensional structure ?'* is discussed. It is not sufficient just to denote a single 0-dimensional edge (i.e. the corner of a cell), because in this case someone has no knowledge about edges with more dimensions. And when the focus is set to a single D-dimensional edge (i.e. a complete cell), this gives us no information about components with less dimensions, either.

Consequently a more adequate type of addresses is introduced to be used in conjunction with cellular hypergraphs. In figure 1 the sample address t^2_{sample} is emphasised: The edge $e^0 = \{f\}^0$ is accentuated with a bigger dot, the bold line segment denotes the edge $e^1 = \{f, g\}^1$ and the shaded area denotes the edge $e^2 = \{a, b, c, d, e, f, g\}^2$. The same address has been emphasised in figure 2 where the boxes symbolizing the edges have been drawn using thicker lines.

For writing down an address, the notation of sets will be used. The sample address can be written down as a list of three edges $t^2_{sample} = (e^0, e^1, e^2) = (\{f\}^0, \{f, g\}^1, \{a, b, c, d, e, f, g\}^2)$. This can be abbreviated in the form $t_{sample} = (f|g|a, b, c, d, e)$ where only the differences between the edges are listed.

In general (for arbitrary dimensions D) each address t^D in g^D consists of a list of edges $t^d = (e^0, e^1, \ldots, e^D)$. In order to form a legal address, the edges in t^d need to conform with the three conditions

T1: $\forall e^d \in t^D : e^d \in g^D$, every edge $e^d \in t^D$ must be contained in g^D.

T2: $\forall d; 0 \leq d \leq D : |\{e^{d'} \in t^D \mid d' = d\}| = 1$, for every number of dimensions d there exists exactly one edge $e^d \in t^D$.

T3: $\forall e^d, e^{d+1} \in t^D; 0 \leq d < D \Rightarrow e^d \subset e^{d+1}$, every edge with more dimensions covers those with less dimensions completely.

It is obvious that a legal address must be located within the graph (condition T1). And it's clear that an address shall contain enough but as little as possible information about a certain location, and this is ensured via condition T2. Finally condition T3 expresses that an address can only make sense if the individual edges of the address are related to each other in a specific way. The edges of the sample address which are visualized in figure 2 illustrate this relationship between the edges with different dimensions. It can be observed that the sample address t^2_{sample} obeys to condition T3 because $e^0 \subset e^1 \subset e^2$.

Now it is discussed how *relative addressing* can be performed using the interpretation of the term *address* given above. Let there be given some address t^D pointing to the graph g^D and a given number of dimensions $d; 0 \leq d \leq D$. Then the relative addressing step $\hat{t}^D = R^d(t^D, g^D)$ has a single and definite address $\hat{t}^D \subset g^D$ as a result. The two addresses t^D and $\hat{t}^D$ are identical, except for the single d-dimensional edge:

$\forall e^{d'} \in \hat{t}^D; d' \neq d : e^{d'} \in t^D$

Performing a relative addressing step in dimension d means, that the d-dimensional edge e^d contained in t^D is replaced by the only possible alternative edge $\hat{e}^d$ taken from the graph g^D, such that the new address conforms to condition (T3). In order to describe the effect of relative addressing steps formally, we need to refer to the sets of edges $U(e^d, g^D) := \{e^{d+1} \in g^D | e^d \subset e^{d+1}\}$ (if $d < D$) and $L(e^d, g^D) := \{e^{d+1} \in g^D | e^d \subset e^{d+1}\}$ (if $d > 0$), which describe the relationships between the edges with different dimensions. Then the set $\hat{E}^d$ of all possible alternative edges can be evaluated as follows:

$\hat{E}^d = U(e^{d-1}) \cap L(e^{d+1}) \backslash \{e^d\}$ if $0 < d < D$

$\hat{E}^d = U(e^{d-1}) \backslash \{e^d\}$ if $d = D$

$\hat{E}^d = L(e^{d+1}) \backslash \{e^d\}$ if $d = 0$

The edge $\hat{e}^d$ is selected from $\hat{E}^d$ as follows:

$\hat{e}^d \in \hat{E}^d$ *if* $E^d \neq \emptyset$

$\hat{e}^d := e^d \in t^D$ *if* $E^d = \emptyset$ *(only possible if* $d = D$*)*

The analysis of the condition G4 and the condition T3 from above leads to the proof that there always exists exactly one alternative edge $\hat{e}^d$. Only in the case $d = D$ such an alternative address may be missing (if the address t^D denotes an component that participates in the *envelope* of the graph g^D (for the definition of the *envelope*, please refer to [3]). In this case the original edge $\hat{e}^d := e^d$ is selected and the operator $R^d(\ldots)$ has the effect of the neutral function.

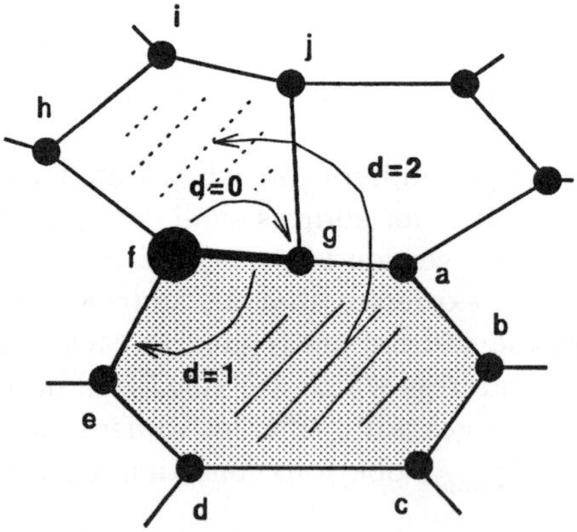

Figure 3: Visualization of relative addressing steps with different dimensions $d \in \{0, 1, 2\}$.

Figure 3 illustrates the effect of a relative addressing step for all possible dimensions $d = 0 \ldots 2$. The arrows denote how the focus to one of the edges of the address changes. The effect of the operator $R^d(\ldots)$ can be interpreted as follows:

$R^2(t^2_{sample}) = (f\|g\|h, i, j)$:	Neighbouring cell, same wall segment
$R^1(t^2_{sample}) = (f\|e\|a, b, c, d, g)$:	Neighbouring wall segment, same cell
$R^0(t^2_{sample}) = (g\|f\|a, b, c, d, e)$:	Alternative end point of actual wall segment

5 Component types and component addresses

The type of addresses that is described above is used for performing relative addressing within cellular hypergraphs. For addressing the *components* of a CHG model like cell bodies or cell walls a different type of addresses is required: The *component addresses.*

For instance it may be desired to associate a specific substance s_2 with the wall segments. This shall be done with respect to the polarity of the walls. From the perspective of a cell each wall has an inner as well as an outer layer. And both layers may have a different state. Of course, an address contains enough information for unequivocally denoting specific components. For instance, when a wall segment shall be denoted, an address that points to this wall contains information about the wall itself (the 1-dimensional edge) as well as the *owner* of the wall (the 2-dimensional edge which denotes the cell body). Additional edges like some 0-dimensional end point of the wall segment do not deliver any additional useful identification and shall be omitted.

The address of a component c^D is derived from some address t^D if specific edges are filtered out from t^D according to their dimensions. Each imaginable filter operation defines a *component type* $C_i = (L_i, \Sigma_i)$. The main part of the description of a component type consist of a list of dimension values L_i. In the case of 2-dimensional models of biological systems the component types C_{body} and C_{wall} with the lists $L_{body} = \{2\}$ and $L_{wall} = \{1, 2\}$ are used. When a component address c^D is computed from an address t^D according to some component type C_i, this is expressed with the formula:

$$c^D = C_i(t^D) \Leftrightarrow \Big(\forall d; 0 \leq d \leq D:\ e^d \in c^D \Leftrightarrow\ \big(e^d \in t^D \wedge d \in L_i\big)\Big)$$

After the term *component addresses* has been defined, it is possible to introduce the notation for denoting the *state values* which are stored within the components of a CHG model.

With each component type C_i a list of symbolic variables $\Sigma_i = \{s_1, s_2, \ldots\}$ with their domain $dom(s_j) = \{\sigma_{j,1}, \sigma_{j,2}, \ldots\}$ is associated. If the variable s_j of a component c^D shall be accessed this is written in the form $s_j(c^D)$. It is assumed that different component types use different identifiers for the variables, such that ambiguities can not occure. If there is given an address t^D and the variable s_j stored in components of type C_i shall be accessed, the following abbreviation can be used: $s_j(t^D) := s_j(C_i(t^D))$.

6 Expressions for writing the local transition function

In the previous sections those terms have been introduced which allow us to address state values and to perform elementary relative addressing steps. This section introduces some more complex expressions which will be used to write down the local transition function of a cellular model with inhomogeneous structure. Towards the end of this section it is demonstrated how these expressions can be used to formulate the local transition function of the example presented at the beginning of this article. And it will be demonstrated how loop expressions can be added to the local transition function such that the global transition function of the cellular model is achieved.

The list of proposed expressions is summarized in the following. It may be complemented in dependence on future experiences:

1. **Address expressions**

   (a) **Address constants.** Any constant set of addresses $T^D = \{t_i^D \mid t_i^D \subset g^D\}$ may be contained in an address expression.

   (b) **Elementary relative addressing steps.** An relative addressing step μ^d denotes a valid expression M. It's effect on a set of addresses computes as $\mu^d(T^D) := \bigcup_{t^D \in T^D} R^d(t^D)$.

   (c) **Complex addressing description.** Any *sequence*, *union*, *intersection* or *complementary* of addressing steps forms a legal address expression: $M = M_1 M_2 \ldots$, $M = (M_1 \cup M_2 \cup \ldots)$, $M = (M_1 \cap M_2 \cap \ldots)$, $M' = \neg M$. If a sequence is evaluated, the addressing steps are applied from right to left, so that the two expressions $M_1 M_2(t^D)$ and $M_1(M_2(t^D))$ are equivalent.

   (d) **Repeating of an addressing expression.** Some addressing sequence may be repeated k times or for an *infinite* number of times: $M' = [M]^k$ or $M' = [M]^*$ with $[M]^1 := M$; $[M]^k := (M^k \cup [M]^{k-1})$ and $[M]^* := lim_{k \to \infty} [M]^k$.

2. **Access to state values**

   (a) **Constant symbols.** Any constants $\sigma \in dom(s(C_i))$ of some component type C_i and some variable $s_j \in \Sigma(C_i)$ may be used within an expression.

   (b) **Status variables.** The state values of a variable with name s_j that is stored within the components of type C_i can be addressed via an expression of the form $s_j(T^D) := \bigcup_{t^D \in T^D} s_j(C_i(t^D))$. The result consists of a *set* of state values.

(c) **Functions.** Functions $F_x(\ldots)$ are defined in order to summarize a set of state values. For instance this function may denote the *sum* $F_\Sigma(\ldots)$, the *product* $F_\Pi(\ldots)$ or the *average* $F_{avg}(\ldots)$ of its arguments. If there is given a set of addresses T^D and some variable s, each of the addresses $t^D \in T^D$ defines a state value $s(t^D)$. The expression $F_x(s(T^D)) = F(s(t_1^D), s(t_2^D), \ldots)$ is used for addressing all of the state values and for combining them according to $F_x(\ldots)$.

3. **Commands**

(a) **Assignment commands.** A new value may be assigned to a state variable s_i if the notation of a *command* $s_i(t^d) \leftarrow < expression >$ is used.

(b) **Parallel and sequential commands.** It is possible to use brackets so that the parallel execution (brackets $\{\ldots\}$ or sequential execution (brackets $< \ldots >$) of a number of commands can be expressed.

(c) **Loop expressions.** A command of the form $\lambda(M(T^D) \to t^d, S(t^d))$ has the effect of a *loop*. When this command is executed, in a first step the result of the address description $\hat{T}^D = M(T^D)$ is computed. Then each address $\hat{t}^D \in \hat{T}^D$ is extracted from the result and input as an argument into the command sequence $S(t^d)$.

Furthermore it shall be allowed to perform the usual arithmetic operations like *addition*, *subtraction* etc. .

Using these terminology, the rules for the example presented in section 2 can be written down very compactly. Let there be given the specific address $t^2 = t^2_{sample}$ that is located in g^2_{sample}, then the local transition function reads:

r_1: $s_2(t^2) \leftarrow s_1(\mu^2(t^2)^{(1)}) \cdot (s_2^{max} - s_2(\mu^1(t^2)^{(2a)}) - s_2(\mu^1\mu^0(t^2)^{(2b)}))$.

r_2: $s_3(t^2) \leftarrow F_{avg}(s_2([\mu^1\mu^0(t^2)]^*))^{(3)}$

These rules can be applied to any wall segments and cells within g^2_{sample}. This means that different rules for the different directions (like in the sample rules for a CA in section 2) need not to be written down separately.

The components of the formulas given above have been labeled with indices *(1)*... *(3)*. They have the following meaning (please compare with the CHG g^2_{sample} and the sample address $t^2 = (g|h|a, b, c, d, e, f)$):

(1) This expression has the result $\mu^2(t^2) = (f|g|h, i, j)$. The resulting address is used for accessing the variable s_1, and as a consequence its value is taken from the outer part of the actual wall segment.

(2a/b) The expressions $\mu^1(t^2) = (f|e|a, b, c, d, g)$ and $\mu^1\mu^0(t^2) = (g|a|b, c, d, e, f)$ address the two neighbouring wall segments (clockwise and counterclockwise) within the same cell. The addressed values of s_2 are used for computing the difference between the values stored in neighbouring wall segments.

(3) In each iteration the expression $\mu^1\mu^0(\ldots)$ advances the actual address to the neighbouring wall segment that is located in clockwise direction. Since this expression is bracketed $[\ldots]^*$ it is repeated until all wall segments of the actual cell have been inspected. The expression $F_{avg}(s_2(\ldots))$ is used for averaging all state values s_2 that are stored within the wall segments.

In order to perform the computations for all cells of g^D_{sample} and all of their wall segments, all addresses where the local transition function shall be applied must be listed systematically. In the terminology of cellular automata, this is called the *global transition function* of the automaton.

Let there be given any address $\hat{t}^2 \in g^2_{sample}$. Then the set of valid addresses T^2 can be computed based on the expression: $T^2 := [\,\mu^2[\mu^1\mu^0(\hat{t}^2)]^* \cup \hat{t}^2\,]^*$. In this case the term $[\mu^1\mu^0(\ldots)]^*$ is used for listing all wall segments of the current cell. If the relative addressing step μ^2 is applied to this set of addresses, all neighbouring cells are listed. If this process is iterated until no further addresses can be determined (effect of the outermost brackets $[\ldots]^*$), all imaginable addresses of the graph are maintained.

Finally, a loop expression is used for initiating the different calculation steps for the whole structure. The resulting global transition function reads:

$$
\begin{array}{l}
\lambda([[\mu^2[\mu^1\mu^0(\hat{t}^2)]^* \cup \hat{t}^2\,] \rightarrow t^2, \\
\quad \{ \\
\qquad s_2(t^2) \leftarrow s_1(\mu^2(t^2)) \cdot (s_2^{max} - s_2(\mu^1(t^2)) - s_2(\mu^1\mu^0(t^2))), \\
\qquad s_3(t^2) \leftarrow F_{avg}(s_2([\mu^1\mu^0(t^2)]^*)) \\
\quad \} \\
)
\end{array}
$$

7 Summary

This paper has introduced fundamentals which allow us to describe how information is processed within an inhomogeneous cellular structure. The presented method is based on the novel *cellular hypergraphs.* In order to store state values and to perform relative addressing, the terms *address*, *component* and *relative addressing* have been introduced.

A versatile method for performing *relative addressing* has been presented that allows us to describe how to access data associated with the components of a CHG. More complex formulas can be written down using the presented schema for loop expressions and for concatenating single addressing steps. This schema gives us the ability to formulate the local transition function as well as the global transition function of a cellular model compactly.

Future activities will concentrate on high level programming languages for dynamic and inhomogeneous cellular structures, which will be based on the methods presented in this paper. This will require investigations on how cellular hypergraphs and replacement systems can be implemented efficiently on parallel computer architectures.

Acknowledgments

This paper reflects results from our project supported by the DFG foundation under the title *Hardware-unterstütztes Simulationssystem für die graphenbasierte und dreidimensionale zellulare Verarbeitung.*

References

1. T. Toffoli and N. Margolus. *Cellular Automata Machines.* MIT Press, 1987.
2. B. Alberts, D. Bray, J. Lewis, M. Raff, K. Roberts, and J.D. Watson. *Molecular Biology of THE CELL, 2nd Ed.* Garland Pub. Inc., 1989.
3. P. Hartmann. Parallel Replacement Systems on Geometric Hypergraphs: A Mathematical Tool for Handling Dynamic Geometric Sceneries. *Proceedings of the VIth International Workshop on Parallel Processing by Cellular Automata and Arrays, (PARCELLA), Potsdam*, 1994.
4. P. Hartmann. Efficient Subgraph Matching within Cellular Hypergraphs. *Proceedings of the Second International Conference on Developments in Language Theory, Magdeburg, Germany*, 1995.
5. P. Hartmann. Parallel and Distributed Processing of Cellular Hypergraphs. *Proceedings of Third International Conference on Parallel Computing Technologies (PaCT-95), St. Petersburg, Russia*, 1995.

GENOMIC CELLULAR AUTOMATA TRANSPOSED ON SILICON: EXPERIMENTS IN SYNTHETIC LIFE

P. MARCHAL, P. NUSSBAUM, C. PIGUET, S. DURAND
CSEM Centre Suisse d'Electronique et de Microtechnique SA, Jaquet-Droz 1
CH-2007 Neuchâtel,
email: <name>@csemne.ch

D. MANGE, E. SANCHEZ, A. STAUFFER, G. TEMPESTI
Laboratoire de Systèmes Logiques, Swiss Institute of Technology, Ecublens,
CH-1015 Lausanne
email: <name>@di.epfl.ch

1 Introduction

Artificial Life is a new scientific field of investigation which has gained acceptance over the past few years. Its definition has itself evolved from "the study of artificial systems which exhibit lifelike behaviour", as proposed in 87 [1], to "the study of biological phenomena by attempting to synthesize them within alternative media (hardware, software, wetware)", as defined more recently [2]. As previous classifications still remain unclear [3,4], we propose the following definitions ofr the three main axes of investigation of Artificial Life:

- Alternative Life (sometimes called Synthetic Biology [5]) is concerned with the research undertaken by biologists attempting to create new biological arrangements of molecules [6] or artificial nucleotides [7]. The related medium is wetware

- Virtual Life is concentrated on computer simulations. In this field of research, scientists study cellular automaton [8], animats [9], virtual robots in virtual worlds [10]. The related medium is software

- Synthetic Life addresses, at the same time, the extraction of the concepts underlying biology and their application to any hardware medium (electronic, mechanic, optic, ..) [11,12].

This paper is concerned with this third aspect of Artificial Life. The aim is neither to define virtual machines nor new nucleotides but to realized functional electronics endowed with lifelike properties such as cloning, differentiation, reproduction and healing. Synthetic Life, as described in this article, is based on the following general hypotheses:

1. The cell deals exclusively with a flow of information; the physical material (usually a silicon substrate) and the energy (power supply) are given a priori

2. The physical space is two-dimensional and as large as desired

3. The physical space is homogeneous, that is, comprised of identical cells, all of which have the same internal architecture and the same connections with their neighbors; only the "state" of a cell (the combination of the values in its memories) can distinguish it from its neighbors

4. The reproduction is asexual: the daughter automaton is identical to the mother automaton.

The possibility of achieving self-reproduction in a universal computing machine has been demonstrated by von Neumann in the 40's [13]. Unfortunately, notwithstanding the improvements and the simplifications to which it has been subjected, it has not yet been possible to simulate, and much less to physically implement, von Neumann's automaton in its entirety. Only recently, simulations of a part of it, i.e., restricted to the universal constructor (less than a fifth of the complete machine), have been achieved [14]. Two decades later, Langton [15] and then his successors (Byl [16], Reggia et al. [17]) developed self-reproducing automata which are much simpler and which have been simulated in their entirety. These virtual machines, however, lack any computing ability and are capable exclusively of self-reproduction.

Recent advances in molecular biology, a better understanding in embryology and genetics, as well as rapid development of programmable devices [18,19] have led us to propose a Genomic Cellular Automaton Transposed in Silicon (GCAT in Silicon). This multi-cellular architecture is comprised of cells, each one containing a copy of the genome that describes the whole system.

Section 2 describes the stack of CA layers composing the mathematical model underlying the GCA. The internal behaviour of each "proto-cell", as well as the relations describing the binding of "proto-cells" in a GCA cell, are given in mathematical format; the corresponding hardware is also presented. Section 3 describes the synthetic counterparts of lifelike properties of living beings such as cloning, differentiation, healing and reproduction. Section 4 illustrates the embryological development (method of construction), self-reproduction and self-repair of a synthetic organism: a reversible counter. Finally we conclude by discussing cellular vs molecular approach to self-reproduction.

2 GCA Architecture: Stacked Layers of "Proto-Cells"

Instead of describing the GCA cell as a whole, it seems easier to describe each of its components (proto-cells) separately. To do so, we have laid down proto-cells on separate Cellular Automaton (CA) layers for which we describe the behaviour (transition function) and the interaction with other layers. As depicted in Fig. 1, the top layer is used to compute the local coordinates (X,Y) of each proto-cell, where X is the horizontal coordinate and Y the vertical coordinate. The second layer is equipped with random access memory (RAM); it is used to store the genomic information describing the synthetic organism. The local coordinates provided by the top layer enable to select the local gene corresponding to that location. This local gene is used to control the bottom layer which contains the functional proto-cell. This proto-cell is a programmable CA, i.e., neighborhood and transition functions are imposed by the value of the local gene.

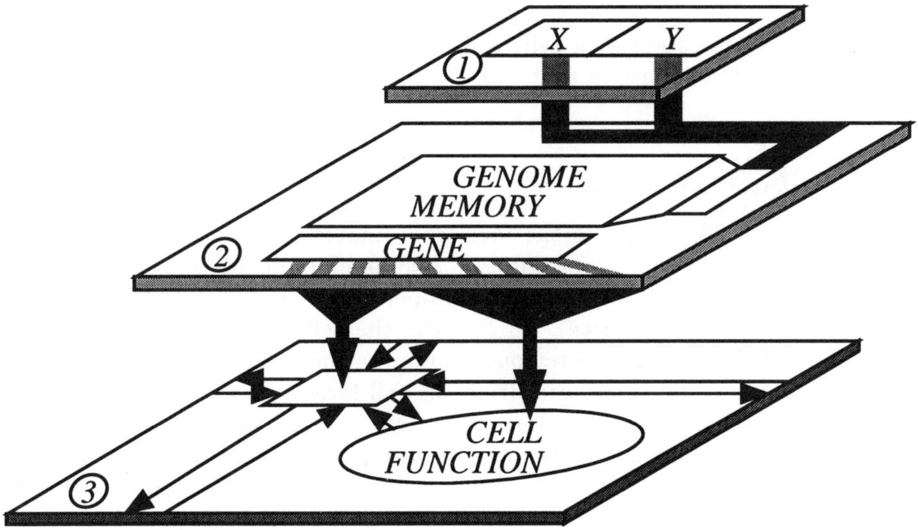

Figure 1: Architecture of the GCA cell: block scheme of the top three layers

2.1 Preliminary Definitions

This section is aimed at providing precise definitions of the cellular spaces used to describe the proto-cells gathered in a GCA cell. Codd's notations [20] have been used throughout this chapter. Let I denote the set of integers. To obtain a cellular space, we associate with the set $I \times I$:

1. The neighborhood function $g : I \times I \rightarrow 2^{I \times I}$, defined by

$$g(\alpha) = \{\alpha + \delta_1, \alpha + \delta_2, \ldots, \alpha + \delta_n\} \text{ for all } \alpha \in I \times I$$

   where $\delta_i (i = 1, 2, \ldots, n) \in I \times I$ is fixed, n being the number of neighbors

2. The finite automaton (V, v_0, f), where V is the set of *cellular states*, v_0 is a particular element of V called the *quiescent state*, and f is the local *transition function* from n-tuples of elements of V into V. The function f is subject to the restriction,

$$f(v_0, v_0, v_0, \ldots, v_0) = v_0$$

3. The *neighborhood state function* $h^t : I \times I \rightarrow V^n$ is defined at a given time t by

$$h^t(\alpha) = (v^t(\alpha), v^t(\alpha + \delta_2), \ldots, v^t(\alpha = \delta_n))$$

   The cellular state of a cell α at time $t + 1$ can now be related to the neighborhood state of that cell at time t by:

$$v^{t+1}(\alpha) = f(h^t(\alpha)).$$

2.2 Top Layer CA: Coordinate Computation

The top layer CA computes the local coordinates of each cell as a function of the neighboring cells' coordinates. A jumper indicates the neighbor cell chosen to perform computations. The jumper may be in 2 states: H (horizontal source) or V (vertical source). In H-state, the cell computes its coordinates from the cell located in the previous column on the same line. In the V-state, the cell computes its coordinates from the cell located in the same column on the line underneath. Let L_1, g_1, h_1, f_1 denote respectively the cellular space, neighborhood function, neighborhood state function and transition function defining the top layer CA: CA_1. Let C, W and S denote respectively the Centre cell, the West and the South neighbor. Then, CA_1 is given by:

$$\begin{aligned} L_1 &= \{I \times I\} \\ g_1(C) &= \{C, W, S\} \\ h_1^t(C) &= \{\text{JUMPER}^t, X^t(W), Y^t(W), X^t(S), Y^t(S)\} \end{aligned}$$

Note that CA_1 is quasi-stationary: as long as the origin of the coordinates is fixed, the coordinates of each cell remain constant. Fig. 2 depicts the block scheme of CA_1 and the algorithm describing its transition function.

The transition function f_1 is given by :
if $JUMPER^t(C) = H$
then (* compute cell address from West *)
$X^{t+1}(C) = X^t(W) + 1$
$Y^{t+1}(C) = Y^t(W)$
else (* compute cell address from South*)
$X^{t+1}(C) = X^t(S)$
$Y^{t+1}(C) = Y^t(S) + 1$
endif

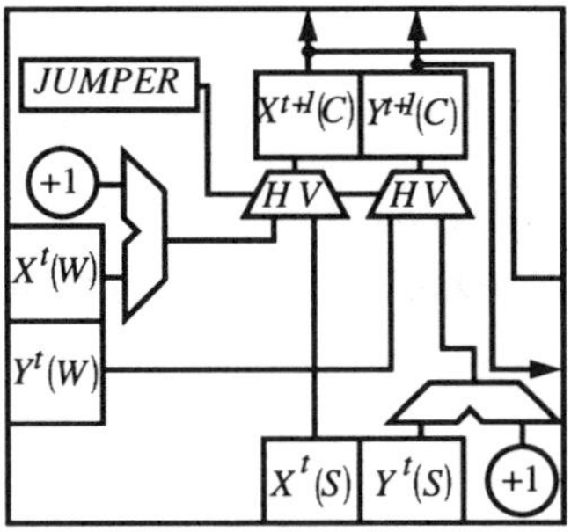

Figure 2: Top layer CA: CA_1 - Coordinates computations

2.3 Mid-Layer CA: Gene Selection out of the Genome Memory

The mid-layer CA selects, out of the genome memory, the functional gene of each cell as a function of the local coordinates of the cell (determined by the top layer CA). The memory is called genome memory since it stores a sequence of instructions matching the concept of genotype.

Let L_2, g_2, h_2, f_2 denote respectively the cellular space, neighborhood function, neighborhood state function and transition function composing the second layer. Let C, C/CA_1 denote respectively the Centre cell and the Centre cell of the top layer CA: CA_1. Then, CA_2 is given by:

$$\begin{aligned} L_2 &= \{I \times I\} \\ g_2(C) &= \{C, C/CA_1\} \\ h_2^t(C) &= \{\text{GENOME MEMORY}^t, X^t(C/CA_1), Y^t(C/CA_1)\} \end{aligned}$$

Fig. 3 depicts the block scheme of CA_2 and the algorithm describing its transition function. We can note that, as long as the CA_1 is quasi-stationary (coordinates of each cell remain constant), CA_2 is also quasi-stationary, i.e., the gene selected at each time step is always the same.

Before selecting any gene, the genome memory must be filled with appropriate genetic data (provided externally). This storage, realized prior to any CA activation, is performed sequentially from either the West or the South neighboring cell (as a function of the jumper described in CA_1). Let L_2, g_2', h_2', f_2' denote respectively the cellular space, neighborhood function, neighborhood state function and transition function composing the second layer during the storage of the genome memory. Let C, W and S denote respectively the Centre cell, the West and the South neighbor. Then, CA_2' is

The transition function f_2 is given by :

$GENE^{t+1}(C) :=$
$GENOME\ MEMORY^{t+1}(X/CA_1, Y/CA_2)$

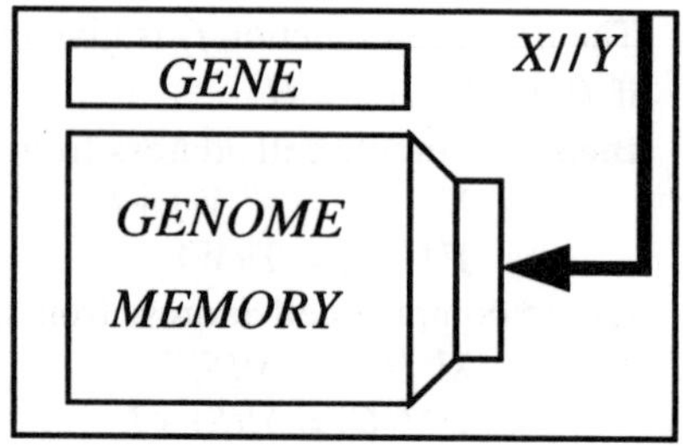

Figure 3: Mid-layer CA: CA_2 - Local gene selection

given by:

$$
\begin{aligned}
L'_2 &= \{I \times I\} \\
g'_2(C) &= \{C, W, S\} \\
h'^{t}_2(C) &= \{\text{GENOME MEMORY}^t(C), \text{JUMPER}_t(C), n, m, \\
&\quad \text{SIPO}_t(C), \text{IN}_t(W), \text{IN}_t(S)\}
\end{aligned}
$$

Fig. 4 shows the block scheme of CA'_2 as well as the algorithm describing its transition function.

The transition function f'_2 is given by :
if $JUMPER^t(C) = H$
then (* load genome memory from West *)
  for i:=1 **to** n **do**
    for j:=1 **to** m **do**
    $SIPO^{t'+1}(C) = SHIFT(SIPO^{t'}//(IN^{t'}(W)))$
    $GENOME\ MEMORY^{\tau+1}(C) = SIPO^{\tau}$
else (* load genome memory from South*)
  for i:=1 **to** n **do**
    for j:=1 **to** m **do**
    $SIPO^{t'+1}(C) = SHIFT(SIPO^{t'}//(IN^{t'}(S)))$
    $GENOME\ MEMORY^{\tau+1}(C) = SIPO^{\tau}$
endif

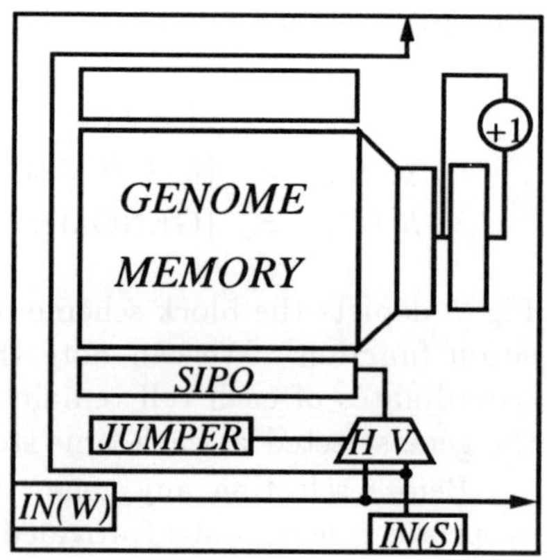

Figure 4: Mid-layer CA: CA'_2 - loading of the genome memory

Note on Synchronous CA: CAs are by definition synchronous machines, that is to say, synchronized by a single clock. A transient phase is nevertheless needed in order to load the genome memory appropriately.

2.4 Bottom Layer CA: What Functional Proto-Cell ?

Let L_3 denote the cellular space composing the third layer. The third layer is a programmable CA: the functional gene, selected by the mid-layer CA, programs both the transition function (functionality of the cell) and its neighborhood function (related neighbors). The functions g_3, h_3 and f_3 are fully programmed by the local gene provided by CA_2. This layer should specify the content of the functional proto-cell, i.e., the logical function of a tile: an "atom" of universal machine. We have considered Shannon's point of view. Shannon [21] has shown that any n-variable logic function can be implemented with a finite number of multiplexers, each multiplexer realizing the universal one-variable logic function:

$$f(z) = a_0 \bullet \overline{z} + a_1 \bullet z \tag{1}$$

where z is the control variable, a_0 and a_1 are (n-1)-variable logic functions, " $-$ ", " $\bullet$ " and " $+$ " are respectively the Boolean NOT, AND and OR functions.

As depicted in Fig. 5, the one-variable multiplexer has a direct output $f(z)$ which implements the combinational function (Eq. 1) and a delayed output $q(z)$ which implements the sequential operation by means of a D flip-flop; the presence of this flip-flop is necessary to implement both finite state machines and shift registers needed for any universal Turing machine.

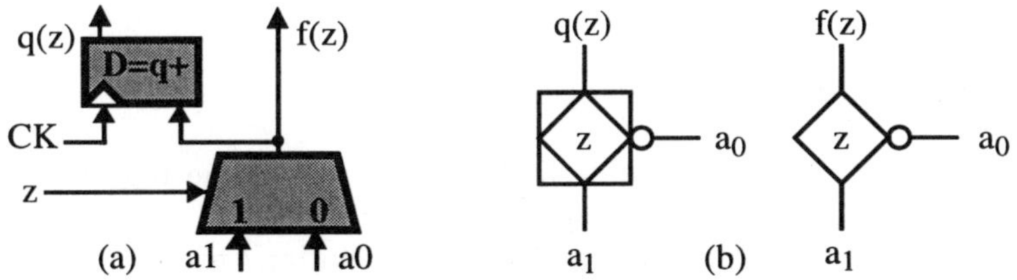

Figure 5: An "atom" of the functional proto-cell (a) 1-variable multiplexer with combinational and sequential outputs (b) logical representations

The functional proto-cell, shown in Fig. 6, is controlled by a 20-bit word $GENE[19:0]$ depicted in Fig. 7. Bit 0 of the gene selects the multiplexer control variable (z). Each input (a_1, resp. a_0) is selected from the neighboring cells through two MUX(8:1) controlled by 3-bits (*LEFT* [2:0], resp. *RIGHT* [2:0]). Eight more bits (*N* [1:0], *S* [1:0], *W* [1:0], *E* [1:0]) are used to control the interconnection buses.

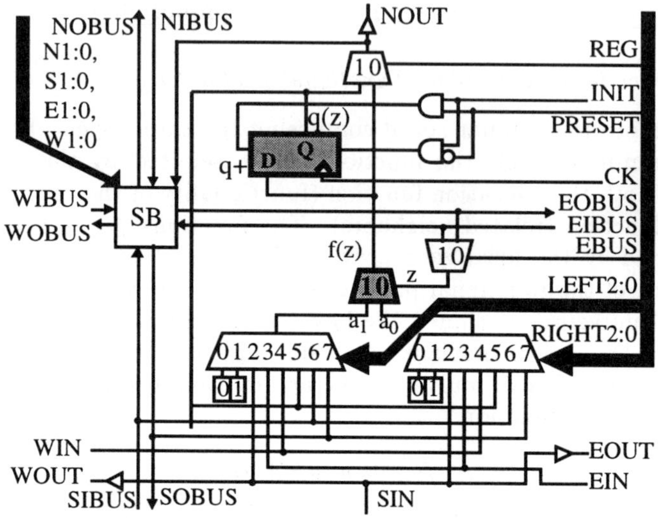

Figure 6: Bottom layer CA: CA_3 - programmable transition and neighborhood functions

0,LEFT2:0	0,RIGHT2:0	N1:0,S1:0	E1:0,W1:0	0,PRESET,REG,EBUS
19 … 16	15 … 12	11 … 8	7 … 4	3 … 0

Figure 7: The 20-bit word GENE

3 Emergence of Lifelike Properties

As mentioned by Morris [22], some lifelike properties come together with Genomic Cellular Architecture. Some are pretty obvious (cloning, differentiation), while others emerge from the Genomic Cellular Automaton (self-repair, self-reconfiguration). This section investigates these lifelike properties.

3.1 Cell Reproduction (Cloning)

Cell reproduction or cloning is a biological process consisting in duplicating exactly the content of a cell. The cell cloning process can easily be performed by a slight modification in the address computation. If top layer CA: CA_1 (coordinate computations) produces a unique pattern of coordinates ($X = 1 \rightarrow 1 \rightarrow 1$ and $Y = 3 \rightarrow 3 \rightarrow 3$ for instance). The gene located in the 13th word of the genome memory will be selected in all cells.

3.2 Cell Differentiation (Organism Realization)

Apart from the simplest forms of life, constituted by isolated cells, all higher organisms, like ourselves, are similar to cellular cities in which groups of cells fulfil specialized functions. Cell differentiation is a biological process consisting in interpreting a different gene as a function of the location (and time). As mentionned above, the purpose of CA_1 is to induce cell differentiation by appropriate coordinates computation.

3.3 System/Organism Differentiation

Let us consider a logical system (synthetic organism) as a rectangular block of m by n cells. Producing several systems or organisms on the same cellular space implies downloading several genomes from several zygote cells. In biology, the zygote cell is the first cell of an organism from which all cells have been copied and differentiated. In our approach the zygote cell is the origin of local coordinates, that receives the genetic data provided externally. As soon as each zygote has been loaded, the different organisms may grow and differentiate at the cell level, given a sufficient amount of available cells.

3.4 Healing (Local Self-Repair or Cicatrization)

In order to be able to self-repair the system should, first of all, be capable of detecting its own faults. We have addressed this point by using Built-In-Self-Test (BIST) techniques [23,24] to design the different layers of proto-cells. A supplementary fourth layer (proto-cell) has been side mounted. This fourth layer checks the correctness of the 3 other layers. As soon as a GCA cell is faulty, it should be removed from the processing space. In silicon, cicatrization cannot bring any material to replace damaged cells. In our approach, this can be done by letting the faulty cell be transparent with respect to local address computation. So, the function that took place in the faulty cell is now performed by the next non-faulty cell.

3.5 System/Organism Healing (Global Self-Repair)

Let us consider a logical system (synthetic organism) to be repaired. As the same pattern of coordinates produces the same pattern of genes, global self-repair can be easily performed if the address computation produces several occurrences of the basic pattern of coordinates ($X = 1 \rightarrow 2 \rightarrow 3$ and $Y = 1 \rightarrow 2 \rightarrow 3$); for instance, the repetition of the horizontal coordinates pattern, i.e. the production of the pattern $X = 1 \rightarrow 2 \rightarrow 3 \rightarrow 1 \rightarrow 2 \rightarrow 3$, produces one copy of the original "mother organism". Cell differentiation, using a modulo adder (modulo 3 in the previous example), induces self-reproduction. This technique can be processed in any direction (X or Y).

4 Method of Construction: an Example

As an example, let us consider the realization of a modulo-4 up-down counter, defined by the following sequences:

$$\text{for } M = 0 \quad : \quad Q1, Q0 = 00 \rightarrow 01 \rightarrow 10 \rightarrow 11 \rightarrow 00 \ (\textit{counting up}) \qquad (2)$$

$$\text{for } M = 1 \quad : \quad Q1, Q0 = 00 \rightarrow 11 \rightarrow 10 \rightarrow 01 \rightarrow 00 \ (\textit{counting down}). \qquad (3)$$

A binary decision diagram [25–27] is a mode of representation of finite state machines based on the symbols of Fig. 5. The up-down counter described by relations (2) and (3) can be represented by the binary decision diagrams of Fig. 8. It is then evident that a multiplexer element, whose operation has been shown in Fig. 5, can implement anyone of the nodes of the diagrams of Fig. 8.

The complete determination of the 9 genes of the up-down counter can be undertaken starting from the detailed characteristics of the multiplexer element described in [28], and finally leads to the diagram of Fig. 9, which depicts the local genes in their respective locations.

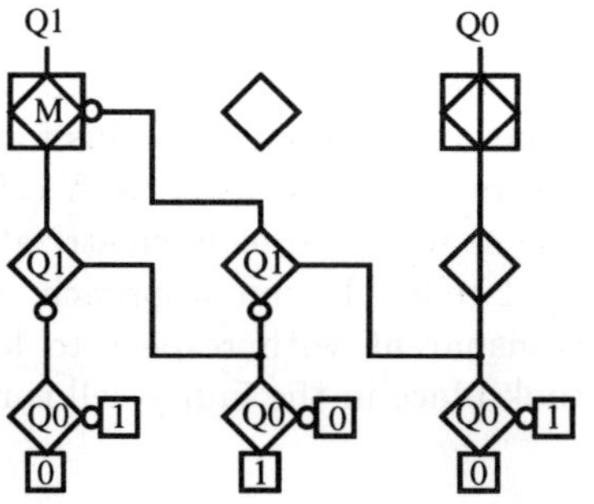

Figure 8: BDD of the reversible counter

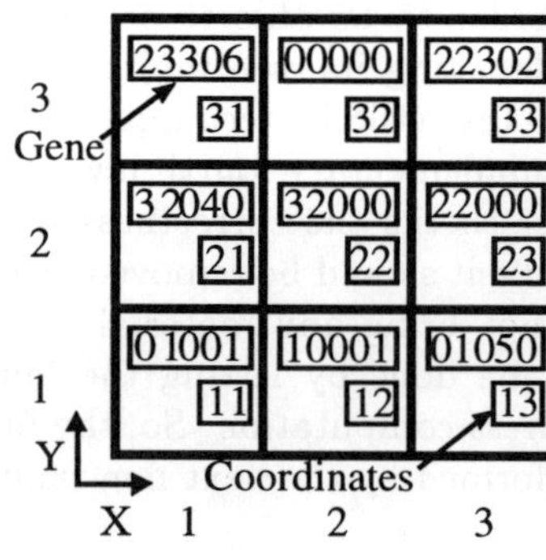

Figure 9: System organization

In the example of the up-down counter, assume that an element is faulty in any column (X = 2 in Fig. 10). Should such a fault be automatically detected using BIST (built-in self-test) techniques, then, given sufficient spare parts (a spare column), the faulty cell is able to deactivate the column to which it belongs, to bypass the data coming from the left neighboring column and to obtain, by a global shift to the right, a faultfree implementation of the original device, as depicted in Fig. 10.

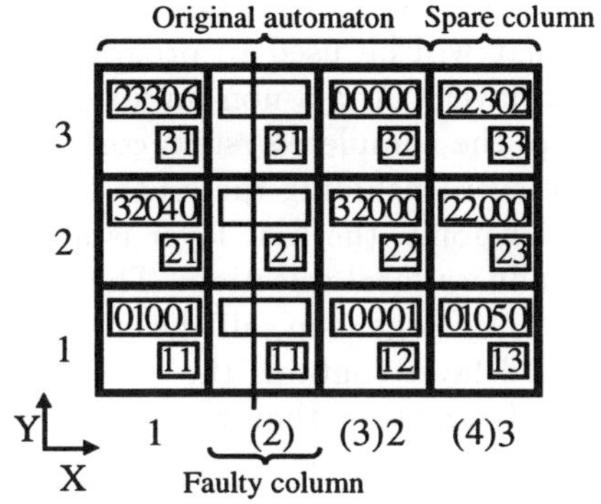

Figure 10: Example of self-repair

5 Conclusion

5.1 Back to von Neumann

In biology, the living cell - small compartment bounded by a membrane and filled with a concentrated solution of chemical - is the smallest part of a being containing the complete blueprint of the being: the "genome". On the basis of this definition, we clearly see that von Neumann's automaton is a uni-cellular organism, in a biological sense, since it contains a single copy of the genome. Each processing element of the automaton is thus a part of a uni-cellular organism, or, i.e., a "molecule". Von Neumann's automaton is therefore a "molecular automaton". In contrast, the GCA architecture is a cellular (in the biological sense) automaton, since each of its elements contains a copy of the genome (nine copies in the case of the up-down counter of Fig. 9). Approximately as many multiplexer elements are needed as there are nodes in the binary decision diagram(s) describing the organism to be implemented (see Fig. 8). This gives a rough idea of the size required for the memory. Nevertheless, the simulation and the physical implementation of the element itself are easily realizable: using a programmable cellular network ("field-programmable gate array") and a RAM, we have implemented a prototype of the GCA cell, which we have called "biodule" [12]. Two versions have been designed in VLSI and are still at the foundry. Using a number of these biodules, we have successfully implemented the self-reproduction and self-repair processes.

5.2 Specialist Viewpoints

From a mathematical point of view, the top layer CA offers a simple means to select 1-out-of-n transition functions. The mid-layer CA enables the set of transition functions that will be used to program the bottom layer to be stored. Finally, the third layer CA is a non-uniform cellular automaton. The complexity emerging from the complete system comes from the feedback loop present at the proto-cell level (including the fouth layer).

From a biologist's viewpoint, the first layer is a simple means to initiate differentiation, since it induces local selection. The second layer CA contains the genotype of the synthetic organism, all the instructions necessary for its implementation. The third layer contains the essence of the phenotype since it determines the physical structure that results from the transcription the genotype instructions.

From the engineer's point of view, the complete system may be viewed as a new kind of FPGA. The bottom layer implements a dedicated FPGAs, the second layer stores the complete field program of all the cells in use in the FPGA. Conventional FPGAs do not offer such a size of memory per cell, but this strategy offers quicker reconfiguration time. By providing a logical address for each fault-free cell, the first layer enables the function description in terms of logical addresses and not physical addresses as done in conventional FPGAs.

5.3 Future Developments

Future developments consider three different axes: (a) increase the complexity at the hierarchical level, increasing the number of levels for instance; (b) increase the complexity of the cell, leading to an increase of functional complexity. Attempts have been made to embed functional blocks much more complex than a simple test tile as done presently; (c) increase the complexity at the biological level, coming closer to biology (for instance closer to the immune system for fault detection, closer to cells and tissues for signal propagation, and so on).

References

1. C. G. Langton: Artificial Life, Proc. of Santa Fe Institute Studies in the Sciences of Complexity, 1-47 (1989).
2. C. G. Langton: Artificial Life II, Proc. of Santa Fe Institute Studies in the Sciences of Complexity, xiii-xviii (1992).
3. H.H. Pattee: Artificial Life, Proc. of Santa Fe Institute Studies in the Sciences of Complexity, 63-77 (1989).

4. S. Harnad; Artificial Life III, Proc. of Santa Fe Institute Studies in the Sciences of Complexity, 539-552 (1994).
5. M. Zeleny, G.J. Klir, and K.D. Hufford: Artificial Life, Proc. of Santa Fe Institute Studies in the Sciences of Complexity, 125-139 (1989)
6. M. Eigen: Steps towards Life, (1992).
7. G. North: Nature 345: 576-578, (1990).
8. H. Gutowitz: Cellular Automata, MIT/North Holland (1991).
9. J-A Meyer, S.W. Wilson: From Animals to Animats, MIT Press (1991).
10. S. Forrest: Emergent Computation, MIT/North Holland (1991).
11. X. Arreguit, E.A. Vittoz: PerAc, IEEE - Comp. Soc. Press, 170-180 (1994).
12. P. Marchal, C. Piguet, D. Mange, A. Stauffer, S. Durand: in Artificial Life IV, MIT Press, 365-370 (1994).
13. J. von Neumann: The Theory of Self-Reproducing Automata, Univ. of Illinois Press (1966).
14. J. Signorini: Springer Proceedings in Physics 46, Berlin, 57 (1990).
15. C. G. Langton: Physica D 10, 135 (1984).
16. J. Byl: Physica D 34, 295 (1989).
17. J. A. Reggia, S. L. Armentrout, H.H. Chou, Y. Peng: Science 259, 1282 (1993).
18. S.D. Brown, R.J. Francis, J. Rose, Z. G. Vranesic: Field-Programmable Gate Arrays; Kluwer Academic Publishers (1992).
19. W. Moore, W. Luk: FPGAs; Abingdon (1991).
20. E.F. Codd: Cellular Automata; Academic Press (1968).
21. C. E. Shannon: Bell System Technical Journal 28, 59 (1949).
22. H. C. Morris: Artificial Life, Proc. of Santa Fe Institute Studies in the Sciences of Complexity, 369-395 (1989).
23. E. J. McCluskey: Logic Design Principles with Emphasis on Testable Semicustom Circuits Prentice-Hall, Englewood Cliffs, (1986);
24. S. Durand and C. Piguet: in 2nd International ACM/SIGDA Workshop on Field-Programmable Gate Arrays, Berkeley, Calif. (1994).
25. S. B. Akers: IEEE Transactions on Computers C-27, 509 (1978).
26. R. E. Bryant: ACM Computing Surveys 24, 293 (1992);
27. D. Mange: Microprogrammed Systems: an Introduction to Firmware Theory, Chapman & Hall (1992).
28. S. Durand, A. Stauffer, D. Mange: Biodule: an Introduction to Digital Biology , Swiss Federal Institute of Technology at Lausanne (1994).

COHESIVE INTERACTION IN BIOMOLECULES AND THEIR ORGANIZATIONS AS ENERGY CONSUMERS

KOICHIRO MATSUNO
Department of BioEngineering
Nagaoka University of Technology
Nagaoka 940-21, Japan

1 Introduction

Describing emergent phenomena as facing biological organizations succumbs inescapably to internal confrontation between two different types of agent. One is the author of description, and the other is an agent experiencing emergent novelty. Any objective description requires the author to remain external to the description itself, while emergent phenomena in biology internalize those agents experiencing novelties. A serious question that may come up at this point is whether or not the author of description and the agent experiencing emergent novelty could be identical.

If the author and the experiencing agent are one and the same as facing emergent phenomena, both the description and the descriptive object to be experienced will remain external to the act of describing and experiencing. External description prohibiting itself from referring to its author certainly satisfies the condition necessary for being objective and perspicuous in its own right.[1] Likewise, the presence of the object to be experienced externally assumes that the experiencing agent performs measurement externally and that perspicuous objectivity of the outcome is its natural consequence.

External description of external measurement, however, suffers from its methodological peculiarity of seeking the origin of emergent novelty solely outside what has been experienced and described. Measurement external to an object exhibiting an emergent novelty has nothing to do with the process of giving rise to the very emergent phenomenon. The present externality of emergent novelty is totally vulnerable to what cannot be described and experienced externally. Only the consequence of emergent phenomena is visible to external description of external measurement. If it is intended to decipher how emergent phenomena could come into being, one would have to go beyond external description of external measurement.

We shall in the present article try to establish a descriptive scheme that can cope with emergent novelty occurring in biological organizations and the underlying cohesiveness being responsible for their coming into being. In order to accomplish the present objective, let us first examine how external description of external measurement would proceed in its more details.

2 External Description of External Measurement

Emergent phenomena require both duration in time and variations acting upon it. If measurement is external in the sense that an object out there exists exactly as it is as measured externally and if the consequence of the measurement is describable without any reference to the author, the notion of duration in time can be sanctioned without any recourse to variations to be imposed. Galilean inertia just happens to be the case. However, the notion of duration met in evolutionary processes at least during the last 3.8 billion years on the earth is quite different from that of Galilean inertia. Molecular replication as a common denominator of the emergent phenomena called life on the earth makes it inevitable to be accompanied by the process of generating variations or making products successively without any interruptions.[2]

Variable duration is inherent to molecular replication, and is totally foreign to invariant duration unique to Galilean inertia.[3] External description of external measurement would make the occurrence of molecular replication no more than an extremely unlikely event paraphrased as a frozen accident. In fact, quantum mechanics understood as external description of external measurement cannot raise such a replicating molecule by itself.[4] Evolutionary emergence of replicating molecules presumes a unilateral sequence of synthetic molecular reactions in which some of the products re-enter into an arbitrary stage in the whole reactions as further reactants, while every reaction or interaction conceivable within external measurement of quantum mechanics is bilateral as admitting both associative and dissociative reactions everywhere at the same time.

Invariant duration cannot serve as an evolutionary precursor of variable duration because of their intrinsic incommensurability. The same difficulty also applies to the notion of natural selection understood as differential retention of slightly modified heritable traits.[5] Underlying natural selection is also variable duration. In particular, if one raises a question of how natural selection could come into being, it would not be feasible any more to have recourse to an onset of molecular replication because both are grounded upon the same notion of variable duration. The present impasse may suggest to us either that the question asking the evolutionary onset of variable duration would be in itself wrongly formulated or that external description of external measurement by itself be methodologically incompetent in coping with such question.

At this point, it is worth pointing out that the relationship between description and measurement is not restricted to the case that both are external. One cannot deny the historical fact such that external description of external measurement has greatly contributed to the promulgation of objective and rigorous sciences over the last few hundred years. Nevertheless, the externality of both description and measurement is no more than merely one frame of reference conceived from and applied to the pair. Measurement in its rudimentary form presumes our sense perception, and description is framed upon our natural languages serving as the infra-structure.

Externalization of both description and measurement is simply a consequence of applying a form of abstraction to both of them by eliminating any agential capacity from the object of description and experience. Otherwise, describing experiences remains internal in the sense that the one who measures and describes the experience of its own is constantly measured and described by others internally as in ordinary conversations in everyday life. The present internal chain of description and measurement has neither its beginning nor end. The agents of both description and measurement are internalized in their mutual acts. This way of looking into the interplay between the two suggests an occurrence of internal description of internal measurement as an opposite extreme case.

3 Internal Description of Internal Measurement

In ordinary conversations in everyday life, everybody as an agent of measuring and describing is constantly measured and described by others. This is exactly the problem area which traditional rigorous sciences have shrewdly avoided up until now. It is of course legitimate to observe that rigorous sciences cannot derive from ordinary conversations that constantly have recourse to the speakers. That rigorous sciences cannot be brought about from mere ordinary conversations rests upon the indefiniteness of the viewpoint of each internal description attempted there. What is unique to internal description is the recursiveness of letting itself be an object to further be measured by others internally within the framework of on-going ordinary conversations. Henceforth, indefinite continuation of ordinary conversations is found to be within indefinite succession of internal description of internal measurement.

The context of ordinary conversations, however, constantly varies with their progression. There is no universality nor a priori invariance in the context in the making. The variability of the context in an unprecedented manner is an undeniable empirical fact in ordinary conversations. Curiously enough, however, ordinary conversations, if properly abstracted, can happen to give birth to objective description and rigorous sciences. In order to objectify ordinary conversations in everyday life, it would again be required to externalize either description or measurement. If both description and measurement are externalized as in traditional rigorous sciences, a cost would inevitably be incurred. No descriptive enterprise coping with historicity and evolutionary processes could be expected therefrom. If experiencing and describing emergent phenomena are intended beyond those intended by internal description of internal measurement, either description or measurement would again have to be externalized.

4 Internal Description of External Measurement

If the external object exists as perceived by our sense organs, this would guarantee the occurrence of experiences upon which rigorous empirical sciences could stand. However, experiences by themselves cannot result in objective and rigorous sciences, the latter of which depend entirely upon how these experiences could be described. Internal description of internal measurement as practiced in ordinary conversations in everyday life certainly provides an infra-structure upon which rigorous sciences could be constructed. What is required for this construction is an abstraction. The present application of an abstraction, though a necessary condition for deriving objectivity therefrom, may induce further reverberations. In fact, if an abstraction of eliminating any agential capacity of measurement from an object to be measured is taken in the form of external measurement, the measured object will turn out also to be an object to be described externally. External measurement now necessitates external description in order to put its intended objectivity into a definite descriptive form.

Nevertheless, if internal description of external measurement is attempted as in the case that there may be allowed many descriptive perspectives internally while maintaining a unique external measurement of the whole, a certain pathological situation would come up. A schizoid separation of the authorship would inevitably follow between the externalist and the internalist stance. The internalist authorship cannot set its descriptive perspective without recourse to the arbitrary internalist stance it assumes, while the externalist counterpart grounds itself upon the invariant stance that guarantees the object to be externalized. Internal description of external measurement would simply be a misnomer at best or a methodology to be avoided earnestly. These considerations come to leave external description of internal measurement as an only likely choice for coping with emergent phenomena to be experienced and to be described.

5 External Description of Internal Measurement

Objective description that has to be observed in any case is external in keeping itself from referring to its author. In spite of the present externality of description, measurement remains internal as our sense perception is. Of course, rigorous sciences based upon the stipulation asking the object of sense perception to stand alone externally as perceived necessitate external description, but the latter does not necessitate external measurement. External description is simply a linguistic pre-condition for aiming at objectivity of intended description.

In fact, the agency of measurement is not monopolized by our sense organs. Any material bodies interacting among themselves can serve as the agents performing measurement.[3,6] In view of the fact that measurement is a most prevailing form of interaction distinguishing between before and after the events and that nothing propagates faster than light, measurement is synonymous with the presence of matter. Measurement is thus internalized within interacting material bodies of any kind endophysically.[7] The problem that now comes up is how to externally describe the contribution of internal measurement.

One of the necessary conditions for external description is the observation of the principle of the excluded middle. Since external description is synonymous with the consequence of description attempted externally, the principle of the excluded middle guarantees the global consistency of description in the product. On the other hand, internal measurement in the making is local both in space and in time. In order to approach the global consistency of external description from internal measurement, a certain condition has to be imposed internally upon the progression of internal measurement. That is that internal measurement proceeds so as to fulfill the principle of the excluded middle in the effect. Internal measurement for the sake of the principle of the excluded middle is a consequence of the global consistency of external description. Put differently, if the global consistency of description in the product fails, internal measurement for the sake of the principle would also fail. The present reflection now raises a serious question on the likelihood of objective description to be externalized. The principle of the excluded middle is a consequence of descriptive objectivity, and not vice versa.

In order to justify the occurrence of external description, something that could remain invariant and objective globally has to be sought in the product of internal measurement. Although each internal measurement is local, the globality of the product makes itself be an object to be measured externally.[8] As far as the consequence of internal measurement is concerned, external measurement can safely intervene there without causing any conflicts with the former. If there is found an invariant quantity within the consequence of internal measurement, that quantity can also be identified by external measurement. Energy to be conserved just happens to be such a quantity, thanks to Robert Mayer. Observation of the conservation of energy by external measurement also confirms the progression of internal measurement for the sake of the conservation in the effect. The present interplay between external and internal measurement as facing the conservation of energy is to provide a means for legitimizing the objectivity of external description of internal measurement. This is owing to the presence of an invariant quantity being independent of descriptive artifacts. What makes external description of internal measurement meet the principle of the excluded middle happens to be the conservation of energy.

Fulfilling the conservation of energy through internal measurement is a local process. The local activity for the sake of the conservation of energy on a global scale lets each interacting material body be an energy consumer acting toward its outside from the inside as wanting or yawning for energy resources in order to compensate the preceding energy consumption in whatever way.[9] The occurrence of energy consumers is thus unique exclusively to external description of internal measurement,[10] and can be totally dismissed in the framework of external description of external measurement. There is no room of agential capacity in the object that is supposedly measured externally.

The significance of energy consumer for emergent phenomena is multifarious. Evolutionary emergence of the phenomenon called life at about 3.8 billion years ago on the earth happened to be an instance of raising a very primitive form of energy consumers. In addition, energy consumers exert a specific form of cohesive force toward their energy resources.[9,11] Compared to various mechanical forces imposed externally in the form of fixed boundary conditions, cohesive forces unique to energy consumers act from its inside toward the outside. Above all, energy consumers carry with themselves the capacity of variable duration as constantly wanting energy resources while being exploited by others at the same time.[12] External description of internal measurement that can come up with energy consumers installs in itself the capacity of variable duration as a necessary condition for emergent phenomena.

Energy consumers are accompanied by energy dissipation whenever and wherever. Energy dissipation is just one instance of describing externally what is being measured internally. What has been measured internally can in turn be measured externally in the completed record at the global scale, for instance, in mechanistic terms of generalized fluxes and their conjugate generalized forces.[13] Nonetheless, cohesive forces yielding energy dissipation are more than what mechanistic forces would literally imply. It is the operation of measurement proceeding internally which signifies the genuine implication of cohesive forces underlying the emergence of energy dissipation in general and energy consumers in particular.

6 Cell Motility: An Example

6.1 Perpetual Disequilibrium between Energy and Momentum Conservation

Recently, a considerable amount of experimental efforts for identifying the physical nature of molecular motors involved in muscle contraction has been accumulated.[14-19] What we may require at this point might be to identify each physical process constituting the operation of molecular motors in their motile organization from the perspective of internal measurement and cohesive interaction.

When it hydrolyzes an ATP molecule, a myosin molecule contacting an actin filament can release the energy stored in the ATP and transform it into the energy of excitations activated along the actin filament. The resultant momentum of the generated excitations measured in the coordinate system resting on the myosin molecule should vanish because the ATP molecule as an energy source remains stationary with respect to the coordinate. A most simple form of excitations is a pair of excitations whose energies are e(**k**) and e(-**k**), in which e(**k**) is the energy of an excitation carrying momentum **k** with e(**k**)=e(-**k**). An example of a likely pair of excitations is an exciton as a bounded particle-hole pair excitation activated along the actin filament.[20]

On the other hand, quantum mechanics of such a pair-excitation tells us that the uncertainty principle still imparts an uncertainty in the momentum by the amount of Δp_x to the excitation spread spatially over the linear extent of Δx along the x-direction in parallel to the filamental longitudinal axis, with the constraint

$$\Delta p_x \Delta x \sim \hbar ,$$

in which $\hbar$ is Planck's constant divided by 2π. The uncertainty Δp_x in momentum cannot however survive indefinitely, otherwise momentum conservation would be violated.[21] In order to restore momentum conservation, another excitation carrying momentum $-\Delta p_x$ with its energy δe has to be generated spontaneously. The source of such spontaneous excitation again should be due to the uncertainty principle in the form of

$$\delta e \delta t \sim \hbar$$

if the time required for the momentum compensation is δt.

The uncertainty δe in energy, however, has to be compensated similarly, otherwise energy conservation would be violated.[3,6] If the source of energy compensation is supplied externally as in the form of ATP molecules to be hydrolyzed, the restoration of energy conservation could certainly be accomplished by annihilating the excitation carrying energy δe while filling the energy deficiency caused by the operation of the energy-time uncertainty principle. Unless the energy source supplied externally is available, there would be no such compensation. Nonetheless, the restoration due to annihilating the excitation carrying the momentum $-\Delta p_x$ would again come to violate momentum conservation by the amount of Δp_x if noting is accompanied. Fulfilling momentum conservation would lead to violating energy conservation if nothing is accompanied and vise versa.

Disequilibrium between energy and momentum conservation thus perpetuates.[3] Equilibration for the sake of eliminating disequilibrium causing violation of conservation laws constantly yields further disequilibration.[8] What is constantly occurring is indefinite sequential alternation between otherwise violating energy conservation for the sake of fulfilling momentum conservation and violating momentum conservation for the sake of energy conservation. Of course, the net result to be observed in the record is both energy and momentum conservation.

The present perpetual disequilibrium between energy and momentum conservation can be facilitated so long as the energy compensation from ATP molecules is available. When the energy e_{ATP} is released from a single ATP molecule uniformly over the time interval t_{ATP}, the compensation for energy conservation requires

$$\delta e t_{ATP}/\delta t \sim e_{ATP} .$$

In addition, the fluctuation Δp_x in momentum is related to the force f acting there through the relationship of the impulse

$$\Delta p_x = f\delta t ,$$

and the displacement Δx is related to the sliding velocity v of the actin filament against the fixed myosin molecule as

$$\Delta x = v\delta t .$$

The relationship connecting the force f to the sliding velocity v now reduces to

$$fv \sim h/(\delta t)^2 \sim e_{ATP}/t_{ATP} .$$

An experimental confirmation of the generation of motile force due to perpetual disequilibrium between energy and momentum conservation can be obtained by examining each measured value of the force $f \sim 2\times10^{-7}$dyne,[22,23] the sliding velocity $v \sim 4\times10^{-4}$cm/s,[16] the energy released from an ATP molecule $e_{ATP} \sim 10^{-12}$erg,[14] and the turn-on time of ATP to be hydrolyzed in an actomyosin complex $t_{ATP} \sim 10^{-2}$s.[16,23] These measured values certainly satisfy an approximate equality $fv \sim e_{ATP}/t_{ATP} \sim 10^{-10}$erg/s at room temperature.

Motile force due to perpetual disequilibrium between energy and momentum conservation in the actomyosin complex is cohesive in that both an actin filament and a myosin molecule are attracting each other. The physical origin of the attractiveness is within fulfilling conservation laws altogether from the inside,[10] while the attractive activity does not fail in constantly leaving further causes for the similar activity in the subsequent stage. Sliding movement of the actin filament, whether or not accompanied by power stroke of a swinging movement of the head of a myosin molecule,[16,24] is simply a derivative of the underlying attractive interaction between those components comprising the motile organization. The attractive interaction is sufficiently long-ranged both in space and in time because conservation laws are not localized in their extent.

In particular, cohesive forces acting in the actomyosin system are direct manifestation of a most fundamental attribute of quantum mechanics, namely, an interplay between the uncertainty principles on energy and on momentum. Measurement of energy (or momentum) subject to its conservation would necessarily come to yield the uncertainty principle on momentum (or energy) violating its conservation if left alone. Since both energy and momentum have to be conserved in the long run, measurement has to be internalized in the sense that any measurement having come to terms with conservation of energy internally induces another measurement going to observe conservation of momentum also internally and vice versa, interchangeably in sequence indefinitely. Internal measurement observing conservation laws in the record constantly passes the operation of the uncertainty principle onto the subsequent stage, while inducing further internal measurement.[25]

The quantum coherence underlying the generation of motile force in the actomyosin system is thus seen due not to applying the uncertainty principle only once, but to coherent alternation of the uncertainty principles accompanied by measurements proceeding internally. Sliding movement of an actin filament on a myosin molecule in the presence of ATP is just manifestation of the quantum phenomenon on how linear momentum could be conserved coherently both in space and in time. Since the uncertainty principle presumes measurement in one way or another, the coherent sliding movement of an actin filament pumped up by the interplay of the uncertainty principles on energy and momentum demonstrates an occurrence of the quantum coherence mediated by an alternated sequence of measurements proceeding internally.

6.2 Quantized Dissipation

The interplay between cohesive interaction and internal measurement in cell motility is, however, subtle in the face of quantum mechanics. Measurement is ubiquitous in the biological realm as in the form of signal detection, and at the same time fundamentally irreversible in making distinction between before and after the events. In contrast, the problem of measurement is intrinsic to quantum mechanics. This simplified overview raises a substantial question on how measurement, irreversibility and quantum mechanics could be related especially with regard to naturally occurring dissipative structures, including biological ones. We now address this question from the viewpoint that measurement is ubiquitous internally and updated at every finite time interval.[3,6]

Let us suppose that an arbitrary material system, whether a molecule or a molecular organization, is involved in measuring its outside at every time interval of δt. Each internal measurement is accompanied with the energy δe specified by the energy-time uncertainty principle

$$\delta e \delta t \sim \hbar .$$

Since each measurement by the measuring system would disturb the outside to be measured energetically by the amount of δe at every time interval δt, the energy flow w from the measuring system towards its outside turns out to satisfy

$$w \sim \delta e/\delta t \sim \hbar/(\delta t)^2 .$$

Implicit in the energy dissipation is the presence of energy sources to be dissipated on the part of the measuring system.

As the time interval of successive internal measurement decreases, the energy flow of dissipation would increase. Conversely, the time interval of internal measurement cannot be made arbitrarily small, otherwise there would be no energy sources to meet the associated dissipation.

One of the most ubiquitous sources of energy is gravitation. If the gravitation on the surface of the earth is the case, the gravitational force f_g acting upon the most stable fundamental unit of mass, that is, a hydrogen atom leads to $f_g \sim 10^{-21}$erg/cm based upon the atomic mass unit of $\sim 10^{-24}$g and the gravitational acceleration on the surface of the earth $\sim 980 cm/s^2$. Since the gravitational force is mediated by graviton propagating at light velocity c,[26] the hydrogen atom measuring the earth as a whole dissipates energy at the rate

$$w_g \sim f_g c \sim 3.0 \times 10^{-11} erg/s .$$

The energy flow of dissipation due to internal measurement through the gravitation on the earth cannot become less than w_g because the hydrogen atom is the minimum mass unit available there. Of course, the hydrogen atom pulling the earth is pulled back by the earth. The dissipated energy is constantly compensated by the earth pulling the hydrogen atom.[21,26]

On the other hand, quantum mechanics of the hydrogen atom measuring the earth yields that the measurement would proceed at every time interval of δt_g given by

$$\delta t_g \sim (\hbar / w_g)^{1/2} \sim 5.8 \times 10^{-9} s$$

because the energy flow of dissipation due to internal measurement updated at every interval of δt_g yields $w_g \sim \hbar / (\delta t_g)^2$. However, the internal measurement accompanied with energy dissipation caused by gravitation may be taken over if other internal measurement to be updated at the time interval smaller than δt_g would appear.

As a matter of fact, various naturally occurring dissipative structures on the earth are not exclusively of gravitational origin. Electrostatic and magnetic interaction are the likely possibility. In order to see the likelihood of natural dissipative structures that could occur through other than gravitational interaction, it would be asked to examine any possibility for internal measurement to be updated at the time interval smaller than δt_g.

One such example of internal measurement facilitated by electrostatic interaction is available from an actin-myosin complex in the presence of ATP molecules as a fundamental functional unit of muscle contraction. Electrostatic interactions involved in the hydrolysis of an ATP molecule at the actomyosin complex are certainly responsible for both the generation of force and the occurrence of the sliding movement of the actin filament. In view of the observed fact that the generated force f_{AM} is about 2×10^{-7}dyne per myosin head[22,23] and the sliding velocity of the actin filament v_{AM} is about 4×10^{-4}cm/s,[6] the rate of energy dissipation w_{AM} is found to be

$$w_{AM} \sim f_{AM} v_{AM} \sim 8 \times 10^{-11} erg/s .$$

The time interval δt_{AM} of successive internal measurement due to the electrostatic interactions leads to

$$\delta t_{AM} \sim 3.6 \times 10^{-9} s ,$$

which is certainly smaller than $\delta t_g \sim 5.8 \times 10^{-9}$s.

It is an undeniable fact that electrostatic and magnetic interaction are ubiquitous and far stronger than gravitational interaction under normal conditions realizable on the earth. Nonetheless, gravitational interaction can exert a significant influence upon the occurrence of internal measurement, or dissipative structures. Electrostatic interaction, though far stronger than gravitational one, is quite short-ranged in comparison because of the imposed condition of electric neutrality.

On the other hand, if internal measurement accompanied by its inevitable energy dissipation is to occur, both the source and the sink of energy would have to be prepared. Otherwise, internal measurement could not take place. If there is neither the source nor the sink to be prepared externally, this stipulation would result in that the time interval for measurement proceeding internally would effectively diverge, that is, quantum mechanics of a unitary transformation accompanied by no measurement. It is at this point where gravitational interaction enters.[26] Although it is very weak compared to electrostatic interaction, gravitational one can facilitate internal measurement by preparing both the energy source and sink for its completion in a self-consistent manner.[21]

Electrostatic interaction can take over gravitational interaction in framing internal measurement and the accompanied energy dissipation only when the rate of facilitated dissipation is quantized and the unit is made greater than the minimum dissipation rate specified by the gravitational interaction. Quantum mechanics of internal measurement,[3,6] which is necessarily irreversible, can materialize dissipative structures by means of electrostatic and magnetic interaction only when the associated dissipation is maintained above a certain minimum level.

Prerequisite to the appearance of dissipative structures through electrostatic and magnetic interaction is quantization of dissipation in the face of gravity. One implication of the quantization of dissipation is the homeostatic capacity of dissipation in maintaining its rate above the minimum quantum level. Unless such homeostatic capacity of approaching energy sources spontaneously from the inside is available, an indefinite sustenance of dissipation could not hold as facing and tolerating adverse conditions exerted temporarily from the outside. In fact, heterotrophic activity associated with biological phenomena ever since the origin of life on the earth about 3.8 billion years ago witnesses the case that dissipation is quantized in its rate.

7 Concluding Remarks

Emergent phenomena associated with biological organizations are methodologically a subtle issue. If one employs external description of external measurement which has been proved in the endeavor for practicing objective and rigorous sciences, a great difficulty would come up with regard to the notion of variable duration. If it is asked, for instance, how natural selection could have originated, external description of external measurement practiced under the name of neo-Darwinism cannot cope with the question.[27] This is because the descriptive scheme is already based upon the notion of variable duration identified as natural selection. Questioning the origin of natural selection is methodologically discouraged or even prohibited there.

On the other hand, if one notes that external description of external measurement is not the only means for attaining descriptive objectivity of our experiences, external description of internal measurement will come up as an alternative that should deserve due attention. As a matter of fact, these two methodologies are by no means antagonistic with each other. External description of internal measurement duly transforms into external description of external measurement in the product because the consequence of internal measurement is externalized in the record and be globally measured externally. Emergent phenomena we experience in biology urges us to recognize that external description of internal measurement is a legitimate methodology for facing emergent novelty as much as external description of external measurement for invariant consistency.

In conclusion, external description of internal measurement is capable for identifying the nature of cohesive forces acting within biomolecules, organelles, cells, organs, tissues, organisms and even among their populations. Cell motility is certainly a concrete case exhibiting an interplay between internal measurement and cohesive interaction within the legitimate framework of quantum-mechanical phenomena.

References

1. G. Farre in *Advances in Synergetics Vol. 1*, ed. G. Lasker & G. Farre (IIAS, Ontario, 1994).
2. B.-O. Küppers, *Information and the Origin of Life* (MIT Press, Cambridge Mass., 1990).
3. K. Matsuno, *BioSystems* **17**, 179 (1985).
4. E. P. Wigner, *Symmetries and Reflections* (MIT Press, Cambridge Mass., 1970).
5. D. J. Depew and B. H. Weber, *Darwinism Evolving* (Bradford/MIT Press, Cambridge Mass., 1994).
6. K. Matsuno, *Protobiology: Physical Basis of Biology* (CRC Press, Boca Raton Florida, 1989).
7. O. E. Rössler in *Real Brains, Artificial Mind*, ed. J. L. Casti & A. Karlquist (Elsevier, New York, 1987).
8. P.-Y. Gunji, *BioSystems* **35**, 33 (1995).
9. K. Matsuno, in *The Origin and Evolution of the Cell*, ed. H. Hartman & K. Matsuno (World Scientific, Singapore, 1992).
10. K. Matsuno, *J. Theor. Biol.* **173**, 137 (1995).
11. R. C. Paton, *Biol. & Philos.* **7**, 279 (1992).
12. P. C. Marijuan, *BioSystems* **25**, 259 (1991).
13. I. Prigogine, *Introduction to Thermodynamics of Irreversible Processes* (Interscience, New York, 1969).
14. R. Cooke, *CRC Crit. Rev. Biochem.* **21**, 53 (1985).
15. H. E. Huxley, *J. Biol. Chem.* **265**, 8347 (1990).

16. Y. Harada, K. Sakurada, T. Aoki, D. D. Thomas, and T. Ynanagida, *J. Mol. Biol.* **216**, 49 (1990).
17. T. Q. P. Uyeda, H. M. Warrick, and J. A. Spudich, *Nature* **352**, 307 (1991).
18. K. Matsuno, and H. Honda, *CRC Crit. Rev. Biomed. Eng.* **18**, 311 (1991).
19. A. Winkelmann, L. Bourdieu, A. Otto, F. Kinose, and A. Libchabei, *Biophys. J.* **68**, 2444 (1995).
20. C. W. E. McClare, *J. Theor. Biol.* **35**, 569 (1971).
21. M. Conrad, *Phys. Lett. A* **152**, 245 (1991).
22. A. Kishino, and T. Yanagida, *Nature* **334**, 74 (1988).
23. J. Finer, R. M. Simmons, and J. A. Spudich, *Nature* **368**, 113 (1994).
24. A. F. Huxley, and R. M. Simmons, *Nature* **223**, 533 (1971).
25. K. Matsuno, *BioSystems* **35**, 209 (1995).
26. R. Penrose, *The Emperor's New Mind* (Oxford University Press, New York, 1989).
27. S. N. Salthe, *Development and Evolution* (Bradford/MIT Press, Cambridge Mass., 1993).

BIOENERGETICS AND BIOCOMMUNICATION

Mae-Wan Ho
Bioelectrodynamics Laboratory, Open University, Walton Hall, Milton Keynes, MK7 6AA, U.K.

1 Introduction

Organisms are so enigmatic from the thermodynamic point of view that Lord Kelvin, co-inventor of the Second Law of thermodynamics, specifically excluded them from its dominion (Ehrenberg, 1967), while Schrödinger (1944) suggested they feed upon "negative entropy" to free themselves from all the entropy they cannot help producing.

Lord Kelvin was impressed with how organisms seem to have energy at will, whenever and wherever required, and in a perfectly coordinated way. That is at once the problem of bioenergetics - how organisms can have energy so readily - and of biocommunication - how the energy mobilizing activities are organized as a whole. Similarly, Schrödinger alluded to the ability of organisms to use the energy they feed on to build up and maintain their dynamic organization. The intuition of both physicists is that energy and organization are intimately linked.

Schrödinger was reprimanded, by Linus Pauling and others, for using the term 'negative entropy', which does not correspond to any rigorous thermodynamic entity (Gnaiger, 1994). However, the idea that open systems can "self-organize" under energy flow became more concrete in the discovery of *dissipative structures* (Prigogine, 1967) that depend on the flow and dissipation of energy, such as the Bénard convection cells and the laser. In both cases, energy input results in a phase transition to global dynamic order in which all the molecules or atoms in the system move coherently. From these and other considerations, I have identified Schrödinger's "negative entropy" as "stored mobilizable energy in a space-time structured system" (Ho, 1994b, 1995a). In this essay, I show how stored mobilizable energy effectively frees the organism from thermodynamic constraints so that it is poised for rapid and specific intercommunication. In the ideal, the organism is a quantum superposition of coherent activities with instantaneous (nonlocal) noiseless intercommunication throughout the system.

2 Energy storage frees the organism from thermodynamic constraints

2.1 Energy storage and mobilization in living systems

The key to understanding the thermodynamics of the living system is not energy flow or energy dissipation, but energy *storage* under energy flow (Fig. 1). Energy flow is of no consequence unless the energy is trapped and stored within the system where it circulates before being dissipated. A reproducing life cycle, i.e., an organism, arises when the loop of circulating energy closes. At that point, we have a life cycle within which the stored energy is mobilized, remaining stored as it is mobilized, and coupled to the energy flow.

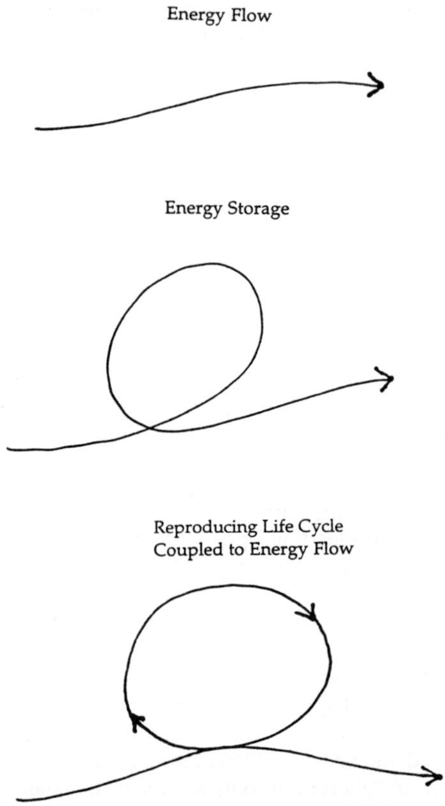

Figure 1. Energy flow, energy storage and the reproducing life-cycle.

The life-cycle is a highly differentiated space-time structure, the predominant modes being themselves cycles spanning many order of magnitudes of space-times all coupled together, from the local and fast (or slow) to the global and slow (or fast) (Ho, 1993;1994; 1995a). An intuitive representation is given in Figure 2.

Cycles are indeed ubiquitous in the living system, appearing in their most familiar forms as biological rhythms. Because the cycles are coupled together, energy is effectively stored and mobilized over all space-time according to the relaxation times (and volumes) of the processes involved. So, organisms can take advantage of two different ways to mobilize energy with maximum efficiency - nonequilbrium transfer in which stored energy is transferred before it is thermalized, and quasi-equilibrium transfer, for which the free energy change approaches zero according to conventional thermodynamic considerations (McClare, 1971; Ho, 1993; see also Lumry, 1991 for evidence that enzymes act to conserve free energy as free energy transducers).

As all the space-time modes are coupled together, energy input into any mode can be readily delocalized over all modes, and conversely, energy from all modes can become concentrated into any mode. In other words, energy coupling in the living system is *symmetrical* (see Ho, 1993, 1994b, 1995a,b).

Figure 2. The many-fold cycles of life coupled to energy flow.

2.2 *Onsager's reciprocity relationship and linearity of coupled flows*

Symmetrical coupling and cyclical flows are both aspects of the living system predicted from the thermodynamics of the steady state: Onsager's (1931) reciprocity relationship, and Morowitz' (1968) theorem. Onsager's reciprocity relationship states that for a system of many coupled linear flows under conjugate forces,

$$\mathbf{J}_i + \Sigma_k L_{ik} \mathbf{X}_k \tag{1}$$

where $\mathbf{J}_i$ is the flow of the ith process (i = 1, 2, 3.....n), $\mathbf{X}_k$ is the kth thermodynamic force (k = 1, 2, 3,.....n), and L_{ik} are the proportionality coefficients (where i = k) and coupling coefficients (where i ≠ k), the couplings for which the $\mathbf{X}_k$s are invariant with time reversal (i.e., velocity reversal) will be symmetrical; in other words,

$$L_{ik} = L_{ki} \tag{2}$$

so long as the **J**s and the **X**s satisfy $T\theta = \Sigma \mathbf{J}_i \mathbf{X}_i$ where θ is the rate of entropy increase per unit volume (I thank Denbigh (personal communication) for this formulation).

Morowitz' theorem states that the flow of energy through the system from a source to a sink will lead to at least one cycle in the system at steady state, *provided that the energy is trapped and stored within the system* (italics mine). This important theorem captures a key aspect of the steady state, and also implies that the steady state - at which global balance is maintained - may harbour nonlinear processes (see Ho, 1993).

Onsager's reciprocity relationship has been extended to the far from equilibrium regime by Rothschild *et al* (1980) for multi-enzyme systems and more recently, by Sewell (1991) for infinite quantum systems. However, the validity and the theoretical basis for the extension of Onsager's reciprocity relationship to biological systems are still under debate (Westerhof and van Dam, 1987). Cortassa *et al* (1991) show that while linear nonequilibrium thermodynamics can describe an autocatalytic system, the matrix of phenomenological coefficients is nonsymmetric. They conclude therefore, that it is the symmetry property (Onsager's reciprocity relationship) and not the linearity of the flow-force relations in the near equilibrium domain that precludes oscillations; and conversely, a system with oscillations cannot at the same time satisfy the symmetry property.

I believe some form of Onsager's reciprocity relationship does hold in living systems if only to account for the ready mobilization of energy on the one hand - why we can have energy at will - and on the other hand, for the linear relationships between steady-state flows and conjugate thermodynamic forces outside the range of equilibrium actually observed in many biological systems (Berry *et al*, 1987, and references therein).

According to Rothschild *et al* (1980), linearity in biological processes can arise in enzymes operating near a multidimensional inflection point far away from thermodynamic equilibrium, if some of the rate constants are linked. That is realistic for living systems which are now known to have highly organized flows in the cytoplasmic matrix (Welch, 1985, and references therein). In common with Rothschild *et al* (1981), Sewell shows how Onsager's reciprocity relationship applies to locally linearized *combinations* of forces and flows, which nonetheless behave globally in nonlinear fashion. That is particularly relevant to the living system, where nested compartments and microcompartments ensure that many processes may be operating locally at thermodynamic equilibrium even though the system or subsystem as a whole is far away from equilibrium (Ho, 1995a). Furthermore, as each process is ultimately connected to every other in the metabolic net through catenations of space and time, even if truly symmetrical couplings are localized to a limited number of metabolic/energy transducing junctions, the effects will eventually be shared or delocalized throughout the system, so that symmetry will apply to appropriate *combinations* of forces and flows over a sufficiently macroscopic space-time scale (Sewell, 1991). That is perhaps the most important consideration. As real processes take time, Onsager's reciprocity relationship cannot be true for an arbitrarily short instant, but must apply at a sufficiently macroscopic time interval when overall balance holds.

To summarize, nonlinearity does not preclude symmetry on the appropriate scale, and local linearity does not exclude the possibility for self-organization at a more global level. Hence, the contention that oscillations typical of self-organized systems is incompatible with symmetry properties (Cortassa *et al*, 1991) may be irrelevant when the entire system or sub-system of *balanced* flows and forces is taken into account. This will become clear as we consider the origins of the thermodynamics of the steady state.

2.3 Thermodynamics of the steady state vs thermodynamics of organized complexity

Denbigh (1951) defines the steady state as one in which "the macroscopic parameters, such as temperature, pressure and composition, have time independent values at every point of the system, despite the occurrence of a dissipative process."[1] That is too restrictive to apply to the living system, which has coupled processes spanning the whole gamut of relaxation times and volumes (Ho, 1993). A less restrictive formulation - one consistent with a "thermodynamics of organized complexity" (Ho, 1994a) - might be to define the living system, to first approximation, as a *dynamic equilibrium in which the macroscopic parameters, such as temperature, pressure and composition have time-independent values despite the occurrence of dissipative processes.* The present formulation omits the phrase, "at every point of the system". Microscopic homogeneity is not crucial for the formulation of any thermodynamic state, as the thermodynamic parameters are *macroscopic* entities quite independent of the *microscopic* interpretation (Ho, 1993). Like the principle of microscopic reversibility, it is extraneous to the phenomenological laws of thermodynamics (Denbigh, 1951).

The first incursion into the thermodynamics of the steady state was W. Thomson's (Lord Kelvin) treatment of the thermoelectric effect (see Denbigh, 1951). This involves a circuit in which heat is absorbed and rejected at two junctions (the Peltier heat), and in addition, heat is absorbed and given off (the Thomson heat) due to current flows between two parts of the same metal at different temperatures. Both of these heat effects are reversible, in that they change sign when the direction of the current is reversed. On the other hand, there are two other effects which are not reversible: heat conduction along the wires and dissipation due to the resistance. It is thus impossible to devise a reversible thermoelectric circuit even in principle. Nevertheless, Thomson took the step of assuming that, at steady state, those heat effects that are reversible, i.e., the Peltier heat and Thomson heat balance each other so that no net entropy is generated,

$$\Delta S_P + \Delta S_T = 0$$

On that basis, he derived the well-known relations between the Peltier and Thomson heats and the temperature coefficient of the electromotive force. It was a bold new departure in the application of the Second Law, but one which was subsequently justified by experimental evidence.

Very similar methods were used later by Helmholz in his treatment of the electromotive force and transport in the concentration cell, where he states clearly that the two irreversible process in the cell, heating and diffusion, are to be disregarded and the Second law to be applied to those parts of the total process which are reversible. Most modern accounts of this system follow the same procedure. A virtual flow of current is supposed to take place across the liquid junction, resulting in a displacement of the ions. The process is taken to be reversible and to generate no net entropy. The justification, according to Guggenheim (cited in Denbigh, 1951), is that the two processes, diffusion and flow of current across the junction, "take place at rates which vary according to different laws" when the composition gradient across the boundary is altered, and so it seems reasonable to suppose that the two processes are merely superposed, and that the one may be ignored when considering the other. Thus, the steady state is treated *as if*

---

[1] Denbigh, 1951, p. 3

there were no dissipative processes, and it is this assumption which is later validated by Onsager's reciprocity relationship.

2.4 *The living system is free from immediate thermodynamic constraints*

In the same spirit, I propose to treat the living system as a superposition of dissipative irreversible processes and non-dissipative processes, so that Onsager's reciprocity relationship applies only to the latter. In other words, it applies to coupled processes for which the net entropy production is balanced or zero,

$$\Sigma_k \, \Delta S_k = 0 \tag{3}$$

This will include most living processes because of the ubiquity of coupled *cycles*, for which the *net* entropy production balances out to zero. The principle applies in the smallest unit cycle in the living system - enzyme catalysis - on which all energy transduction in the living system is absolutely dependent. Over the past 30 years, Lumry and his coworkers (see Lumry, 1991) have shown convincingly how the flexible enzyme molecule balances out entropy with enthalpy to conserve free energy during catalysis. The organism is, in effect, a closed, self-sufficient energetic domain of cyclic non-dissipative processes coupled to the dissipative processes (Ho, 1995b). In the formalism of conventional thermodynamics, the life cycle, or more precisely, the living system in dynamic equilibrium, consists of all cyclic processes for which the net entropy change is zero, coupled to dissipative processes necessary to keep it going, for which the net entropy change is greater than zero (Fig. 3).

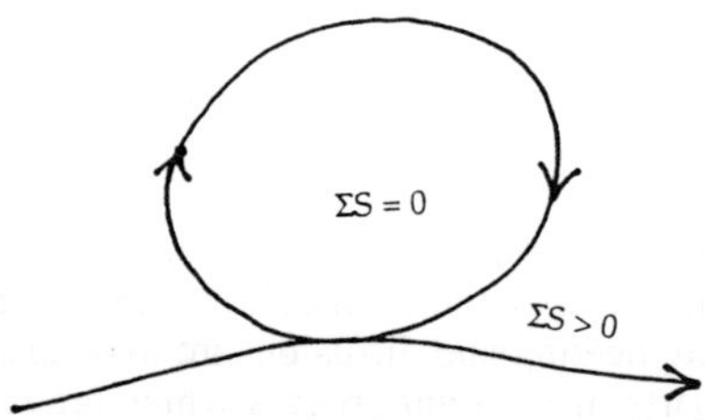

Figure 3. The organism frees itself from the constraints of energy conservation and the second law of thermodynamics.

Consequently, the organism is free from the immediate constraints of energy conservation - the First Law - *and* the Second Law of thermodynamics. *There is always energy available within the system, which is mobilized at close to maximum efficiency and over all space-time modes.* This in turn creates the conditions for rapid, sensitive and specific intercommunication throughout the system.

3 The exquisite sensitivity of organisms

3.1 *Energy self-sufficiency and sensitivity*

One distinguishing feature of the living system is its exquisite sensitivity to weak signals. For example, the eye can detect single photons falling on the retina, and the presence of several molecules of pheromones in the air is sufficient to attract male insects to their mates. That exquisite sensitivity applies to all levels of 'information processing' in the organism, and is the direct consequence of its energy self-sufficiency. No part of the system has to be pushed or pulled into action, nor be subjected to mechanical regulation and control. Instead, coordinated action of all the parts depends on rapid *intercommunication* throughout the system. The organism is a system of "excitable media" (see Goodwin, 1994,1995), or excitable cells and tissues poised to respond specifically and disproportionately to weak signals because the large amount of energy stored can amplify weak signals into macroscopic actions. It is by virtue of its energy self-sufficiency, therefore, that an organism is a *sentient* being - a system of sensitive parts all set to intercommunicate, to respond and to act appropriately as a whole to any contingency.

3.2 *The polychromatic whole*

Evidence for constant intercommunication throughout the living system may already exist in the physiological literature. I refer to 'deterministic chaos' which has been used to describe many living functions from the complex, locally unpredictable behaviour of ant colonies (Goodwin, 1994) to unrepeatable patterns of brain activities (Freeman, 1995). A different understanding of the complex activity spectrum of the healthy state is that it is polychromatic (Ho, 1995d), approaching 'white' in the ideal, in which all the modes of stored energy are equally represented. It corresponds to the so-called $f(\lambda) = \text{const.}$ rule that Fritz Popp (1986) has generalized from the spectrum of light or "biophotons" found to be emitted from all living systems. I have proposed that this polychromatic ideal distribution of stored energy is the state towards which all open systems capable of energy storage naturally evolve (Ho, 1994b). It is a state of both maximum and minimum entropy - maximum because energy is equally distributed over all space-time modes, and minimum because the modes are coupled together to give, in effect, a single degree of freedom (Popp, 1986; Ho, 1993). In a system with no impedance to energy mobilization, all the modes are intercommunicating and hence all frequencies are represented. But when coupling is imperfect, or when the subsystem, say, the heart, or the brain, is not communicating properly, it falls back on its own modes, leading to impoverishment of its activity spectrum. The living system is necessarily a polychromatic whole, it is full of variegated complexity that nevertheless cohere into a singular being, and that is the ultimate problem of biocommunication that needs to be addressed.

4 The intercommunicating whole

Recent advances in biochemistry, cell biology and genetics are giving us a concrete picture of the organism as an interconnected, intercommunicating whole. It is becoming increasingly clear that living organization cannot be understood in terms of mechanistic controls, nor of endless processings of genetic information.

4.1 A molecular democracy of distributed control

Henrik Kacser (1987) was among the first to realize that in a network, especially one as complicated as the metabolic network, it is unrealistic to think that there could be special enzymes controlling the flow of metabolites under all circumstances. He and a colleague pioneered *metabolic control analysis* to discover how the network is actually regulated. After more than 20 years of investigation by many biochemists and cell biologists, it is now generally acknowledged that so-called 'control' is invariably distributed over many enzymes (and metabolites) in the network, and moreover, the distribution of control differs under different conditions. The metabolic network turns out to be a "molecular democracy" of distributed control.

4.2 Long-range energy continua in cells and tissues

Studies over the past 25 years have also revealed that energy mobilization in living systems is achieved by protein or enzyme molecules acting as "flexible molecular energy machines" (Ho, 1995a and references therein) transferring energy directly from the point of release to the point of utilization, without thermalization or dissipation. These direct energy transfers are carried out in collective modes extending from the molecular to the macroscopic domain. The flow of metabolites is channeled coherently at the molecular level, directly from one enzyme to the next in sequence, in multi-enzyme complexes (Welch and Clegg, 1987). At the same time, high voltage electron microscopy and other physical measurement techniques reveal that the cell is more like a 'solid state' than the 'bag of dissolved enzymes' that generations of biochemists had previously supposed (Clegg, 1984). Not only are almost all enzymes bound to an intricate "microtrabecular lattice", but a large proportion of metabolites as well as water molecules are also structured on the enormous surfaces available. Aqueous channels may be involved in the active transport of solutes within the cell in the way that the blood stream transports metabolites and chemical messengers within the organism (Wheatley and Clegg, 1991).

As Welch and Berry (1985) propose, the whole cell is linked by "long-range energy continua" of mechanical interactions, electric and eletrochemical fluxes and in particular, proton currents that form a "protoneural network", whereby metabolism is regulated instantly and down to minute detail. Cells are in turn interconnected by electrical and other cytoplasmic junctions. And there is increasing evidence that cells and tissues are also linked by electromagnetic phonons and photons (see Popp, Li and Gu, 1992; Ho, 1993; Ho, Popp and Warnke, 1994). As I shall show later, the cell (as well as organism) is not so much a "solid state" as liquid crystalline. Living systems, therefore, possess just the conditions favouring the rapid propagation of influences or 'information' in all directions, which are naturally gated in cascades (see Ho, 1993) by the relaxation space-times of the processes involved. These are precisely the conditions that can yield linear flow force relationships in a system globally far from thermodynamic equilibrium (Berry *et al*, 1987). Global phase transitions may often take place, which can be initiated at any point within the system or subsystem. Abrupt, phase-transition like changes in the electrical activities of whole areas of the brain are indeed frequently observed in simultaneous recordings with a large array of electrodes (Freeman, 1995), for which no definite centre(s) of origin can be identified.

4.3 Organism and environment - a mutual partnership

Biology today remains dominated by the genetic paradigm. The genome is seen as the repository of genetic information controlling the development of the organism, but otherwise insulated from the environment, and passed on unchanged to the next generation except for rare random mutations. The much publicized Human Genome Project is being promoted on that basis (Ho, 1995e). The genetic paradigm has already been fatally undermined at least ten years ago, when a plethora of 'fluid genome' processes were first discovered, and many more have come to light since. These processes destabilize and alter genes and genomes in the course of development, some of the genetic changes are so well correlated with the environment that they are referred to as "directed mutations". Many of the genetic changes are passed on to the next generation. As I pointed out at the time, heredity can no longer be seen to reside solely in the DNA passed on from one generation to the next. Instead, the stability and repeatability of development - which we recognize as heredity - is distributed in the whole gamut of dynamic feedback interrelationships between organism and environment from the socioecological to the genetic. All of these may leave imprints that are passed on to subsequent generations: as cultural traditions or artefacts, maternal or cytoplasmic effects, gene expression states, as well as genetic (DNA sequence) changes (see Ho, 1986;1996).

4.4 The distributed organic whole

Thus, the essence of the organic whole is that it is *distributed* throughout its constituent parts, with no centre of control, no governors, no hierarchical levels of line-managers or regulators processing information down the line of command. Instead, pervasive, moment to moment intercommunication throughout the system renders part and whole, local and global completely indistinguishable. The existing mechanistic framework is most inadequate in coming to grips with the organic whole. In the next Section, I shall present an alternative frame-work based on *coherence*, in particular, on *quantum* coherence.

5 The organism as an autonomous coherent whole

5.1 The coherence of organisms

I mentioned earlier that the living system is necessarily a polychromatic whole - a variegated complexity that nevertheless cohere into a singular being. The *wholeness* of the organism is the ultimate problem of biocommunication: how to account for the continuity that encompasses the activities of elementary particles and atoms, molecules and cells, tissues and organs all the way to the organism itself (see Joseph Needham,

1935) The problem has never been adequately addressed until Herbert Fröhlich (1968; 1980) presented the first detailed theory of *coherence.* He argued that as organisms are made up of strongly dipolar molecules packed rather densely together (c.f. the 'solid state' cell), electric and elastic forces will constantly interact. Metabolic pumping will excite macromolecules such as proteins and nucleic acids as well as cellular membranes (which typically have an enormous electric field of some 10^7V/m across them). These will start to vibrate and eventually build up into collective modes, or *coherent excitations*, of both phonons and photons extending over macroscopic distances within, and perhaps also outside, the organism.

The emission of electromagnetic radiation from coherent lattice vibrations in a solid-state semi-conductor has recently been experimentally demonstrated for the first time (Dekorsy *et al*, 1995). The possibility that organisms may use electro-magnetic radiations to communicate between cells was already entertained by Soviet biologist Gurwitsch (1925) early this century.This hypothesis was revived by Popp and his coworkers in the late 1970s, and there is now a large and rapidly growing literature on "biophotons" believed to be emitted from a coherent photon field (or energy storage field) within the living system (see Popp, Li and Gu, 1992).

In collaboration with Fritz Popp, we have found that a single, brief exposure of synchronously developing early fruitfly embryos to white light results in the re-emission of relatively intense and prolonged flashes of light, some tens of minutes and even hours after the light exposure (Ho *et al*, 1992b). The phenomenon is reminiscent of phase-correlated collective emission, or *superradiance*, in atomic systems, although the time-scale is orders of magnitude longer, perhaps in keeping with the coherence times of organisms. For phase-correlation to build up over the entire population, one must assume that each embryo has a *collective* phase of all its activities, in other words, each embryo must be considered a highly (quantum) coherent domain, despite its multiplicity of activities (Ho, Zhou and Haffegee, 1995).

During the same period of early development, exposure of the embryos to weak static magnetic fields also cause characteristic global transformation of the normal segmental body pattern to helical configurations in the larvae emerging 24 hours later (Ho *et al*, 1992a). As the energies involved are several order of magnitude below the thermal threshold, we conclude that there can be no effect unless the external field is acting on a coherent domain where charges are moving in phase, or where magnetically sensitive liquid crystals are undergoing phase alignment globally (Ho, *et al*, 1994). Liquid crystals may indeed be the material basis of many, if not all aspects of biological organization (Ho *et al*, 1995).

5.2 Organisms as polyphasic liquid crystals

Liquid crystals are phases of matter between the solid and the liquid states, hence the term, *meso*phases (DeGennes, 1974). Liquid crystalline mesophases possess long range orientational order, and often also varying degrees of translational order. In contrast to solid crystals, liquid crystals are mobile and flexible, and above all, highly responsive. They undergo rapid changes in orientation or phase transitions when exposed to electric and magnetic fields (Blinov, 1983) or to changes in temperature, pressure, pH, hydration, and concentrations of inorganic ions (Collings, 1990; Knight, 1993). These properties are ideal for organisms (Gray, 1993; Knight, 1993). Liquid crystals in organisms include all the major constituents of the organism: the amphiphilic lipids of cellular membranes, the DNA in chromosomes, all proteins, especially cytoskeletal proteins, muscle proteins, collagens and proteoglycans of connective tissues. These adopt a multiplicity of mesophases that may be crucial for biological structure and function at all levels of organization (Ho *et al*, 1995) from channeling metabolites in the cell to pattern determination and the coordinated locomotion of whole organisms.

The importance of liquid crystals for living organization was recognized by Joseph Needham (1935) among others. He suggested that living systems actually *are* liquid crystals, and that many liquid crystalline mesophases may exist in the cell although they cannot then be detected. Indeed, there has been no direct evidence that extensive liquid crystalline mesophases exist in living organisms or in the cytoplasm until our recent discovery of a noninvasive optical technique (Ho and Lawrence, 1993; Ho and Saunders, 1994; Newton, Haffegee and Ho, 1995). This enables us to obtain high resolution and high contrast coloured images of live organisms based on visualizing just the kind of coherent liquid crystalline mesophases which Needham and others had predicted.

The technique amplifies small birefringences typical of biological liquid crystals, enabling us to see the whole living organism down to the phase alignment of the molecules that make up its tissues. Brilliant interference colours are generated, specific for each tissue, dependent on the birefringence of the molecules and their degree of coherent phase alignment. The colours are generated even as the molecules in the tissues are moving about, busily transforming energy. That is possible because visible light vibrates much faster than the molecules can move, so the tissues will appear indistinguishable from static crystals to the light passing through so long as the movements of the constituent molecules are sufficiently coherent. With this imaging technique, one can see that the organism is *thick with activities at every level, coordinated in a continuum from the macroscopic to the molecular.* And that is what the coherence of the organism entails.

These images also bring out another aspect of the wholeness of the organism: all organisms, from protozoa to vertebrates without exception, are polarized along the anteroposterior axis, so that all the colours in different parts of the body are maximum when the anteroposterior axis is appropriately aligned, and they change in concert as the organism is rotated from that position. The anteroposterior axis is the *optical axis* of the whole organism, which is, in effect, a single (uniaxial) crystal. This leaves us in little doubt that the organism is a singular whole, despite the diverse multiplicity and polychromatic nature of its constituent parts.

The tissues not only maintain their crystalline order when they are actively transforming energy, the degree of order seems to *depend* on energy transformation, in that the more active and energetic the organism, the more intensely colorful it is, implying that the molecular motions are all the more coherent (Ho and Saunders, 1994). The coherence of the organism is closely tied up with its energetic status, as argued in the beginning of this essay: energy and organization are intimately linked. The coherent whole is full of energy - it is a *vibrant* coherent whole.

5.3 *Quantum coherence in living organisms*

The above considerations and observations convince me that the wholeness of organisms is only fully captured by *quantum* coherence (Ho, 1993). An intuitive way to understand quantum coherence is to think of the 'I' that each and every one of us experience of our own being. We know that our body is a multiplicity of organs and tissues, composed of many billions of cells and astronomical numbers of molecules of many different kinds, all capable of working autonomously, and yet somehow cohering into the singular being of our private experience. That is just the stuff of quantum coherence. Quantum coherence does not mean that everybody or every element of the system must be doing the same thing all the time, it is more akin to a grand ballet, or better yet, a very large jazz band where everyone is doing his or her own thing while being perfectly in step and in tune with the whole.

A quantum coherent system maximizes both global cohesion and local freedom (Ho, 1993). This property, technically referred to as *factorizability*, enables the body to be performing all sorts of different *coordinated* functions simultaneously (Ho, 1995b). It also enables *instantaneous (nonlocal) and noiseless intercommunication to take place throughout the system* (Ho 1995f). As I am writing, my digestive system is working independently, my metabolism busily transforming chemical energy in all my cells, putting some away in the longer term stores of fat and glycogen, while converting most of it into readily utilizable forms such as ATP. Similarly, my muscles are keeping in tone and allowing me to work the keyboard, while, hopefully, my neurons are firing in wonderfully coherent patterns in my brain. Nevertheless, if the telephone should ring in the middle of all this, I would turn to pick it up without hesitation.

The importance of factorizability is evoked by the movie character, Dr. Strangelove, portrayed by Peter Sellers as a megalomaniac scientist who wanted to rule the world. He was a wheelchair-bound paraplegiac, who could not speak without raising his arm in the manner of a Nazi salute. That is just the symptom of the loss of factorizability which is the hallmark of quantum coherence.

The coherent organism is, in the ideal, a quantum superposition of activities - organized according to their characteristic space-times - each itself coherent, so that it can couple coherently to the rest (Ho, 1995b). It is, in effect a vast array of Fröhlich systems all coupled together. This picture is fully consistent with the earlier proposal that the organism stores energy over all space-time domains each intercommunicating (or coupled) with the rest. It is also consistent with Onsager's reciprocity relationship or symmetrical coupling between all energy modes. Furthermore, quantum superposition enables the system to maximize its potential degrees of freedom so that the single degree of freedom required for coherent action can be *instantaneously* accessed.

The main implication of quantum coherence for living organization is that, in maximizing both local freedom and global intercommunication, the organism is in a very real sense completely free. Nothing is in control, and yet everything is in control. Thus, it is the failure to transcend the mechanistic framework that makes people persist in enquiring which parts are in control, or issuing instructions or information. These questions are meaningless when one understands what it is to be a coherent, organic whole. An organic whole is an *entangled* whole, where part and whole, global and local are so thoroughly implicated as to be indistinguishable, and where each part is as much in control as it is sensitive and responsive. The challenge for us all is to rethink information processing in the context of the coherent organic whole.

Acknowledgments
I am very grateful to Kenneth Denbigh, Geoffrey Sewell, Rickey Welch, Koichiro Matsuno, Roy Cuthbertson, Sonia Cortassa and Miguel Aon for instructive and helpful comments.

References

Blinov, L.M. (1983). *Electro-optical and Magneto-optical Principles of Liquid Crystals,* John Wiley and Sons, London.
Clegg, J.S. (1984). *Am. J. Physiol.* **246**, R133-151.
Collings, P.J. (1990). *Liquid Crystals, Nature's Delicate Phase of Matter*, Princeton University Press, Princeton.
Dekorsy, T., Auer, H., Waschke, C., Bakker, H.J., Roskos, H.G. and Kurz, H. (1995). *Physical Rev. Letters* **74**, 738-741.
Denbigh, D. (1951). *The Thermodynamics of the Steady State*, Methuen & Co., Ltd., New York.
De Gennes, P.G. (1974). *The Physics of Liquid Crystals*, Clarendon Press, Oxford.
Ehrenberg, W. (1967). *Scient. Am.* **217**, 103-110.
Freeman, W.J. (1995). *Societies of Brains. A Study in the Neuroscience of Love and Hate*, Lawrence Erlbaum Associates, Hove.
Fröhlich, H. (1968). *Int. J. Quant. Chem.* **2**, 641-649.
Fröhlich, H. (1980). *Adv. Electronics and Electron. Phys.* **53**, 85-152.
Gnaiger, E. (1994). *Modern Trends in BioThermoKinetics* **3**, 62-70.
Goodwin, B.C. (1994). *How the Leopard Changed Its Spots: The Evolution of Complexity*, Weidenfeld and Nicolson, London.
Goodwin, B.C. (1995). In *Biocommunication, S327 Living Processes*, pp.183-230, Open University Press, Milton Keynes.
Gray, G. (1993).British Association for the Advancement of Science, Chemistry Session: Molecular Self-Assembly in Science and Life, Sept. 1, Keele.
Gurwitsch, A.G. (1925). *Bot. Gaz.* **80**, 224-226.
Ho, M.W. (1986). *Rivista di Biologia* **79**, 407-447.
Ho. M.W. (1993). *The Rainbow and the Worm: The Physics of Organisms*, World Scientific, Singapore.
Ho, M.W. (1994a). In *New Metaphysical Foundations of Modern Science* (W.Harman and J. Clark, eds.), pp. 179-213, Institute of Noetic Sciences, Sausalito.
Ho, M.W. (1994b). *Modern Trends in BioThermoKinetics* **3**, 50-61.
Ho, M.W. ed. (1995a) *Bioenergetics, S327 Living Processes, An Open University Third Level Science Course,* Open University Press, Milton Keynes.
Ho, M.W. (1995b). *Neural Network World* **5**, 733-750.
Ho, M.W. (1995c). In *Encyclopedia of Comparative Psychology* (G. Greenberg and M. Haraway, eds.), Garland Publishing, New York.
Ho, M.W. (1995d). *J. Consciousness Studies* (submitted).

Ho, M.W. (1995e). *Soundings* **1**, 77-98.
Ho, M.W. (`1995f). (in preparation).
Ho, M.W. (1996) *Annal. Human Genetics* **60**: 81-84,
Ho, M.W., French, A., Haffegee, J. and Saunders, P.T. (1994). In *Bioelectrodynamics and Biocommunication* (M.W. Ho, F.A. Popp, and U. Warnke, eds.) pp.195-207, World Scientific, Singapore.
Ho, M.W. and Fox, eds. (1988).*Evolutionary Processes and Metaphors* , Wiley, London.
Ho, M.W., Haffegee, J., Newton, R. and Ross, S. (1995). (submitted)
Ho, M.W. and Lawrence, M. (1993). *Microscopy and Analysis*, **September**, 26.
Ho, M.W., Popp, F.A. and Warnke, eds. (1994). *Bioelectrodynamics and Biocommunications*, World Scientific, Singapore.
Ho, M.W. and Saunders, P.T. (1994).In *Bioelectromagnetism and Biocommunication* (M.W. Ho, F.A. Popp and U. Warnke, eds.) pp213-225, World Scientific, Singapore.
Ho, M.W., Stone, T.A., Jerman, I., Bolton, J., Bolton, H., Goodwin, B.C. , Saunders, P.T. and Robertson, F. (1992a). *Physics in medicine and Biology* **37**, 1171-1179.
Ho, M.W. Xu, X., Ross, S. and Saunders, P.T. (1992b). In *Advances in Biophotons Research* (F.A. Popp, K.H. Li and Q. Gu, eds.), pp. 287-306, World Scientific, Singapore.
Ho, M.W., Zhou, Y.M. and Haffegee, J. (1995). In *Physics in Biology* (L. Trainor and C. Lumsden, eds.), Academic Press (in press).
Kacser, H. (1987). In *The Organization of Cell Metabolism* (G.R. Welch and J.S. Clegg, eds.), Plenum Publishing Corporation, New York.
Knight, D. (1993). British Association for the Advancement of Science, Chemistry Session: Molecular Self-Assembly in Science and Life, Sept. 1, Keele.
McClare, C.W.F. (1971). *J. theor. Biol.* **30**, 1-34.
Needham, J. (1936). *Order and Life*, MIT Press, Cambridge, Mass.
Newton, R., Haffegee, J. and Ho, M.W. (1995). *J. Microscopy* **180**, 127-130.
Popp, F.A. (1986). In *Disequilibrium and Self-Organization* (C.W. Kilmister, ed.), p.207, Reidel, Dordrecht.
Popp, F.A., Li, K.H. and Q. Gu, eds. (1992). *Recent Advances in Biophoton Research and its Applications*, World Scientific, Singapore.
Prigogine, I. (1967). *Introduction to Thermodynamics of Irreversible Processes*, John Wiley & Sons, New York.
Schrödinger, E. (1944). *What is Life?* Cambridge University Press, Cambridge.
Sewell, G.L. (1991). In *Large-Scale Molecular Systems: Quantum and Stochastic Aspects* (W.Gans, A. Blumen and A Amann eds.), pp. 1-48, Plenum Press, New York.
Welch, G.R. and Berry, M.N. (1985). In *Organized Multienzyme Systems* (G.R. Welch, ed.), Academic Press, New York.
Welch, G.R. and Clegg, J.S. eds. (1987). *The Organization of Cell Metabolism*, Plenum Press, New York.
Wheatley, D. and Clegg, J.S. (1991). *Am. Zool.* **31**, 504-513.

PARAMETER ESTIMATION FOR A DIFFUSION APPROXIMATION TO STEIN'S MODEL

SATISH IYENGAR
Department of Mathematics & Statistics
University of Pittsburgh
Pittsburgh, PA 15208 USA

1 Introduction and Summary

The activity of neurons can be studied at several levels. Sometimes the intracellular subthreshold activity can be monitored. In many cases, however, only the extracellular region is accessible, so the neuron's successive firings, or spike train, are the available data. These data are used, for instance, to provide a quantitative characterization of the neuron in terms of indices such as the mean and variance of the interspike intervals (ISIs), and their autocorrelation. The effects of different experimental conditions can then be inferred from changes in these indices.

Such uses, however, are crude in the sense that they do not relate the data or indices to any known (or hypothesized) details of the electrophysiology or anatomy of the neuron. When such information is gathered from experiments, more realistic mathematical models for neural activity have been proposed. For example, Lapicque's model neuron, [13], behaves like an RC circuit: if $V(t)$ is the neuron's membrane depolarization at time t, then

$$C\frac{dV}{dt} + \frac{V}{R} = I, \tag{1}$$

where C is the capacitance, R is the resistance, and I is the input current. The quantity RC is the membrane time constant, so that in the absence of input the membrane potential decays as $V(0)e^{-t/RC}$. The current comes from synaptic inputs (or artificially, using electrodes) impinging upon the neuron. This equation is valid only as long as $V(t)$ is below a (possibly time-varying) threshold; when the threshold is reached, the neuron fires, and the membrane potential is reset to some lower value. Stein, [12], later modified this deterministic model by regarding the input to the neuron as the superposition of Poisson processes of excitatory and inhibitory signals, so that $V(t)$ becomes a stochastic process. When the number of inputs is large, this process is well approximated by the diffusion process first studied by Ornstein and Uhlenbeck, [10].

The use of diffusion models for subthreshold neural activity has a long history. The earliest example in the neuroscience context seems be due to Gerstein and Mandelbrot, [4], who used Brownian motion, or the Wiener process with drift as a model for $V(t)$; its deterministic analog is also given by (1), but without the resistance term. When neural firing corresponds to the first time that the process reaches a constant threshold, this model leads to the inverse Gaussian distribution for the firing time. However, the inverse Gaussian distribution is inappropriate in the present context, because it ignores the decay of the membrane potential between inputs. Thus, a number of more realistic diffusion models have been proposed since that time: see [13].

The main difficulty with the study of first passage times of more realistic diffusion processes is that they are not analytically tractable. The inverse Gaussian has a known density function that readily allows inference, such as maximum likelihood estimation and hypothesis testing. And certain generalized inverse Gaussian distributions, [2], with densities proportional to $t^{\lambda-1}\exp(-\alpha t - \beta/t)$, arise as the first passage times for diffusions, but these diffusions have not yet been used in the neuroscience context. For Stein's and other more realistic models, though, it is hard to estimate parameters from available data to fit models to data, to judge the goodness of fit, or to compare various models.

In this paper, we study the use of the Ornstein-Uhlenbeck process (OUP) approximation, and discuss the problem of estimating the parameters of the underlying process when the observed data are the ISIs. We assume that the firing threshold is a constant, and that each time the neuron fires, it is instantly reset to its resting potential; that is, we ignore the refractory period. The spike train then forms a renewal process. The Laplace transform of the ISI density is known: in fact, for any temporally homogeneous diffusion satisfying the Chapman-Kolmogorov equation, the Laplace transform of the first passage time to a constant boundary can be written in terms of the Laplace transform of the diffusion's transition density. Unfortunately, for the OUP, the first passage time Laplace transform is a complicated expression involving parabolic cylinder functions, so it has not been inverted to yield a tractable density. Consequently, little progress has been made towards developing inferential procedures using first passage time data for this process.

Recently, Ricciardi and Sato, [11], provided useful approximations for the OUP first passage density and moments for certain ranges of parameter values. Their work made it easier to use numerical methods to evaluate the density, so that inference using maximum likelihood estimates is now feasible. Next, in [5], we have used the asymptotic properties of parabolic cylinder functions to study the Laplace transform in detail. In particular, we have shown that the parameters of the OUP are not identifable; rather, certain easily interpreted functions of them are identifiable (that is, can be estimated), and we have shown the consistency, asymptotic normality, and efficiency of solutions to the likelihood equations. We have also written computer programs to evaluate these estimates and their standard errors.

In Section 2, we sketch Stein's model, its diffusion approximation, the OUP, and derive the Laplace transform of the first passage time. In Section 3, we discuss the identifiability of parameters, and state our main theorem about the asymptotic distribution of the maximum likelihood estimates (MLEs), based on the firing times, of those parameters of the underlying process. In Section 4, we describe the computer programs that we wrote to compute the MLEs and their standard errors. We then describe the data that we used: artificial spike trains from simulations of Stein's integrate-and-fire model, and actual spike trains from the retinal ganglia of goldfish. In Section 5, we describe our results, and end with a discussion of our current related work in Section 6.

2 Stein's Model and the OUP Approximation

An account of the neurophysiological background Stein's model and the OUP is given in several sources, perhaps the most careful treatment being that of Ricciardi [10]; we include a sketch of it here. Let the model neuron have $p+q$ synapses, each of which sends excitatory and inhibitory signals to the neuron. Suppose that these incoming signals are independent Poisson processes: let α_k be the rate for the kth excitatory input, with $e_k > 0$ being the magnitude of the excitatory postsynaptic potential ($k = 1, \ldots, p$); similarly, let β_l be the rate for the lth inhibitory input, with $i_l < 0$ being the magnitude of the inhibitory postsynaptic potential ($l = 1, \ldots, q$).

Let $V(t)$ denote the neuron's depolarization at time t, and suppose that the neuron is initially at rest, so that $V(0) = 0$. When a signal arrives at time t, $V(t)$ jumps to either $V(t) + e_k$ or $V(t) + i_l$, depending on the signal's nature. When there is no input, the depolarization decays exponentially with time constant τ towards its resting state: $V(t+h) = V(t)e^{-h/\tau}$ when there is no input during $[t, t+h]$. Thus, $V(t)$ is a continuous time Markov process; ignoring the neuron's firing threshold and maximum hyperpolarization for the moment, its state space is the entire real line.

Denote the transition density of the process $V(t)$ by $f(t, x, y)$, where x is the initial state, and y is the state at time t. The assumptions above imply that $V(t)$ is a time-homogeneous Markov process, whose transition density therefore satisfies the Chapman-Kolmogorov equation [6]:

$$f(t+h,x,y)=\int_{-\infty}^{\infty} f(t,x,u)f(h,u,y)\,du\ . \tag{2}$$

Let $\mu=\sum_k \alpha_k e_k+\sum_l \beta_l i_l$ and $\sigma^2=\sum_k \alpha_k e_k^2+\sum_l \beta_l i_l^2$; when $\mu>(<)\ 0$, the diffusion has a positive (negative) drift, modelling a net excitatory (inhibitory) input. In the present context, a Taylor expansion of (2) shows that $f(t,x,y)$ satisfies the Kolmogorov backward equation,

$$\frac{\partial f}{\partial t}=\left(-\frac{x}{\tau}+\mu\right)\frac{\partial f}{\partial x}+\frac{\sigma^2}{2}\frac{\partial^2 f}{\partial x^2}.$$

with initial condition

$$\lim_{t\downarrow 0} f(t,x,y)=\delta(y-x).$$

Thus, $V(t)$ is an OUP, [6]; the solution to this problem is

$$f(t,x,y)=\frac{1}{\sqrt{2\pi v}}\exp\left[-\frac{1}{2}\frac{(y-m)^2}{v}\right]\ ,$$

where

$$v=v(t,\tau,\sigma^2)=(1-e^{-2t/\tau})\sigma^2\tau/2,$$

and

$$m=m(t,x,\tau,\mu)=xe^{-t/\tau}+\mu\tau(1-e^{-t/\tau}).$$

We now turn to the first passage time distribution. Suppose that $V(0)=x$, and for $y>x$ let $T=\inf\{t>0:V(t)=y\}$, be the first passage time to y, and let $g(t,x,y)$ be the density of T. For any $z>y$ the strong Markov property for the OUP, [6], implies the following convolution:

$$f(t,x,z)=\int_0^t g(s,x,y)f(t-s,y,z)\,ds.$$

Defining the Laplace transform of f,

$$\hat{f}(\lambda,x,y)=\int_0^{\infty} e^{-\lambda t}f(t,x,y)\,dt,$$

and of g similarly, we get

$$\hat{g}(\lambda,x,y)=\frac{\hat{f}(\lambda,x,z)}{\hat{f}(\lambda,y,z)}.$$

A lengthy calculation (which we omit) shows that for $x < y$,

$$\hat{f}(\lambda, x, y) = \Gamma(\lambda\tau)\sqrt{\frac{\tau}{\pi\sigma^2}\frac{e^{(x-\mu\tau)^2/2\sigma^2\tau}}{e^{(y-\mu\tau)^2/2\sigma^2\tau}}}H_{-\lambda\tau}\left(\frac{\sqrt{2}[y-\mu\tau]}{\sigma\sqrt{\tau}}\right)H_{-\lambda\tau}\left(-\frac{\sqrt{2}[x-\mu\tau]}{\sigma\sqrt{\tau}}\right),$$

where

$$\Gamma(x) = \int_0^\infty t^{x-1}e^{-t}\,dt$$

is the gamma function, and

$$H_\nu(x) = \frac{e^{-x^2/4}}{\Gamma(-\nu)}\int_0^\infty t^{-\nu-1}e^{-(xt+t^2/2)}\,dt$$

is the parabolic cylinder function of order $\nu < 0$, [5; the standard notation is $D_\nu(x)$]. Thus, for $x < y$,

$$\hat{g}(\lambda, x, y) = \frac{\exp([x-\mu\tau]^2/2\sigma^2\tau)H_{-\lambda\tau}\left(-[x-\mu\tau]\sqrt{2/\sigma^2\tau}\right)}{\exp([y-\mu\tau]^2/2\sigma^2\tau)H_{-\lambda\tau}\left(-[y-\mu\tau]\sqrt{2/\sigma^2\tau}\right)},$$

The inversion of this Laplace transform to get $g(t, x, y)$ does not seem to be tractable, so Ricciardi and Sato have suggested a several approximations which are useful for numerical computations.

3 Parameter Estimation

Returning to the neurophysiological background, suppose that the resting potential of the neuron is a, its firing threshold is b, τ is its time constant, and μ and σ are the mean rate and standard deviation of the input, respectively. Then $\eta = (a, b, \tau, \mu, \sigma)$ are the parameters of the underlying process. Let T represent the firing time of the neuron; its Laplace transform is

$$Ee^{-\lambda T} = \frac{e^{(a-\mu\tau)^2/2\sigma^2\tau}H_{-\lambda\tau}(-[a-\mu\tau]\sqrt{2/\sigma^2\tau})}{e^{(b-\mu\tau)^2/2\sigma^2\tau}H_{-\lambda\tau}(-[b-\mu\tau]\sqrt{2/\sigma^2\tau})} = \frac{e^{\theta_1^2/2}H_{-\lambda\theta_3}(-\theta_1\sqrt{2})}{e^{\theta_2^2/2}H_{-\lambda\theta_3}(-\theta_2\sqrt{2})} \quad (3)$$

where

$$\theta = (\theta_1, \theta_2, \theta_3)' = \left(\frac{a-\mu\tau}{\sigma\sqrt{\tau}}, \frac{b-\mu\tau}{\sigma\sqrt{\tau}}, \tau\right)'.$$

Note that the Laplace transform is a function of η only through θ; θ_1 and θ_2 are the standardized distances between $\mu\tau$, the limiting mean voltage of $V(t)$, and the resting and firing thresholds, respectively, and θ_3 is the time constant. Henceforth, let $p(t;\theta)$ denote the density of T, and let $\hat{p}(\lambda;\theta)$ be its Laplace transform. We now state without proof several technical results leading to our main theorem. Proofs of these results and theorem are given in [4].

The following asymptotic expansion for $H_\nu(x)$ as $\nu \to \infty$,

$$H_{-\nu}(x) \sim \frac{1}{\sqrt{2}} \exp\left[-\frac{\nu}{2}\log(\nu) + \frac{\nu}{2} - x\sqrt{\nu} + \frac{x}{2(\sqrt{\nu}+\sqrt{\nu-1/2})} - \frac{x^3}{24\sqrt{\nu-1/2}}\right]$$

(see [1]) implies that distinct values of θ yield distinct Laplace transforms; thus θ is identifiable. This expansion implies that standard conditions for the inversion integral ([3], Section 70) applies, so that

$$p(t;\theta) = \frac{1}{2\pi i}\int_{-i\infty}^{i\infty} e^{t\lambda}\hat{p}(\lambda;\theta)\,d\lambda. \tag{4}$$

They also imply analogous expressions for partial derivatives of $p(t;\theta)$ with respect to θ (of all orders): for instance,

$$\frac{\partial}{\partial\theta_1}p(t;\theta) = \frac{1}{2\pi i}\int_{-i\infty}^{i\infty} e^{t\lambda}\frac{\partial}{\partial\theta_1}\hat{p}(\lambda;\theta)\,d\lambda. \tag{5}$$

Now let $\Theta = \{\theta : -\infty < \theta_1 < \theta_2 < \infty, \theta_3 > 0\}$ be the parameter space. The problem then is to estimate θ given interspike interval data $T^{(n)} = (T_1, \ldots, T_n)$. Let

$$l_n(\theta|T^{(n)}) = \sum_{i=1}^{n}\log p(T_i;\theta)$$

be the log-likelihood, and consider the likelihood equations

$$D_\theta l_n(\theta|T^{(n)}) = 0. \tag{6}$$

Our main result, then, is the
Theorem: With probability tending to 1 as $n \to \infty$, there exist solutions $\hat{\theta}_n$ to (6) based on the observations $T^{(n)}$ such that (*i*) $\hat{\theta}_{jn}$ is consistent for θ_j, $j = 1,2,3$, (*ii*) $\sqrt{n}(\hat{\theta}-\theta)$ is asymptotically normal with mean 0 and covariance matrix $I(\theta)^{-1}$, where $I(\theta) = (I_{ij}(\theta)) = (\partial I(\theta)/\partial\theta_i\partial\theta_j)$ is the information matrix, and (*iii*) $\hat{\theta}_{jn}$ is asymptotically efficient.

An existence theorem such as this is, of course, of little use without a method for finding the solutions $\hat{\theta}_n$. In the next Section, we sketch our computational procedures. Our approach is to use an approximation to the moments discussed in [10] to find a good initial value, and then iterate to a solution. When it is known that τ is large, it may be worthwhile to use the inverse Gaussian approximation to get an initial guess. Although it is possible that the log-likelihood $l_n(\theta|T^{(n)})$ is multimodal, in all the cases that we tried we found that it was in fact concave in the regions that we examined, so the optimizing algorithm typically converged in about ten steps.

4 Computation

We briefly describe our computer programs, all of which are available upon request. We did all computations on a Sun Sparcstation 20, using FORTRAN 77, and the IMSL library of routines. To compute the MLEs, we needed an accurate approximation to the density $p(t;\theta)$. For large and small values of t, we used the approximations given in [11]; for other values of t, we numerically inverted the Laplace transform, (4), using the DINLAP routine in IMSL. To find the MLEs, we used the optimization routine DUMCGF; we used rough moment approximations given in [10] to provide starting values for the optimizer. Finally, to evaluate the standard errors, we first calculated the observed Fisher information matrix (of second partial derivatives evaluated at the MLE) using the routine DDERIV, and then inverted the matrix.

To check our programs, we simulated Stein's integrate-and-fire model, and used the resulting data to compare the known values with the estimates. Using a 1-msec time step, we generated Poisson inputs. We set the membrane time constant at 5 and 10 msec, and various values of θ_1 and θ_2. The integrator ran until it reached a specified threshold, at which time the artificial neuron fired, reset to zero, and continued the renewal process; the simulations resulted in about 800 to 1200 spikes.

We also used our programs to study spike trains from the retinal ganglia of goldfish, which were kindly provided to us by Professor M. Levine of the University of Illinois at Chicago. The goldfish were sacrificed, their ganglia removed and placed in an oxygen bath that was maintained for up to six hours. A fuller account of the laboratory methods is contained in Levine, [8], and the references contained therein. We report the results for two units, called f68 and f83: each was recorded for 30 seconds under the same conditions, and resulted in the following ISI summaries, shown in Table 1.

TABLE 1. *Summary of ISIs from 2 goldfish retinal ganglia units.*

unit	spikes	mean	median	s.d.
68	814	36.85	15	39.56
83	1028	33.52	12	33.52

The large coefficients of variation, and the large differences between the means and medians indicate highly skewed ISI histograms, as do the plots given in the next Section.

5 Results

For the simulated data, the MLEs were in good agreement with the true values. Four of the cases are given Table 2.

TABLE 2. *True and estimated parameter values for simulated data. Standard errors are given in parentheses.*

spikes	θ_1	θ_2	θ_3	$\hat{\theta}_1$	$\hat{\theta}_2$	$\hat{\theta}_3$
823	0	1	5	-0.03 (.05)	0.96 (.06)	5.31 (.21)
951	0	1	10	-0.02 (.05)	1.08 (.05)	9.04 (.56)
1006	0	5	5	-0.11 (.07)	5.16 (.05)	5.80 (.49)
1128	0	5	10	-0.02 (.05)	5.03 (.05)	11.05 (.61)

Notice that for this ideal data the standard errors for the time constant, $\tau = \theta_3$ is larger than those for the other two parameters; there appears to be less information in the data about the time constant. Figure 1 depicts a (smoothed) simulated data histogram, along with the fit of the OUP first passage time curve; The fit of the OUP is much better than, for example, the best fitting gamma density, which is commonly fit to ISI histograms.

Next, the parameter estimates for the goldfish cells are given in Table 3.

TABLE 3. *Estimated parameter values for goldfish retinal cells. Standard errors are given in parentheses.*

unit	$\hat{\theta}_1$	$\hat{\theta}_2$	$\hat{\theta}_3$
68	-8.2 (0.6)	-3.5 (.5)	7.1 (1.04)
83	-12.1 (0.3)	-6.3 (0.3)	12.4 (0.9)

The fact that θ_1 and θ_2 are both negative indicates a net excitatory input with limiting mean level, $\mu\tau$, that is above the firing threshold. The standard errors for the time constant are larger than for the other two parameters. Notice that the standard errors here are larger than those for the simulated data sets, while the number of spikes are roughly the same. This is an indication the OUP model may not be the correct model for the ISI data. In Figures 2 and

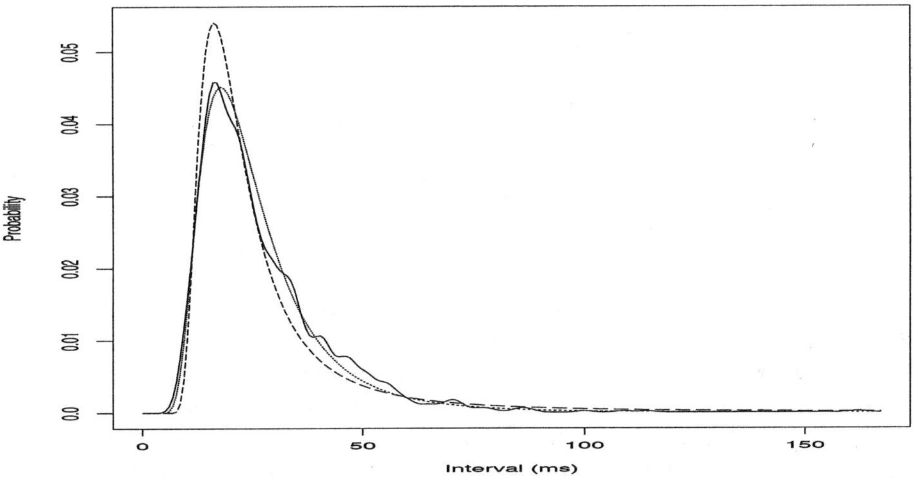

Figure 1: ISI histogram (solid curve) from simulated data with mean firing rate=26.74 and CV=14.70. Fits using on OUP first passage time (dotted curve) and gamma densities (dashed curve).

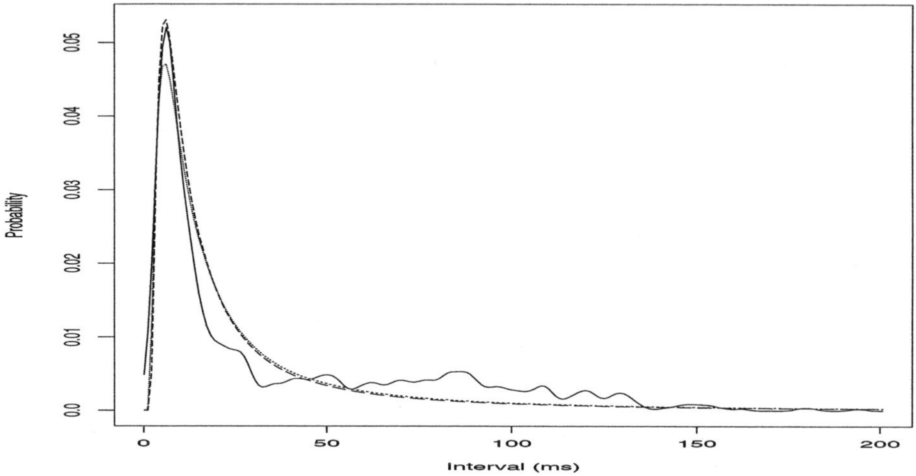

FIgure 2: ISI histogram (solid curve) from goldfish unit 68. Fits using OUP (dotted curve) and gamma (dashed curve) densities.

3, we once again plot the (smoothed) histogram, the OUP fit, and the gamma fit. For unit 68, the gamma density is better than the OUP near the mode, but the overall fit (as measured by the mean squared distance between the two curves) of the OUP is better; and for unit 83, the OUP model is better than the gamma both near the mode and in the tails. However, the plots show that there are modes beyond the 70 msec range that are not accounted for either model.

6 Discussion

In this paper, we have shown the feasibility of the use of a diffusion approximation to Stein's model, using simulated data. The analysis of the goldfish retinal units is less convincing, because of the presence of modes in the range of 50 to 150 msec range. All of the standard curves used to fit ISI histograms (inverse Gaussian, gamma, reciprocal gamma) are all unimodal. The same seems to be true of the OUP first passage time density, although a rigorous proof of that is not yet available. The OUP does have the advantage

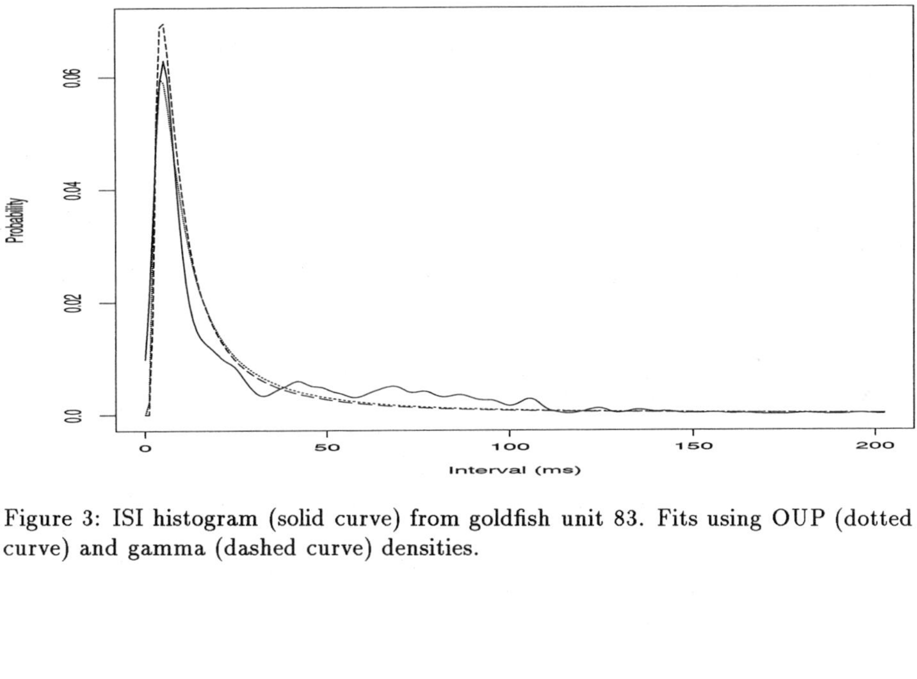

Figure 3: ISI histogram (solid curve) from goldfish unit 83. Fits using OUP (dotted curve) and gamma (dashed curve) densities.

that it is based on an actual physical model, with interpretable parameters. Thus, we plan to generalize th OUP to accomodate those modes. In particular, these modes indicate that the units may effectively have different thresholds at different times, with the threshold governed by a gating mechanism. One promising way of modeling gating is to use mixtures of OUP first passage time densities, with different firing thresholds for the different components of the mixture. We are currently developing the methodology for this more complicated model.

Acknowledgments

I thank Ray Paton for his very generous help in the preparation of this manuscript. This research was supported by the United States Office of Naval Research, grant # N00014-89-J-1496.

References

1. Abramowitz, M. and Stegun, I. A. (1965). *Handbook of Mathematical Functions.* Dover, New York.
2. Barndorff-Nielsen, O., Blaesild, P., and Halgreen, C. (1978). First Hitting Time Models for the Generalized Inverse Gaussian Distribution. *Stochastic Processes and their Applications* 7, 49-54.
3. Churchill, R.V. (1981) *Operational Mathematics.* McGraw-Hill, New Delhi.
4. Gerstein, G., and Mandelbrot, B. (1964) Random Walk Models for the Spike Activity of a Single Neuron. *Biophysical J.* 4, 41-68.
5. Iyengar, S. (1995) Maximum Likelihood Estimation for the Ornstein-Uhlenbeck Process. Technical Report. Statistics Department, University of Pittsburgh. Submitted for publication.
6. Karlin, S. and Taylor, H. (1981) *A Second Course in Stochastic Processes.* Academic Press, New York.
7. Lehmann, E.L. (1983) *Theory of Point Estimation.* Wiley, New York.
8. Levine, M.W. (1992) Modeling the Variability of Firing Rate of Retinal Ganglion Cells. *Mathematical Biosciences.* 112, 225-242.
9. Magnus, W., Oberhettinger, F., and Soni, R.P. (1966) *Formulas and Theorems for the Special Functions of Mathematical Physics.* Third edition. Springer, New York.
10. Ricciardi, L. (1977) *Diffusion Processes and Related Topics in Biology.* Springer. New York.
11. Ricciardi, L. and Sato, S. (1988) First Passage Time Density and Moments of the Ornstein-Uhlenbeck Process. *J. Applied Probability.* 25, 43-57.

12. Stein, R.B. (1965) A Theoretical Analysis of Neuronal Variability. *Biophysical J.* 5, 173-194.
13. Tuckwell, H. (1988) *Introduction to Theoretical Neurobiology.* Volumes 1 and 2. Cambridge University Press, New York.

Dynamical Models of Cellular Systems and Information Processing

"Times flies like an arrow,
fruit flies like bananas".
Marx (Groucho, not Karl)

Complex trajectories from simple feedback circuits

R. THOMAS
Department of Molecular Biology, Université Libre de Bruxelles,
rue des Chevaux 67, B-1640 Rhode-St-Genèse, Belgium

1 Differential systems seen in terms of feedback circuits.

There may have been some ambiguity in the past concerning the precise definition of "interactions" and of "feedback circuits" (should one consider only the regulatory interactions, etc...). Any ambiguity disappears if one simply considers the jacobian matrix of the system and state that variable j exerts a positive (vs negative) action on variable i if element a_{ij} of this matrix has a positive (vs negative) value. Feedback circuits (or more generally unions of disjoint circuits) are identified as sets of non-zero elements of the matrix whose indices i and j are permutations of each other ; in particular, each non-zero diagonal term (a_{ii}) denotes an one-element circuit. A circuit is positive or negative according to the parity of the number of negative elements it comprises.

The typical[a] behavior of simple feedback circuits can be described as follows:

- an one-element circuit (direct autoregulation ; direct autocatalysis if one deals with a positive circuit) generates (in the subspace of the variable considered) a steady state which is attractive or repulsive depending on whether the circuit is negative or positive ;
- a two-element *positive* circuit forces the variables involved in the circuit to choose between two attractors (it generates a saddle point, located on

[a] "Typical" means that the behavior described is found for a wide range of parameter values, such that the interactions which constitute the circuit considered are strong enough and not hindered by other interactions.

a separatrix which divides the space of the variables considered into two attraction basins);

- a two-element *negative* circuit generates a periodic approach to a steady state (a stable focus), but in the presence of an autocatalytic term the focus can be destabilized, thus resulting in a periodic departure from its vicinity ;
- a three-element circuit can generate a saddle-focus which, for a negative circuit, is attractive along one direction and periodically repulsive along a normal surface, and, for a positive circuit, is attractive on a (separatrix) surface and repulsive along a normal direction.

2 Application to the Rössler-type systems.

An admirable (because astonishingly simple) system of differential equations giving rise to deterministic chaos was discovered by Rössler[1]. His system consists of a set of three ordinary differential equations with only one non-linearity. In a form slightly modified by Gaspard and Nicolis[2] in order to have a (convenient) steady state with coordinates (0,0,0), it writes:

$$\begin{aligned} \dot{x} &= \quad - y - z \\ \dot{y} &= x + ay \\ \dot{z} &= bx + xz - cz \end{aligned} \tag{1}$$

(a, b, c are positive coefficients). The jacobian matrix is:

$$\begin{pmatrix} 0 & -1 & -1 \\ 1 & a & 0 \\ b+z & 0 & x-c \end{pmatrix}$$

A relevant aspect of the system is the existence of two[b] unstable steady states (saddle-focuses) ; one is attractive following the z axis and periodically repulsive in x y, the second is periodically attractive in a plane and repulsive along a normal direction[2].

In spite of the remarkable structural simplicity of the system, the exact role of each term in the equations is by no means obvious. With the hope of better understanding this aspect, one may reason in terms of feedback circuits

[b] As forcasted by Nicolis, a single appropriate steady state is sufficient. Using the same type of reasoning as above, I found indeed a variant of this system which generates deterministic chaos with a single steady state (Thomas and Nicolis, in preparation). See also Goldbeter[3] for a more intricate system with a single steady state.

and try to re-build a system which would display the two types of steady states just mentioned.

Using what we know (see the end of section 1) about the properties of feedback circuits, the first steady state can be built as follows. In order to be attractive along the z axis, we need a negative diagonal term in z, in order to be periodic in xy we need a negative circuit in xy, and in order for this focus to be repulsive, we need a diagonal autocatalytic term in x or y. One of a few equivalent possibilities is described by the qualitative jacobian matrix:

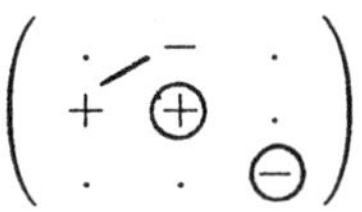

One can easily check[c] that even a linear system of this structure can generate the first type of saddle-focus.

Similarly, in order to generate the second steady state, one can use a negative circuit in xz in order to have a periodic attractivity in plane xz, and a positive circuit of y on itself in order to be repulsive in y; for example:

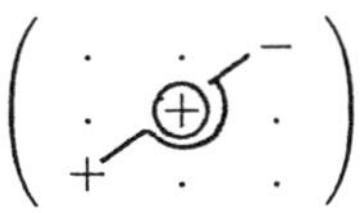

Combining the two matrices, we get:

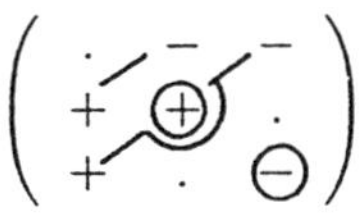

Note that term a_{22} is common to the two matrices and indeed serves two purposes: destabilize the first focus and generate repulsivity along the y axis in the second focus. As a matter of fact, this term represents the only positive circuit in the system, and it is thus responsible for the existence of two distinct steady states.

Comparing this matrix with the jacobian matrix of the Rössler system lead me to ask whether in Rössler's equations there is really a *structural* need

[c]Take, for example, the system
$\dot{x} = -2y$
$\dot{y} = 2x + 0.5y$
$\dot{z} = -10z$
The roots of the characteristic equation are -10 and +0.25 +/- 1.98 i. Starting, say, from (0.1, 0.1, 2), the trajectory has the expected shape.

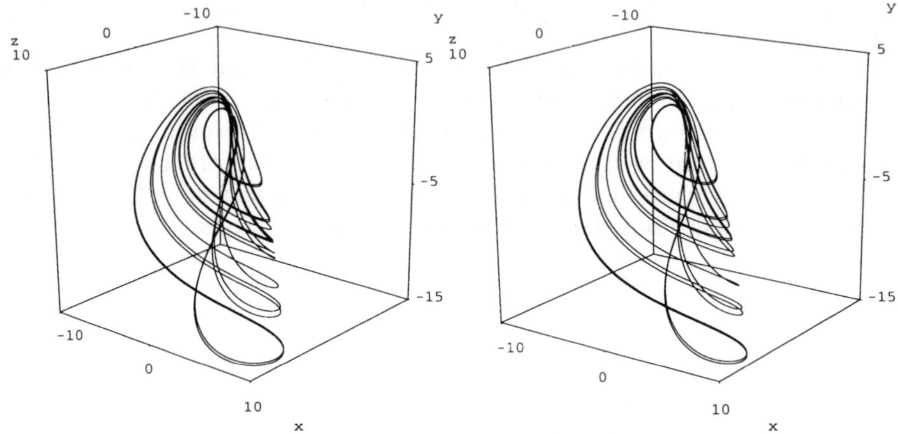

Figure 1: System (2) ($\dot{x} = -y - z, \quad \dot{y} = x + ay, \quad \dot{z} = bx^2 - cz$) with a = 0.385 , b = 0.3, c = 2. Initial state was (.1, .1, .1) but integration was run (step: 0.01) from time 0 to time 100 without recording the trajectory (in order to eliminate transitories), then from time 100 to 200. The chaotic character was checked by determining the Lyapunov exponents.

for an xz term, or whether it is there simply because there must be at least one non-linear term in the differential equations in order to have a complex behavior.

In order to investigate this point, I checked what happens when one deletes the xz term in Rössler's equation and renders another term non-linear. For example:

$$\begin{aligned} \dot{x} &= \quad\quad - y - z \\ \dot{y} &= x + ay \\ \dot{z} &= bx^2 \quad\quad - cz \end{aligned} \tag{2}$$

For a wide range of values of the three parameters, this system displays a deterministic chaotic behavior (Fig. 1) . As a matter of fact, any of a number of nonlinear functions can be used: x^2, x^3, x/(1+x), sin x, tg x, tanh x, ... Note that some of them yield three unstable steady states and two symmetrical chaotic attractors.

Also, the non-linearity can be located elsewhere (for example in position a_{13}) although not anywhere, for reasons which are understood. As expected from the properties of the feedback circuits, one can also permute the signs of the terms forming the negative loops (thus inverting the way of rotation) without loosing the characteristic behavior of the system[d].

The interest of the above is that we now can build systems of an extremely simple structure, which generate deterministic chaos or multiple periodicity, and in which the exact role of each term is understood.

Reference

1. O.E. Rössler, *Phys. Letters* **57A** , 397 (1976)
2. P. Gaspard and G. Nicolis, *J. Stat. Phys.* **31** , 499 (1983)
3. A. Goldbeter, *Biochemical Oscillations and Cellular Rythms, Cambridge University Press (1995)*

Acknowledgments

I wish to thank Professor G. Nicolis for stimulating discussions, Rob De Boer for kindly providing his GRIND Program and Pascale Gyserman who provided me with her program giving Lyapunov exponents and kindly helped me using it.

[d]As can be inferred from the last part of section 1, another way to generate the two complementary types of saddle-foci would consist of using two three-element circuits, one positive, one negative, as, for example:

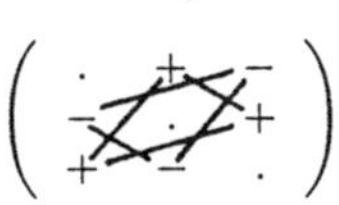

(with one non-linear term in the set of differential equations).

In the absence of diagonal terms, one deals with conservative systems, which display a rich variety of trajectories with multiple periodicity. It remains to see whether deterministic chaos will be found in the corresponding dissipative systems.

RECONSTRUCTION OF EXCITABLE TISSUE PHYSIOLOGY AND PATHOPHYSIOLOGY FROM CELLULAR MODELS

M.R.BOYETT, A.V.HOLDEN, H.ZHANG
Department of Physiology, University of Leeds, Leeds LS2 9JT

1 Introduction

Cardiac tissue is a functional syncytium composed of coupled cells. Thus cardiac tissue can be idealised as a spatially extended, bounded system, modeled by a network of coupled ordinary differential equations, or a partial differential equation.

The anatomical and histological structure of the heart together with the obvious anisotropy in propagation and known electrophysiology of cardiac cells makes accurate modelling computationally prohibitive. However, simple idealizations can provide biophysically realistic results. Here we reconstruct some tissue behaviours from models of cellular excitability. The models of cellular excitability describe membrane currents in terms of conductances, active pumps and intra-and extracellular concentrations, and represent a summary of extensive and on-going series of voltage-clamp investigations on cardiac tissue - see [1]. These models are in the form of stiff, high order nonlinear differential equations, and so require numerical integrations and numerical methods for characterising their bifurcation structure. Cardiac cells are coupled by ohmic conductances, and so the electrical activity of cardiac tissue, which is contractile, anisotropic, heterogenous and has a complicated geometry, can be idealised by systems of coupled ordinary differential equations or partial differential equations. Accurate simulation of propagating activity in cardiac tissue is a computationally intensive problem that is approached using a variety of mathematical and computational methods [2,3].

In reconstructing the electrical behaviour of the heart, there is a danger of being lost in biophysical, physiological, anatomical or clinical details. These details are necessary in the context of cardiac electrophysiology, but in the context of theoretical integrative physiology the heart is an organ whose function is begining to be understood in terms of its molecular biology (contractile proteins and membrane channels), the coupling of membrane potential to subcellular and cellular dynamics, the coupling of cells into a tissue and the coupling of tissues into an organ. Thus the heart is a system where the relations between structural, functional and dynamical hierarchies can be explored in a concrete, quantitative way. Such detailed case studies are necessary to provide firm support for ideas about integrative physiological functions [4]

2 Coupled ordinary differential equations.

The pacemaker of the mammalian heart, the sinoatrial node, is located in the wall of the right atrium. In the rabbit, the action potential is first initiated in the centre of the sinoatrial node and the action potential preferentially propagates in an oblique cranial direction to the atrial muscle of the crista terminalis. Atrial muscle, unlike the sinoatrial node, does not exhibit pacemaker activity and it must be "driven" by the sinoatrial node. The interaction of the sinoatrial node and the atrial muscle is not just one way - there is evidence that the non-pacemaking atrial muscle affects the pacemaker activity of the sinoatrial node. Kodama & Boyett [5] showed the rate of action potentials generated by small preparations of the periphery of the node is faster than that of small preparations from the centre of the node. It is, therefore, paradoxical that the centre and not the periphery of the sinoatrial node is the leading pacemaker site in the intact heart. This paradox can be explained if in the intact heart the pacemaker activity of the periphery of the node is suppressed by the surrounding atrial muscle. Confirmation of this was obtained by Kirchhof *et al.* [6], who showed that when the atrial muscle was cut away from the sinoatrial node, the pacemaker site shifted from the centre to the periphery of the node and there was an increase in the rate of action potentials. Winslow *et al* [7] using a massively parallel computer have modelled the sinoatrial node as a two dimensional cell network. When the sinoatrial node was not connected to an atrial cell network, the action potential was initiated in the periphery of the node, but when it was connected to an atrial cell network the leading pacemaker site was shifted from the periphery towards the centre of the node.

Parasymapthetic nerve fibers in the vagus nerve innervate both the sinoatrial node and the atrial muscle. Stimulation of the parasympathetic nerve fibers causes the release of acetylcholine (ACh) from the nerve endings. The principal action of ACh is to slow the heart rate by decreasing the rate of action potentials generated by the sinoatrial node; we propose it also has a second effect, via the atrial modulation of the pacemaker rate.

We first consider two coupled isopotential cells, one having the properties of an autoryhthmic sinoatrial cell, one the properties of a quiescient atrial cell. Each cell is represented by an ordinary differential equation. As the coupling strength between these two cells is varied, we measure how the cycle length changes. To simulate the effects of vagal stimulation, we include the acetylcholine current in the equations.

2.1 Mathematical model

Each cell takes the form

$$C\frac{dV}{dt} = -I_{tot} \tag{1}$$

where C is the membrane capacity (in μFcm^{-2}), V the membrane potential (in mV), t the time (in s) and I_{tot} the total current (in nA):

$$I_{tot} = i_f + i_k + i_{k1} + i_{to} + i_{b,Na} + i_{b,Ca} + i_p + i_{NaCa} + i_{Na} + i_{Ca,f} + i_{Ca,s} + i_{pulse} + I_{Ach}. \tag{2}$$

For the meaning of the components of I_{tot} and other equations and the related parameters, see [8].

I_{ACh} is described by the Egan-Noble equation [8,9]:

$$I_{ACh} = I_{K,AChm}([K]_i - [K]_o \exp(-VF/RT))/140. \tag{3}$$

2.2 Coupled sinoatrial node cell and atrial cell

The equations for a sinoatrial node cell coupled to an atrial cell are:

$$C_s\frac{dV_s}{dt} = -I_{tot_s} + d \times I_j$$

$$C_a\frac{dV_a}{dt} = -I_{tot_a} - I_j \tag{4}$$

$$I_j = g(V_a - V_s)$$

The coupling between the sinoatrial node and the atrial cell is through the junction current I_j. g is the junctional conductance (in nS) specifying the coupling strength between these two cells. The sub-script s denotes the sinoatrial node, and sub-script a notes the atrial cell. Considering the difference between C_s and C_a, we scale with a parameter d in the equation for sinoatrial node cell. In numerical experiments, we use the Noble-Noble [10] equations as the representation of the ionic mechanism for a sinoatrial cardiac cell, while use the Earm-Noble equations [11] for an atrial cell. We take $d = 50$.

Figure 1(a, b) shows numerical solutions at different values of $I_{K,AChms}$, with the coupling strength fixed: the effect of $I_{K,AChms}$ is to increase the cycle length. Figure 1(c) plots the cycle length of the oscillatory solution

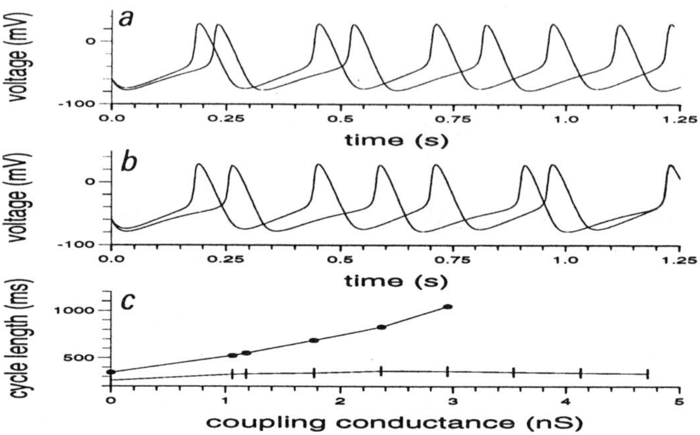

Figure 1. The action potential time series for sinoatrial node with different values of $I_{K,Achms}$ while keep g fixed at 2.35 nS. $I_{K,Achma} = 0$ for all simulations. (a) $I_{K,Achms} = 2.0\ nA$ (b) $I_{K,Achms} = 3.0\ nA$. (c) The cycle length of the sinoatrial cell as a function of coupling conductance with different values for $I_{k,Achms}.I_{K,Achma} = 0$ for all simulation. | for $I_{K,Achms} = 0\ nA$, • for $I_{k,Achms} = 3nA$.

for the sinoatrial cell as a function of coupling strength for different values of $I_{K,AChms}$, while keeping $I_{K,AChma} = 0$. The junction conductance g was varied from 0. to 4.7 nS. The cycle length increases with increase in $I_{K,AChms}$ at all coupling strengths.

2.3 Effect of ACh on the sinoatrial cell.

We simulate how vagal stimulation of the sinoatrial node influences the rate of the activity of the sinoatrial node by increasing $I_{K,AChms}$ in equation (1) while leaving $I_{K,AChma} = 0$.

2.4 Effect of ACh on the atrial cell

Figure 2 illustrates how $I_{K,AChma}$ changes the waveform of numerical solutions for the atrial cell, that have been driven by the endogenous activity of the sinoatrial cell, simulating direct vagal effects on atrial muscle. In addition to the decrease in duration, there is an increase in cycle length of the sinoatrial cell at all values of coupling strength, as shown in fig. 2(b). With an increase in $I_{K,AChma}$ in the equations for the atrial cell, a larger coupling strength is required for the sinoatrial cell to excite the atrial cell. For a fixed coupling strength, the larger the acetylcholine current $I_{K,AChma}$ on the atrial cell, the longer the cycle length for the sinoatrial node cell.

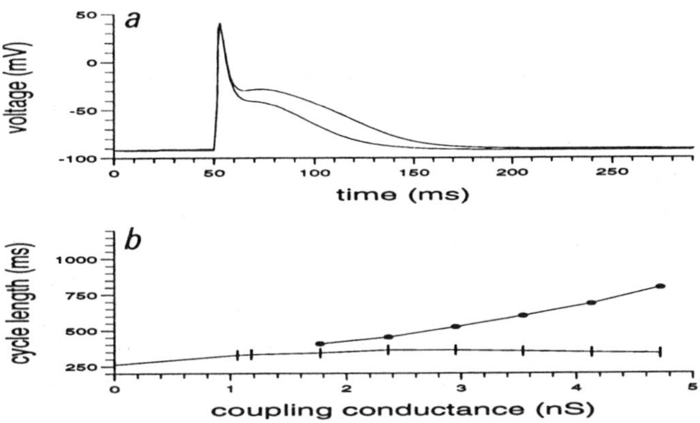

Figure 2. (a). The action potential for an atrial cell with different values of $I_{K,Achma}$ in the Earm-Noble equations. $1 : I_{K,Achma} = 0nA, 2 : I_{k,Achma} = 0.1nA$. The acetylcholine current will result in a hyperpolarization of the membrane and a shortening of the action potential. (b) The cycle length of the activity of the sinoatrial cell as a function of coupling strength with different values for $I_{K,Achma}$. $I_{K,Achms} = 0$ for all simulations. | for $I_{K,Achma} = 0$, • for $I_{K,Achma} = 0.08nA$.

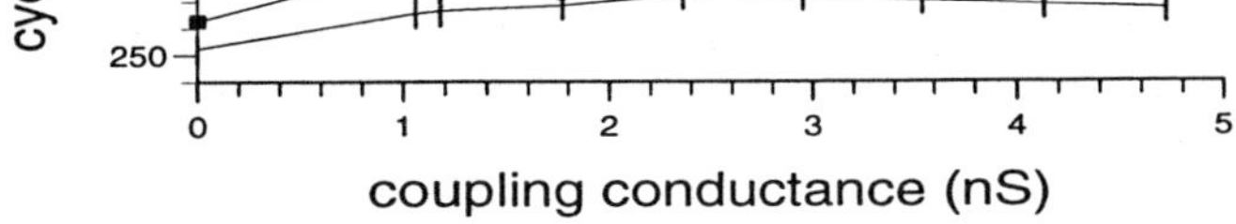

Figure 3. The cycle length of the activity of the sinoatrial cell as a function of coupling conductance with different values for $I_{K,Achms}$ and $I_{K,Achma}$: | for $I_{K,Achms} = 0nA, I_{K,Achma} = 0nA$; • for $I_{K,Achms} = 0nA, I_{K,Achma} = 0.02nA$; ■ for $I_{K,Achms} = 2.0nA, I_{K,Achms} = 0.0nA$; • for $I_{k,Achms} = 2.0nA, I_{k,Achms} = 0.02$ nA.

2.5 Effect of ACh on sinoatrial node cell and atrial cell

When an $I_{K,AChm}$ is added to either the sinoatrial cell or the atrial cell in equation (1), there is an increase in cycle length. Now we study how the cycle length changes with coupling strength if we add an I_{ACh} to both the sinoatrial cell and the atrial cell in equation (1).

Figure 3 plots the cycle length as a function of the junction conductance for different sets of $I_{K,AChms}$ and $I_{K,AChma}$: there is nonlinear interaction between the changes in cycle length produced by $I_{K,AChms}$ and $I_{K,AChma}$. This nonlinear interaction is a result of the nonlinearities of the constant field term in (4) and the nonlinear relation between cycle length and applied current: it is not analogous to the nonlinear interaction between synaptic potentials, that is due to the shunting action of conductance changes.

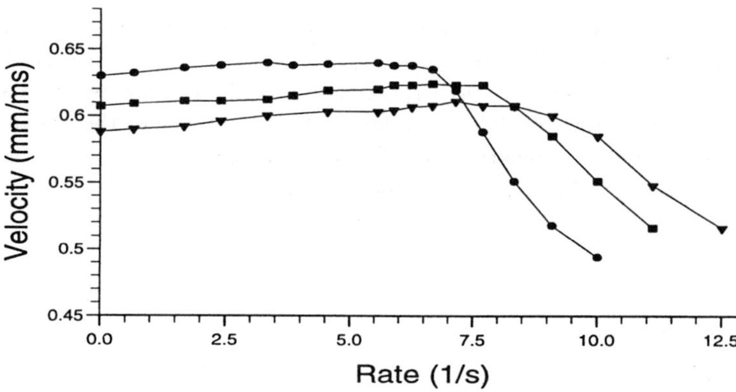

Figure 4: The estimated conduction velocity of the stable travelling wave solution in a $1-D$ model of atrial fiber against the stimulation rate $(1/s)$ with the effect of vagal stimulation. • : $I_{KAChm} = 0nA$; ■ : $I_{K,AChm} = 0.05nA$; ▶ : $I_{K,AChm} = 0.1$ nA. At lower stimulation rate, the effect of vagal stimulation will decrease the conduction velocity, at higher stimulation rate, the effect of vagal stimulation will increase the conduction velocity.

3 One dimensional phenomena

3.1 Mathematical model

A $1-D$ excitable medium model of atrial tissue may be represented by a parabolic partial differential equation of reaction diffusion type based on the Earm-Hilgemann-Noble equations for a single atrial cell [11, 12]:

$$\partial V/\partial t = -I_{tot}(V,\mathbf{u})/C_m + (D\nabla^2 V) \qquad (6)$$
$$\partial \mathbf{u}/\partial t = f(V,\mathbf{u})$$

where $\mathbf{u}$ is a vector of the activation and inactivation gating variables and concentrations that act on $I_{tot}(V,\mathbf{u})$. D is the diffusion coefficient, which is set to $1.25cm^2s^{-1}$ to give the conduction velocity of a solitary planar wave about $0.6ms^{-1}$.

The partial differential equation is solved by the explicit Euler method with a space step $\Delta x = 0.32mm$ and a time step $\Delta t = 0.1$ ms. The space constant of the model is $1mm$, and the space step of 0.32 corresponds to the long axis of about 4 cells, and is sufficiently small for stability of the numerical solution [13].

3.2 The rate dependence of conduction velocity and the effects of ACh

Numerical experiments were carried on a $1-D$ atrial fiber modelled by equation (3.1), with fiber length is $32mm$, discretised as 100 nodes, and with non-flux boundary conditions. The fiber is stimulated by a sequence of periodic pulses at one end with different frequencies.

Figure 4 plots the estimated conduction velocity of a solitary wave along a $1-D$ atrial fiber against the stimulation rate with different values of $I_{K,AChm}$. With $I_{K,AChm} = 0$, when the simulation rate is below 6 beats/s (period over $167ms$), the time interval between two successive propagation wave is long enough for the atrial cells to recover from a previous excitation. The conduction velocity is almost constant for stimulation rate below 6/s. However, further increase of the stimulation rate (beyond $6/s$) causes the reexcitation of atrial cell before it is fully recovered, and so the conduction velocity starts to decrease. When the stimulation rate is further increased beyond $10/s$ ($100ms$ in period) unstable travelling waves will be formed . Above $14/s$ (about $71ms$ in period), one to one response of the excitation to stimulation fails. Re-entrant excitation in a $2-D$ model of atrial tissue using Earm-Hilgemann-Noble model has period $75ms$ [13, 14], which is close to the period of the unstable region for propagation in a $1-D$ atrial fiber.

3.3 Vulnerability in 1-D model of atrial tissue

A premature stimulation applied in a specific time window to the refractory tail of a propagating wave may evoke an excitation wave that travels in the retrograde direction and is blocked in the anterograde direction. This time window is known as the vulnerable window. In a ring, such a unidirectional block in conduction may lead to reentrant excitation. Reentrant excitation is associated with cardiac arrhythmias [15].

The width of the vulnerable window (TW) may be used as an index of its vulnerability [16]. Here we study how ACh changes the width of the vulnerable window and shifts the vulnerable window in time.

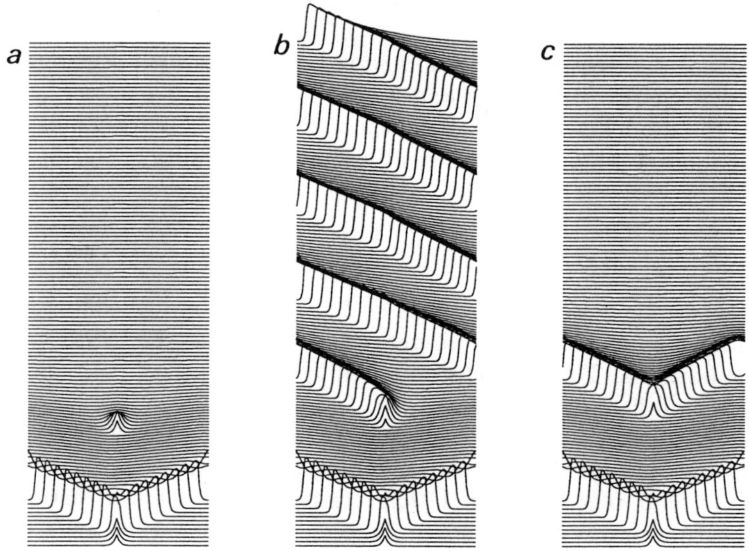

Figure 5 : *A* premature stimulation falling in *TW* may generate disparity in membrane refractoriness, which produces unidirectional block leading to a sustained reentrant excitation in a ring model of atrial tissue. The premature stimulation is applied to node 80 after a time delay Δt after the wave front reaches it. $a)\Delta t = 80ms; b)\Delta t = 88ms; c)\Delta t = 90$ ms.

Numerical experiments are carried out in a $1-D$ ring model (i.e., a $1-D$ cable with periodic boundary conditions) of atrial tissue of circumference $48mm$, discretised by 150 nodes. A conditioning excitation wave is initiated by applying a stimulus at node 0. The initiated waves will propagate towards each other and annhilate on collision. With a stimulus applied at node 80, a time delay Δt after the conditioning excitation wavefront arriving the node 80, it may generate a unidirectional block leading to a sustained reentrant excitation wave if the premature stimulation falls in a critical time window $(T1, T2)$. The time window $(T1, T2)$ is called the vulnerable window of the tissue, and $T2 - T1$ is defined as the width of the vulnerable window. In our numerical experiments the test stimulation is a pulse with duration $0.6ms$

and amplitude of -1.3 nA. This test stimulation is analogous to a premature excitation in rhythmically active cardiac tissue.

To see the effect of the vagal stimulation on TW, we change the value of $I_{K,AChm}$. The result is shown in figure 6(a) where we can see that by increasing $I_{K,AChm}$ the vulnerability of the atrial fiber will decrease, as measured by the decrease of the TW.

ACh will also shift the vulnerable window in time. Increasing $I_{K,AChm}$, the vulnerable window occurs earlier in time. Figure 6(b) plots $\Delta T1$ against the amplitude of the ACh activated K^+ current. Thus ACh decreases the width of the vulnerable window , but advances it in time.

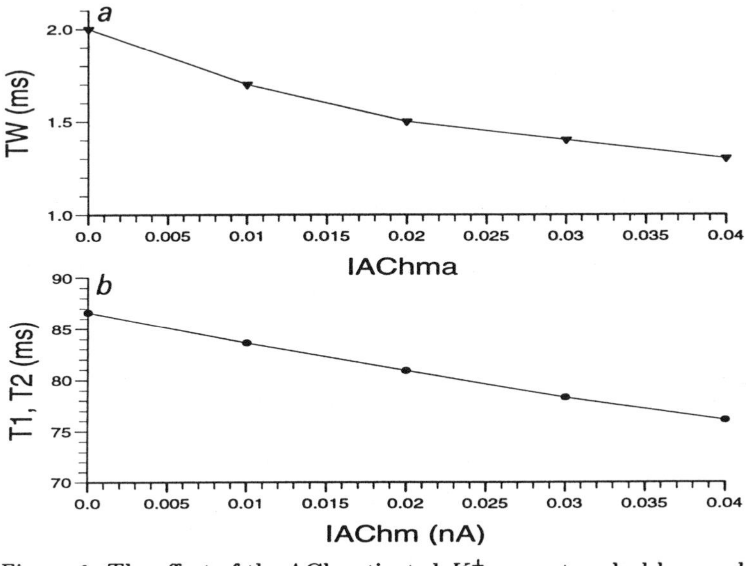

Figure 6: The effect of the ACh activated K^+ current evoked by vagal stimulation on the vulnerable window of the atrial fiber to reentry. a) With $I_{K,AChm} = 0.04nA, TW$ is reduced by 35% . b) With $I_{K,Achm} = 0.04nA$, the vulnerable window is shifted $10.5ms$ in advance.

4 Two dimensional phenomena

Both the generation and modulation of normal pacemaker activity and the generation of ectopic beats are essentially one-dimensional problems in a radial symmetric geometry, in which a circular idealised sino-atrial node is surrounded by an annulus of atrial tissue. If radial symmetry is not present, as in an anatomically accurate model of the sino-atrial node and atrium, or in re-entrant spiral wave activity, a simplification to a one-dimensional problem is not possible. An isolated spiral wave models a re-entrant arrhythmia of the leading circle type [17], in which the same wave of activity propagates around a region of functional block in a homogeneous sheet of tissue.

We have incorporated the EHN model into a planar two dimensional excitable medium, and obtained the characteristics of atrial spiral waves [13]. A two-dimensional excitable medium may be represented by a parabolic partial differential equation of the reaction-diffusion type. Such a system, with appropriate parameter values and a judicious choice of initial conditions can support a single-armed, spiral wave solution that (depending on the parameter values) rotates rigidly around a circular core, or meanders around a cycloidal, linear or irregular trajectory. The initiation and motion of spiral waves, and their relevance to cardiac arrhythmias, is reviewed in [18, 19].

The equations for the membrane current need to be scaled for a two-dimensional homogeneous medium. This could be done by using the cell size and a cylindrical geometry, to obtain the membrane current density at a point and to combine appropriate values for intracellular resistance and intercellular coupling [7, 20] to obtain the diffusion coefficient. This would provide a model of a single layer of cells. A simpler method is to choose the diffusion coefficient D indirectly, from the value that gives an appropriate propagation velocity for a solitary plane wave.

The propagation of a wavefront depends on its curvature, and the time since the preceeding wave; in cardiac tissue the propagation velocity also depends on fibre orientation. Estimates of conduction velocity for the rabbit atrium are $0.3 - 0.8ms^{-1}$; we take $0.62ms^{-1}$ as the conduction velocity of a plane wave in our partial differential equation model. In the coupled ordinary differential equation lattice model of Winslow *et al.* [7, 20] this corresponds to a cell to cell coupling of 1μS. This gives the diffusion coefficient $D = 125mm^2ms^{-1}$. Computations with the atrial model show an aging phenomenon [13]: the tip follows an approximately circular trajectory of $1.2mm$ radius for the first 15 rotations, and then changes to a cyloidal trajectory that can be enclosed by a circle of radius 4 mm.

The period of the computed spiral wave compares with measurements of $85ms$ for the initial cycle length in reentrant tachycardia in the rabbit atrium [17]. Thus the EHN model for a single atrial cell incorporated in an excitable medium approximates the period of re-entry in the rabbit atrium: a quantitative characteristic of an organ pathophysiology has been derived from cellular dynamics, and a behaviour of the atrium can be understood in terms of the behaviour of a two-dimensional excitable medium.

Atrial re-entry as atrial flutter, is not lethal, and is usually self-terminating. ventricular re-entry, as monomorphic ventricular tacycardia, is a more serious clinical problem and is often lethal. Figure 7 illustrates a spiral wave solution for membrane potential for a ventricular model, in which excitability is described by the Oxsoft 3.8 standard guinea-pig ventricle cell model, with a diffusion coefficient of $1.25cm^2s^{-1}$and space and time steps of $0.2mm$ and $0.01ms$. The spiral has a period of $105 - 110ms$, the change in period being due to rotation around a precessing five -lobed orbit, shown in Figure 8.

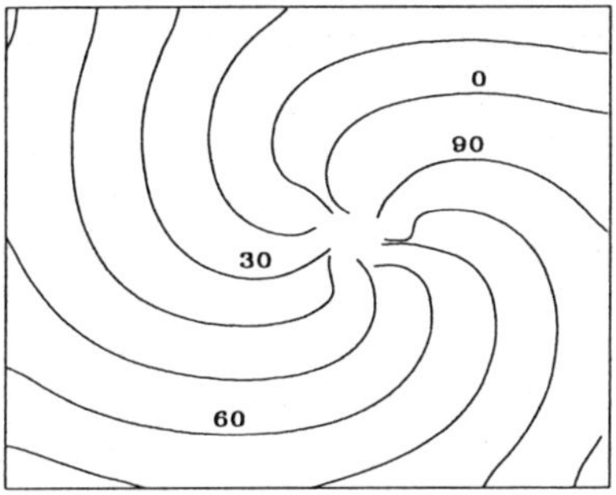

Figure 7: A spiral wave in a $2-D$ ventricular model that has developed after breaking a plane wave half way across the medium and resetting half of the medium to its resting state

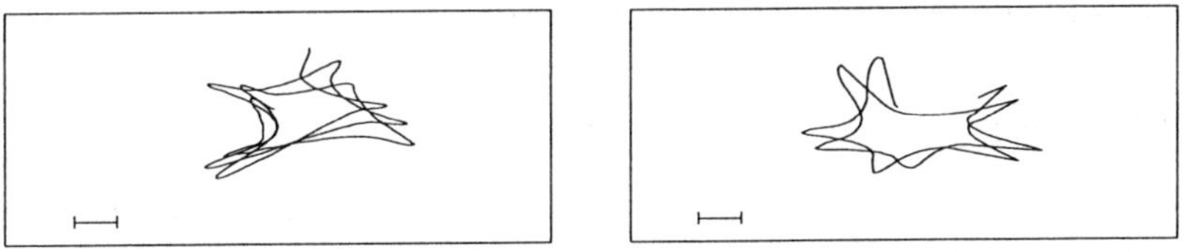

Figure 8 Trajectory of tip of spiral of figure 7 showing precessing orbit. Space scale 1 mm

It would be desirable to eliminate the re-entrant wave if it were to be in atrial or ventricular tissue. A spiral wave can be forced to move by a spatially uniform, time periodic perturbation of appropriate frequency [22]. Resonant drift in the location of a spiral occurs when the frequency of perturbation is the same as the frequency of rotation of the spiral; in effect, each perturbation is applied at the same phase of rotation and so has the same effect on the spiral; since the spiral is stable this effect can only be a displacement in tip position and/or a change in phase of the spiral. This general property of excitable media has been exploited as a means of moving spiral waves to the boundaries of the medium. A boundary can repel a resonantly drifting spiral, and this reflection of the spiral can be overcome by appropriate timing of the repetitive perturbation, and so the spiral can be driven onto the inexcitable boundaries and be extinguished [23]. In principle, resonant drift under feedback control could provide a means of eliminating re-entrant activity in cardiac tissue [24]. This will be practical only if any re-entry is eliminated within a reasonable time, say less than 30s.

Numerical simulations of resonant drift with the atrial model have shown that in all cases of this resonant drift the tip of the spiral approaches a boundary of the medium; as it approaches its averaged trajectory bends, and the spiral is extinguished when the tip is driven to the boundary [14]. The impermeable boundaries model the junction of excitable atrial tissue with inexcitable tissue, say at the junction with the great veins. For the $4 \times 4cm$ medium, the spiral wave was extinguished within $10s$ by repetitive stimuli all less than 10% the threshold for a single defibrillating shock of the same duration.

Cardiac tissue is clearly not isotropic, and the conduction velocity varies 5-fold with orientation, being faster along the axis of atrial fibres, and so a reentrant spiral will not have the shape of the involute of a circle. The effects of this anisotropy, and the possibility of laminar or Y shaped cores, on resonant drift have not been investigated. Further, discretely localised inhomogeneities in the medium, corresponding to say fibrous tissue, may trap or pin spiral waves; this has been seen in thin slices of ventricular tissue [24] and in simple, FitzHugh-Nagumo two dimensional models [25].

5 Conclusions

These computations illustrate that simple coupling of excitable cells in one and two dimensional models leads to the emergence of qualitatively new types of behaviour : travelling waves, vulnerablity to unidirectional conduction block, and re-entrant waves and their control by resonant drift. These behaviours provide a quantitative explanation of some aspects of cardiac function and dysfunction, and provide an example of how physiological functions emerge in organised, spatially extended systems. A tissue physiology is in the process of being constructed from a cellular physiology, which is reaching down into intracellular and molecular biophysics as transmembrane fluxes are coupled with fluxes into intracellular calcium compartments and stores.

Acknowledgements

This work was supported by the Wellcome Trust (042352, 044365).

References

1. Noble, D., *Chaos solitons and fractals* **5** 321 (1995)
2. Holden, A.V., Poole, M.J. and Tucker, J.V., *Chaos solitons and fractals* **5** 691 (1995)
3. Panfilov, A.V. and Holden, A.V. (eds) *The computational biology of the heart.* John Wiley Chichester 1996.
4. Chauvet, G. *Traité de Physiologie Théorique* vols 1,2,3 Masson, Paris (1987,1990); *Theoretical systems in biology. Hierarchical and functional integration* Vol 1,2 Pergamon, Oxford 1995
5. Kodama, I., Boyett, M.R., *Pflugers Arc.* **404**, 214 (1985)
6. Kirchof, C.J.H.J., Bonke, F.I.M., Allessie, M.A. and Lammers, W.J.E.P., *Pflugers Arch.* **410**, 198 (1987)
7. Winslow, R.L., Kimball, A., Varghese, A., Noble, D., *Physica–D***64**, 281(1993)
8. Noble, D., Oxsoft HEART Version 3.8 manual. Oxsoft: Oxford 1990
9. Eagan, T.M., Noble, S.J., *Proc. R. Soc.* **B230**, 315(1987)
10. Noble, D., Noble, S.J., *Proc. R. Soc.* **B222**, 295(1984)
11. Earm, Y.E., Noble, D., *Proc Roy Soc Lond B* : **240**83(1990)
12. Hilgermann, D.W., Noble, D., *Proc Roy Soc Lond B***230**, 163(1987)
13. Holden, A.V., Zhang, H., *J. Theor. Biol.* **175**, 545 (1995).
14. Biktashev, V.N., Holden, A.V., *Proc. Roy. Soc.* **B260**, 211(1995).
15. Zipes, D.P., Jalife, J., eds., *Cardiac Electrophysiology: From Cell to Bedside.* W.B.Sanders, Philadelphia (1995).
16. Starmer, C.F., Biktashev, V. N., Romashko, D.N., Stepanov, M.R., Makarova, O.N., Krinsky, V. I., *Biophysical Journal* **65** 1775 (1993).
17. Allessie, M.A., Bonke, F.I.M., Schopman, F.J.G., **33** 54-62(1973).
18. Holden, A.V., M.S.El Naschie (eds.) (1995) *Chaos, soliton and fractals* **5** 1995.
19. Holden, A.V., Markus, M. & Othmer, H.G. ,eds. (1991) *Nonlinear wave processes in excitable media.* Plenum Press, New York
20. Winslow, R.L., Varghese, A., Noble, D., Adlakha, C., Hoythya, A., *Proc Roy Soc Lon B* : **25**455(1993)
21. Allessie, M.A., Bonke, F.I.M., Schopman, F.J.G., *Circ Res* **39** 168 (1976).
22. Biktashev, V.N., Holden, A.V., *Physics Lett A* **181**, 216 (1993).
23. Biktashev, V.N., Holden, A.V., *J theor Biol* **169** 101-112 (1995)
24. Davidenko, J.M., Pertsov, A.M., Salomonsz, R., Baxter, W., Jalife, J., *Nature* **335**, 349-351 (1992).
25. Vinson, M., Pertsov, A., Jalife, J., *Physica* **72D**, 119 (1994).

ROLES OF THE ION CHANNELS IN REENTRANT CARDIAC ARRHYTHMIA

TERESA REE CHAY
Department of Biological Sciences
University of Pittsburgh
Pittsburgh, Pennsylvania 15260 USA

1 Introduction

The Cardiac Arrhythmia Suppression Trial Team (CAST)[1] has reported that some Class IC drugs (*e.g.*, flecainide and encainide) can cause sudden cardiac death to patients with a history of myocardial infarction. To further investigate the effect of antiarrhythmic drugs, Frame's group used an "adjustable" canine atrial tricuspid ring which was cut and then reconnected electrically by a sensor.[2,3] This sensor senses activation on one side of the cut and paces the other side after an adjustable delay. When the concentration of flecainide was increased i) the tachycardia cycle length was increased more than the effective refractory period, ii) reentry was sustained for a longer period of time (proarrhythmic), and iii) the amplitude of cycle length oscillation was decreased. When d-sotalol was introduced, they found that the action potential duration (APD) increased more than the cycle length and stable reentry became non-sustained (i.e., antiarrhythmic). In the transition region between non-sustained and sustained reentries, cycle length became oscillatory, and the amplitude of oscillation was longer for the ring treated with the drug.

Recently, Chay and Lee[4] have shown that development of reentry and its relation to ectopic foci can be explained theoretically by a bifurcation analysis. Their approach provides remarkable insight into the complex interaction between membrane excitability and conduction properties in the heart. An interesting problem that may be examined with this approach is to find a theoretical explanation for the pro- and anti-arrhythmic effects of the "channel" drugs. Indeed, the bifurcation analysis will allow us to see graphically the roles of the key channels in controlling reentrant arrhythmia in quick and convenient ways.

In this paper, our previous work on smaller ring sizes is extended to the more realistic ring sizes of Frame et al. With this ring model, we then study the effect of Na^+, Ca^{2+} and K^+ channel blockers on reentry and cycle length oscillations. There is no pure drug which can target only one type of the ion channels in the plasma membrane of the cardiac cells. Thus, experimental evidence on how Classes I, II, III, and IV drugs affect reentrant arrhythmia gives no clear insight into the roles of the key channels in reentrant arrhythmia. The motivation of this study is to clarify their roles theoretically. The Beeler-Reuter model[5] is a simple yet a realistic model which mimics the ventricular action potential very closely, and we use this model for our analysis. The synopsis of this work has appeared elsewhere.[6,7]

2 Methods

We use a cable model, where each segment of the cable is represented by a simplified Beeler-Reuter model.[4,8] This model contains five types of current — the sodium current I_{Na}, the calcium current I_{si}, the delayed rectifying K^+ current I_{x1}, the inward rectifying time-independent K^+ current I_{k1}, and the background Na^+ current I_{Nab}. Single cells were arranged in a one-dimensional ring and connected electrically by a coupling constant (g_{gap}). The values of the model's basic parameters are as follows: g_{Na} = 15 mS/cm^2, g_{Nab} = 0.003 mS/cm^2, g_{x1} = 1.0 mS/cm^2, g_{k1} = 1.0 mS/cm^2, g_{si} = 0.09 mS/cm^2, V_{Na} = 40 mV, V_{si} = 80 mV, V_K= -94 mV, g_{gap} = 1 mS/cm^2, C_m = 1 μF/cm^2. Here, g_{x1} and g_{k1} are the multipliers of the delayed rectifying and the inward rectifying potassium currents in the Beeler-Reuter model. The value of g_{gap} is taken to be 1 mS/cm^2, which is about the coupling constant of the cardiac cells.

In the case of the dynamic simulations, our algorithm makes use of the methods of Rush and Larson.[9] In the case of a bifurcation analysis, we used AUTO developed by Doedel.[10] A detailed procedure on how to solve a cable equation and how to use AUTO for the bifurcation analysis in the ring are described in Chay.[11]

3 Bifurcation Analysis

3.1 The Critical Ring Size Which Divides Unstable and Stable Reentry.

The critical ring size (CRS) is defined as the ring size which divides stable reentry from unstable one.[6] Figure 1 shows how the four types of ion channels influence the CRS. Here, the ordinate is the conduction velocity (CV), which is defined as the inverse of the time (in msec) taken for the action potential (AP) to travel from one segment of the ring to the next. The abscissa, on the other hand, is N which is the ring size L divided by the length of a cell Δx. Note that each type of the channel blockers affects the CRS in a different manner.

Along the horizontal portion of the curves, the impulse travels with a constant conduction velocity and thus reentry is stable (see the dynamic representation of stable reentry shown in Fig. 4A). In the region where the curve starts to bend, the action potential encounters a substantial fraction of refractory cells not yet recovered from the previous reentrant wave. In this region, APD and CV oscillate with a complex pattern (see Fig. 4B for oscillatory APD and CV). As the curves become steeper, reentry becomes unstained (Figs. 4C and D for non-sustained reentry).

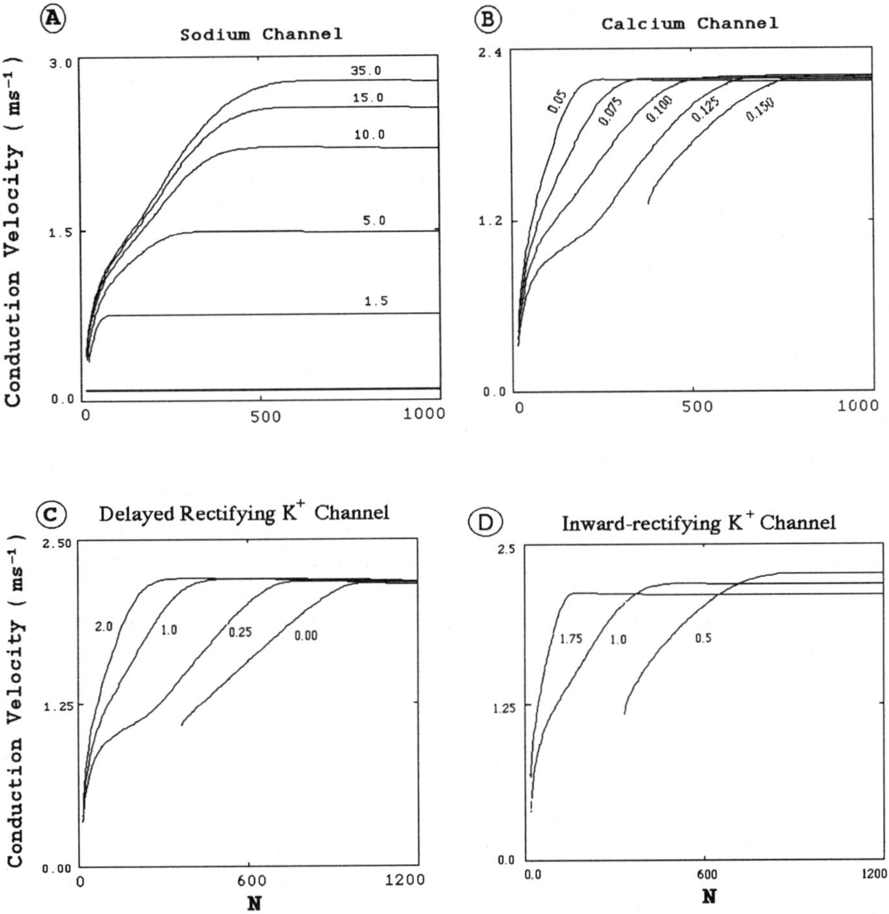

Figure 1: The conduction velocity versus N (i.e., the ring size L divided by the length of the cell Δx) at various values of the conductance (as labeled in the figure). Here, CV is the inverse of the time (in msec) required for the action potential (AP) to travel from segment of the ring to the next and N corresponds to the number of segments in the ring with size L. Frames A-D show the roles of the sodium, calcium, delayed-rectifying potassium and inward-rectifying potassium channels in CV, respectively. To generate the curves in Frames A-D we varied g_{Na}, g_{si}, g_{x1}, and g_{x1}, while fixing other three conductances at their basic values, respectively. Note that as the conductances of sodium and calcium channels decrease, the CRS decreases while the reverse holds for the potassium channels.

The loci of the CRS can be used to find the efficacy of antiarrhythmic drugs — the drugs that shift the CRS to the left are proarrhythmic, while those that shift the CRS to the right are antiarrhytmic. This can be seen from the following examples.

Draw a vertical line along the N = 400 in frame A. When g_{Na} = 35 reentry is non-sustained since the system is at the steeply curved portion. When the g_{Na} is decreased to 15, reentry becomes oscillatory since the system is now in the CRS regime. As the Na^+ conductance decreases (by the action of the drug) to g_{Na} = 5, the oscillatory mode transforms to stable reentry (since the system landed on the horizontal portion). As both g_{Na} and g_{si} decrease, the CRS shifts to the left, and Na^+ and Ca^{2+} channel blockers have the ability to transform nonsustained reentry to sustained reentry (proarrhythmic). From frames C and D, one can explain why the two types of potassium channel blockers can terminate sustained reentry. Take for example N = 600 in frame C. When g_{x1} = 1 reentry is stable. When g_{x1} is decreased to 0.25 (a partial block) or to 0.0 (complete block) the system lands on the curved region. This causes stable reentry to become non-sustained.

We observe in Fig. 1A that when g_{Na}=1.5 mS/cm^2, stable reentry persists in a ring as small as 50Δx in circumference. When g_{Na} is increased to 35 mS/cm^2, a ring size greater than 550Δx is required for reentry to be sustained. A similar effect is found in the case of Ca^{2+} channel blockers. When g_{si}=0.05 mS/cm^2 sustained reentry persists even though the ring size is as small as 100Δx; however, when g_{si} is increased to 0.125 mS/cm^2 a ring size larger than 600Δx is required to sustain the reentry. Although Na^+ channel blockers can decrease the CRS, the Ca^{2+} channel blockers are much more effective in decreasing CRS (which implies the latter blockers may be much more proarrhythmic). In the case of the delayed rectifying K^+ channels, they can also shift the location of CRS but in the opposite direction. When the delayed rectifying K^+ channel is completely blocked (*i.e.*, g_{x1}=0) a ring size greater than 900Δx is required for sustaining the reentry. However, when this current is enhanced (*i.e.*, g_{x1} = 5 mS/cm^2), reentry persists in a ring as small as 100Δx. Thus, these blockers are very antiarrhythmic by shifting the CRS far to the right. In the case of the inward rectifying K^+ channel, blocking of I_{k1} also shifts the CRS to the right, but near blocking of this current gives rise to ectopic beating which is another type of arrhythmia (the result not shown).

In addition, Fig. 1 explains how the channel blockers influence the conduction velocity. Note in frame A that the Na^+ channel blockers have the ability to lower the plateau level. Consequently, for a fixed ring size, the Na^+ channel blockers can slow down the CV considerably (proarrhythmia). On the other hand, both Ca^{2+} and delayed rectifying K^+ channel blockers have little effect on the CV in the region where stable reentry exists. Unlike g_{x1}, decreasing g_{k1} lifts the plateau upward, which implies that CV increases as the I_{k1} channel is blocked. A blocker that lowers the CV curve is proarrhythmic, while a blocker that lifts the curve is antiarrhythmic. Based on this criterion of CV, Na^+ channnel blockers are proarrhythmic while Ca^{2+} and K^+ channel blockers are not.

3.2 How Do Antiarrhythmic Drugs Influence the Reentrant Action Potential?

It would be interesting to obtain information on how the impulse propagates along the ring and how the three channel blockers affect the propagating wave. Figures 2A-C show the reentrant action potential as it propagates along the ring. In frames A-C, there are altogether N=1000 sites in the ring.

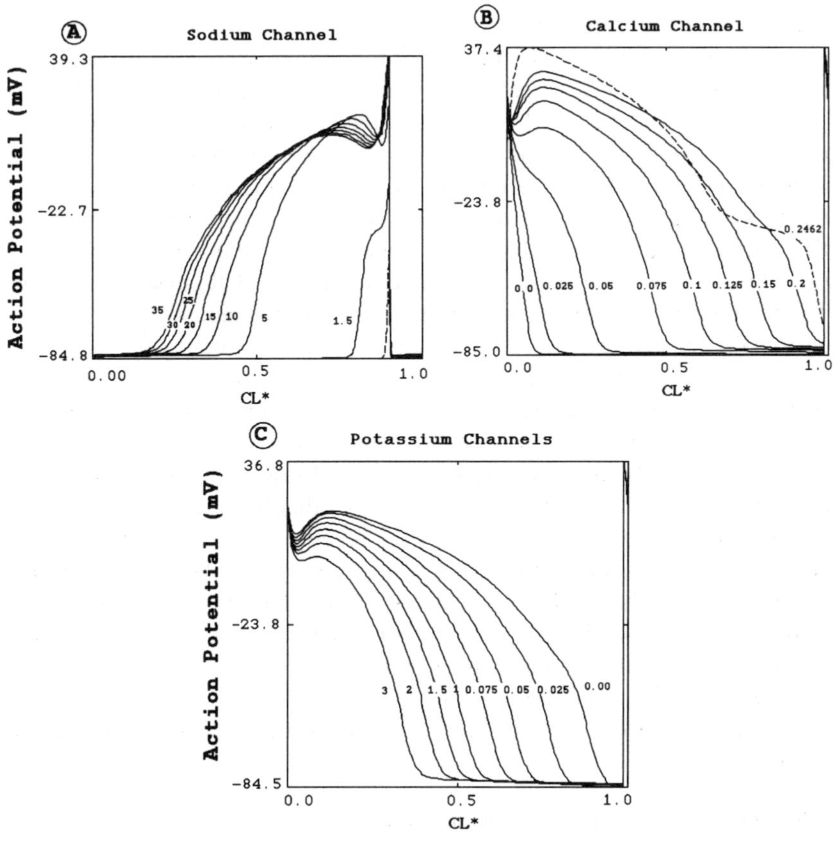

Figure 2: Propagating action potentials along the ring (L=1000Δx) when g_{Na}, g_{si}, and g_{x1} are varied. Here, CL* is the site along the ring divided by the total site (i.e., 1000). The dashed line in frame A is the AP which coexists with other full grown AP's. The dashed line in frame B is the AP at the critical point where ectopic beating exists.

In Fig. 2, the abscissa (labeled CL*) represents the sites along the ring divided by the length of the ring. The dashed line in frame A is the unstable action potential which coexists with other full grown stable action potentials. On the other hand, the dashed line in frame B is the action potential at which ectopic beating arises. In this region, the ring is either in the state of ectopic beating or in the state of quiescence. Above this region, the reentrant wave is no longer sustained. Ectopic beating also arise when the inward rectifying K^+ channel is blocked (the result not shown here).

As shown in this figure, the APD decreases with decreasing g_{Na} and g_{si} and increases with decreasing g_{x1}. In the case of the sodium channel blockers, the APD decreases slightly as g_{Na} decreases from 35 to 10. A drastic decrease in APD occurs as g_{Na} decreases further to 1.5. In the case of the calcium channel blockers, there is a gradual reduction in the APD as g_{si} decreases. The delayed rectifying K^+ channel blockers are effective in changing the APD but in the opposite direction. In summary, both Na^+ and Ca^{2+} channel blockers shorten the APD relative to cycle length, while the K^+ channel blockers lengthen the APD relative to the cycle length. This result is expected from a model (including B-R) that assumes that I_{si} and I_{x1} are responsible for the maintenance of the plateau, while I_{Na} is mainly responsible for generating a brief inward current which activates I_{si}.

In addition to giving information on how the conductance affects the propagating action potential, Fig. 2 also gives information on how the shape of the action potential is affected when the conductance varies. As shown in frame A, a decrease of g_{Na} shortens the initial spike height and lowers the maximum diastolic potential (MDP) by a very slight amount. On the other hand, g_{si} does not affect the spike, but a decrease in g_{si} lowers the plateau level as well as the MDP (see frame B). In the same figure, the AP at the transition point (the dashed line) reveals what is expected for the shape of ectopic beats. Note that this AP consists of two "plateaus," one at about 37 mV and another at about -45mV. The shape of the ectopic beat is very similar to the upper part of AP in that its MPD is the second plateau (i.e., it oscillates between -45 mV and 37 mV). As seen in frame C, g_{x1} does not affect the spike height or MDP. However, the decrease of g_{x1} raises the plateau to a higher level.

3.3 How Do the Drugs Affect the Restitution Curves?

The restitution curve (RC) is a curve which reveals the relation between APD and the preceding diastolic length (DL). Drugs that can lower the RC is proarrhythmic, while drugs that can lift the curve upward is antiarrhythmic. The RC curve also tells us whether g_{Na} can really affect APD or whether the increases in APD are just secondary to the ranges of cycle length involved. The RC thus can be used to find efficacy of the drug.

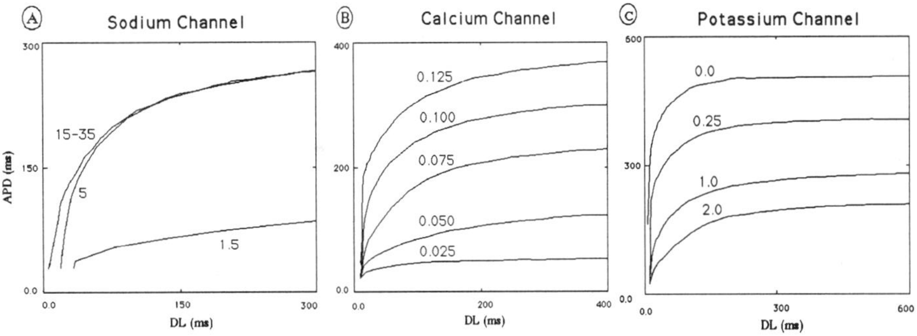

Figure 3: The restitution curves which show the relation between the action potential duration and the diastolic length. Frames A, B, and C reveal how the sodium, calcium, and delayed-rectifying potassium blockers affect the restitution curves, respectively.

To find the effect of channel drugs on the restitution curve, we obtained the RC from the AUTO outputs and displayed them in Figs. 3A-C. Note in frame A that g_{Na} affects only slightly on the RC for $g_{Na} > 15$. The deviation becomes visible only only when g_{Na} becomes less than 5 and DL becomes less than 70 ms. When g_{Na} is reduced drastically to 1.5, the curve deviates significantly from the rest. Thus, the Na^+ channel blockers do not influence the restitution curve as long as $g_{Na} > 5.0$ mS/cm^2. But below this conductance value, the Na^+ channel blockers exert very significant proarrhythmic action. Calcium channel blockers, on the other hand, lower the RC, which implies that these blockers can exert a proarrhythmic effect. The delayed rectifying potassium channel blockers lift the RC upward, which implies that they can exert an antiarrhythmic effect.

4 Simulations

4.1 How Does Stable Reentry Become Non-Sustained As the Ring Size Shrinks?

Although the bifurcation analysis gives information on where CV starts to deviate from the infinite ring behavior, it does not tell us how CV becomes unstable. In the region where the CV curve starts to bend, oscillatory reentry resides. In the region where the curve bends sharply, reentry becomes nonstained. How reentry terminates and why cycle length oscillates as the ring size shrinks can be understood by dynamic simulations (i.e., by solving the cable equation). In this section, using the simulation approach we show rich dynamic structures which reside along the g_{Na}=15 CV curve in Fig. 1, as the size of the ring decreases.

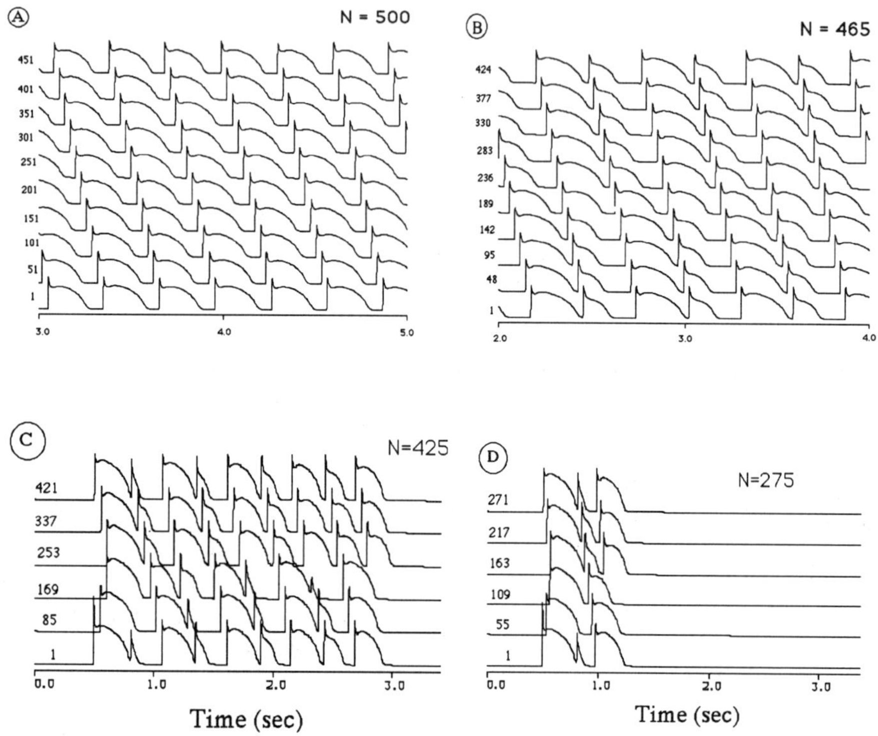

Figure 4: The dynamic simulations which reveal four types of wave along the g_{Na}=15 curve shown in Fig. 1. Reentry was induced by first applying a primary pulse to the first cell and a premature pulse to the 10th cell in the ring which contains 500 (frame A), 465 (frame B), 425 (frame C), and 275 (frame D). Note that stable reentry changes to oscillatory, and then to non-sustained reentry.

Let us examine how stable reentry transforms to nonsustained reentry via oscillatory reentry. Figures 4A-D depict the propagation of the action potential along the ring. These four frames came from the four regions in the g_{Na} = 15 mS/cm^2 curve shown in Fig. 1A — a ring containing 500 cells where CV is on the flat portion, a ring containing 465 cells where the CV is on the curved portion, a ring containing 425 cells where the curve becomes steep, and a ring containing 275 cells where the curve becomes very steep. The numbers listed on the left side of each frame are the sites where the impulse propagation is recorded. Reentry was induced by the following recipe: A primary pulse of strength 0.2 mA/cm^2 and duration of 2 ms was first given to the first cell at 200 milliseconds after the recording. Then, a premature pulse of the same strength and duration was given to the tenth cell when t reached 289 milliseconds after the first impulse passed the tenth cell.

Note that reentry is stable when there is an excitation gap (frame A) but becomes oscillatory when reentry encounters a substantial fraction of refractory cells (frame B). When the tissue becomes very refractory (i.e., no excitation gap), reentry terminates after making several turns (frame C). When the system is farther away from CRS, reentry terminates after making only two turns (frame D). As seen in the lower frames, APD is very heterogeneously distributed (i.e., some are short and some are long). This heterogeneous distribution of ADPs is what causes the termination of reentry, i.e., the termination results when the advancing impulse encounters very refractory tissues.

Figures 5A-D demonstrate how the pattern of oscillations changes, as the ring size shrinks from 500 to 430. Note that when N=500, the APD dampens to a steady state, which implies that reentry becomes stable after the transient oscillatory period (see frame A). In the region of CRS (see frame B), the model exhibits electrical alternans of long-short APDs. When it is observed for a long period of time, however, the alternans disappear and then reappear again. That is, there is a slow wave underlying the fast spikes, which repeats every ~290 turns. This slow wave becomes more visible when N is reduced to 450. This is shown in frame C, where the slow wave repeats about every 55 cycles. When the ring size is reduced to N=430, the cycle repeats every 13th turn (see frame D). As the ring size is decreased to N=420, the CL oscillation terminates after several cycles (the result not shown).

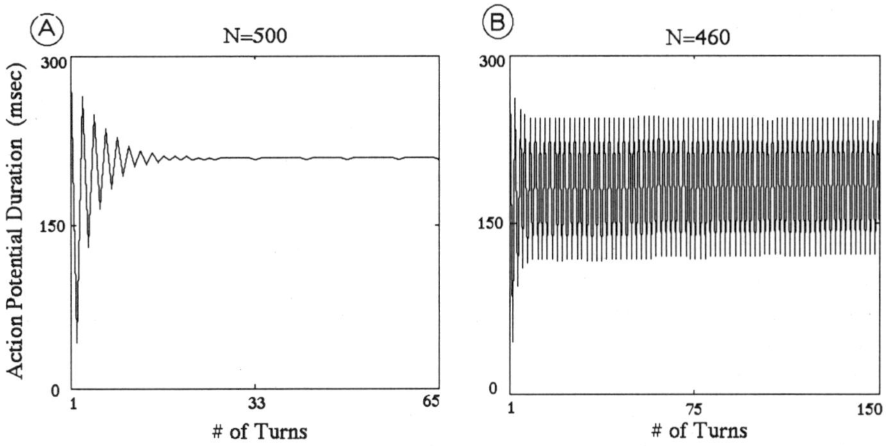

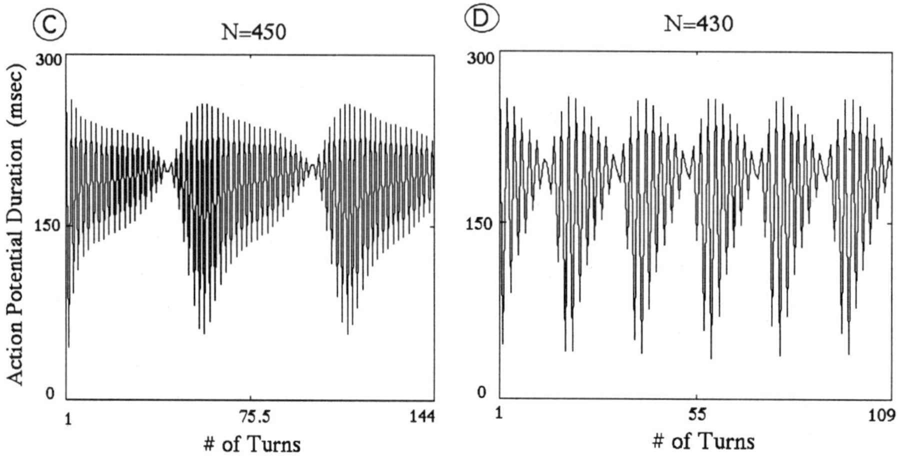

Figure 5: The action potential duration recorded at the first site in the neighborhood of the critical ring size (CRS). Frames A-D were obtained using N=500, 460, 450, 430, respectively.

4.2 What Is the Mechanism Involved in Cycle Length Oscillation?

Figure 6 shows how the incomplete recovery gives rise to the cycle length oscillation. If λ, the wavelength of CL oscillation, is exactly equal to 2L, the CL oscillation is simple (i.e., the APD alternates between long and short APDs). Note in this figure that a tiny AP first appears at the first cell, and it appears again at the 130th cell on the third turn. Since it appears on the third turn (not at the end of the second turn or the end of the third turn), λ must be shorter than 2L. The fact that λ is not equal to nL (where n is a whole number) is the reason for the complex oscillations. This oscillation may even be quasi-periodic.

Our simulation in Fig. 5 shows that as the ring size decreases so does the wavelength. Furthermore, it suggests a form for λ; it must be equal to 2L - f(L) where f(L) is an "inverse" function of L, such that f(L) is almost equal to zero on the right boundary of CRS (i.e., L=460Δx) and L on the left foundary of CRS (i.e., L=430Δx). It has been shown earlier that λ increase with a square of L.[12] Our simulation negates this relationship, i.e., f(L) does not increase with L but decreases with L (see Fig. 5).

Cycle length oscillation observed in the CRS regime is due to two inward currents, I_{Na} and I_{Si}, which have different activation thresholds and kinetics. The activation threshold of I_{Na} is lower than that of I_{Si}, and the calcium channel opens much slower than the sodium channel. Furthermore, only when I_{Na} is strong enough I_{Si} activates fully. I_{Na} becomes strong only if there is a sufficiently large excitation gap. A full activation of I_{si}, in turn, elicits a full grown APD, which causes a decrease in the excitation gap. This decrease induces a weak I_{Na} when the wave front moves to the next cycle. Thus, complex cycle length oscillation results due to varing gap sizes.

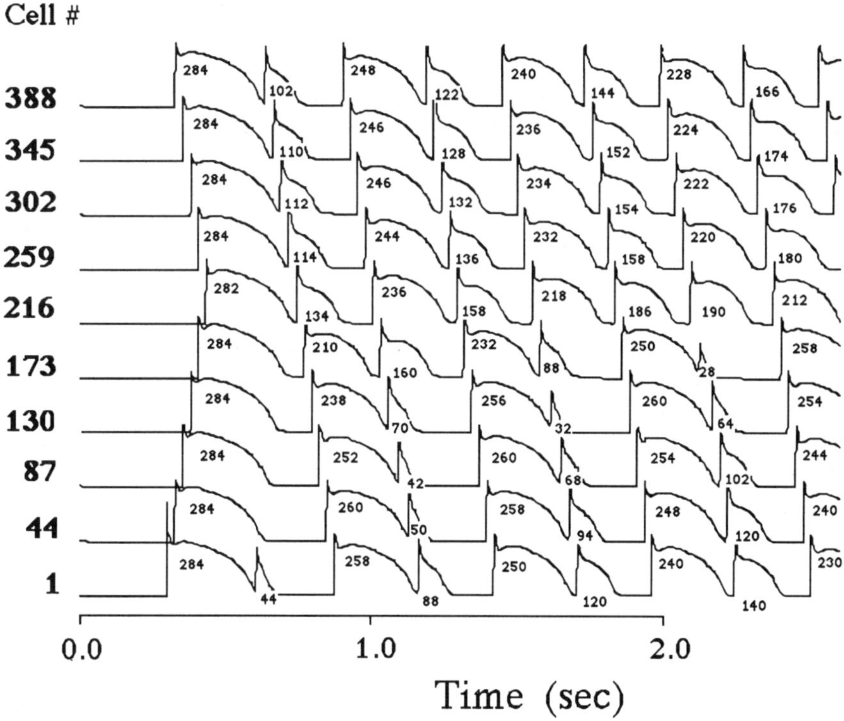

Figure 6: The spatial propagation of the action potentials when the ring size is N=430. The traveling wave was recorded at every 43rd position, starting from the first cell at the bottom trace and ending at the 388th cell on the top. The numbers in the figure are the action potential duration in msec.

Indeed, our simulation indicates that the short-long alternans often seen in the heart tissues,[13] the whole heart,[14] as well as in computer simulation[15] may not be a simple oscillation but rather a complex oscillation where high frequency AP's ride on a slow wave with wavelegth λ. On the right boundary of the CRS, the slow wave moves around the ring so slowly that the CL appears to oscillate with long-and-short alternans (see Fig. 5B). The wavelength λ is a function of the ring size that it decrease quickly with L in the CRS regime.

5 Discussion

Using the Beeler-Reuter model which was modeled for the ventricular myocardium action potential, we examined how the key ion channels in the plasma membrane of cardiac cells may influence reentrant rhythm. The purpose of this examination was to find the effect of various "channel" drugs on their anti- and proarrhythmic actions on reentry. Antiarrhythmic action was measured by the following four means: **i)** Shifting the critical ring size (CRS) to the right, **ii)** an increase in CV, **iii)** an increase in the relative ratio between action potential duration (APD) and cycle length (CL), and **iv)** lifting the restitution curves upward. We found that: **i)** Both sodium and calcium channel blockers shift the CRS to the left, the latter much more effectively than the former. On the other hand, both I_{x1} and I_{k1} blockers shift the CRS the right, the latter more effectively than the former. But, near blocking of I_{k1} gives rise to another type of arrhythmia — ectopic beating. **ii)** The Na^+ channel blockers influence the conduction velocity (CV) much more than any other channel blockers. While the Ca^{2+} channel blockers had little influence on CV, the K^+ channel blockers influenced CV only very weakly. **iii)** As for the APD, both sodium and potassium channel blockers increase APD, but the former only slightly and the latter significantly. When the sodium current is nearly blocked, however, APD decreases drastically. The calcium channel blockers always decrease APD. **iv)** While the Na^+ current has very little effect on the restitution curve, the calcium blockers lower the curve and the delayed rectifying potassium channel blockers lift the curve upward.

The questions arise: How is our observations on the ring applicable to a heart tissue which has a 2-D structure? Why does a drug like flecainide enhance fibrillation only to ischemic patients (but not to healthy ones)? My answer is as follows. Scarred tissues contain islands of rings with various sizes. The scar size may not exceed $400\Delta x$ (i.e., ~12cm) which is the CRS under the drug free condition (otherwise they may be already dead). As long as their sizes are less than 12cm, the impulse that circulates around the rings is terminated (i.e., short lasting tachycardia). However, in the presence of sodium and calcium channel blockers, reentry may become stable since these drugs can shift the CRS considerably to the left (see Figs. 1 A&B). This perhaps explains why it is easier for the patients with an ischemic heart to develop ventricular fibrillation when drugs such as flecainide or encainide are administered. The potassium channel blockers, on the other hand, suppress reentry by shifting the CRS to the right. Thus, this type of drugs may be useful to a ischemic heart patient who is prone to ventricular fibrillation.

As demonstrated in this paper, bifurcation analysis is a very useful way of finding the roles of the ion channels in the cell plasma membrane in reentrant arrhythmias. Compared to the numerical solution, the result of the bifurcation analysis is not only numerically accurate but also computationally very fast. Besides these advantages, it gives a graphical way of understanding the effect of channels on reentrant rhythms. We have demonstrated the usefulness of the bifurcation analysis by constructing Figs. 1-3. With Fig. 1, one can explain why some Class IC drugs give rise to sustained reentry and why Class III drugs can terminate sustained reentry. It will be interesting to apply our bifurcation analysis approach to other cell types such as neural network.[16]

Acknowledgments

This work was supported by National Science Foundation MCB-9411244 and in part by a grant from the Pittsburgh Supercomputing Center through the NIH Division of Research Resources cooperative agreement U41 RR0415.

References

1. The cardiac arrhythmia suppression trial (CAST) investigators, *New. Engl. J. Med.* **321**, 406 (1989) & **327**, 227 (1992).
2. L.H. Frame, E. K. Rhee, H. Fei, and W. Luchetti. *Pace.* **14**, 1728 (1991).
3. L.H. Frame and H. Fei, *Circulation* **80**: II-507, 2016. (1992).
4. T.R. Chay and Y. S. Lee, *J. Theor. Biol.* **155**: 137 (1992).
5. G.W. Beeler and H. Reuter, *J. Physiol.* **268**, 177 (1970).
6. T.R. Chay and T.Y. Chay, *Comput. Cardiol.* pp. 245-248, (1994).
7. T.R. Chay, J. Electrocardiology. **28** Supplement. (1995).
8. T.R. Chay and Y. S. Lee, Biophys. J. **47**, 641 (1985).
9. S. Rush and H. Larsen, IEEE Trans. Biomed. Engin. **25**, 389 (1978).
10. E. Doedel, T*echnical Report in Applied Mathematics*, California Institute of Technology (1986).
11. T.R. Chay, *Intern. J. Bifurc. Chaos* No. 6 (1995).
12. M. Courtemanche, L. Glass, and J.P. Keener, *Phys. Rev. Lett.* **70**, 2182 (1993).
13. L.H. Frame and M.B. Simson, *Circulation* **78**, 1277 (1988).
14. B.D. Nearing, A.H. Huang, and R.L. Verrier, *Science* 437 (1991).
15. W. Quan and Y. Rudy, *Pace* **14**, 1700 (1991).
16. T.R. Chay, Y.S. Lee and Y.S. Fan, *Intern. J. Bifurc. Chaos* **5**, 595 (1995).

THE EFFECTS OF RANDOM VARIATION IN STIMULATION TIMING AND MAGNITUDE ON EXCITABLE AND OSCILLATORY FORMS OF A LUTEINIZING HORMONE PULSE GENERATOR MODEL

J.P.A. FOWERAKER AND D. BROWN

Department of Neurobiology, The Babraham Institute, Babraham, Cambridge, CB2 4AT, UK.

1 Introduction

Luteinizing hormone (LH) secretion from the pituitary gland has the key function in all mammals of mediating neural control of ovulation. In intact animals a surge of LH is the main precursor of ovulation; at other times LH is released in intermittent small pulses, the frequency and amplitude of which are critical for follicular development. The pattern is partly determined by feedback from the ovary; in the ovariectomised (OVX) animal, ovarian feedback is abolished and LH is released in large amplitude pulses. This demonstrates that the intrinsic LH pulse generator is not dependent on feedback from the ovaries, although its action is modified by feedback. The amount and pattern of hypothalamic LH-releasing hormone (LHRH) release into the portal blood circulation in the brain, responsible in turn for LH release from the pituitary gland, is critically controlled by the activity of discrete populations of neurones within the hypothalamus.

It has been difficult to attribute a relative importance to the large number of neurotransmitters and neuromodulators which are involved in the regulation of LHRH neuronal activity. Nonetheless, it is generally acknowledged that the adrenergic neurotransmitters, adrenaline and noradrenaline, and the inhibitory amino acid, gamma-aminobutyric acid (GABA) are important physiological regulators of the pulsatile release of LHRH, and the model proposed by Brown *et al.*[1] includes components related to these. LHRH is released from LHRH cells in the hypothalamus (whose mean electrical activity is represented by v in equations (1) and (2)), and there is evidence that these cells have reciprocal connections with a population of GABA cells[2,3] (mean electrical activity, w), which, possessing oestrogen receptors, mediate oestrogen feedback to the LHRH system in the intact animal. Adrenergic input is also critical to the functioning of this system, in that if this input is blocked, LHRH and LH pulses cease. Pulsatile adrenergic input is modelled as discrete increments in LHRH activity which vary randomly in magnitude and timing: instantaneous, positive perturbations of v, of size Δv_i which follow a lognormal distribution. They arrive at intervals Δt_i, which follow a negative exponential distribution with displaced origin, r, *i.e.* the intervals consist of a constant absolutely refractory period, r, followed by a variable period, which has a negative exponential distribution, mean τ, characteristic of a Poisson process. Both GABA and LHRH activity are also subject to tonic influences of the adrenergic input, a, which can be excitatory or inhibitory. There is evidence of significant interactions between LHRH neurones[4], modelled as a cubic polynomial in v in differential equation (1), and also that local rises in LHRH concentrations induce short term increases in GABA activity and release. This is represented in the model by GABA being subject to linear input from v, its activity declining exponentially in the absence of other influences. There is also assumed to be a one-to-one relationship between pulses of LH and pulses of LHRH, and we restrict our attention to LHRH in the remainder of this paper. A simplified version of the differential equation model of the continuous variables presented by Brown *et al.* used previously[5] is:

$$\frac{dv}{dt} = \gamma[-v(v-\alpha)(v-1)-w+I_{net}] \tag{1}$$

$$\frac{dw}{dt} = v-\beta w \tag{2}$$

This is a form of the Bonhoeffer-van der Pol or FitzHugh-Nagumo model[6,7]. Here I_{net} reflects the interpretation of the tonic adrenergic input, a, by the LHRH/GABA network, and acts in the same way as an applied current in the BVP/FHN model. Initially, $I_{net} = da$ where $a = \mathrm{E}[\Delta v_i]/\mathrm{E}[\Delta t_i]$, d is constant, and E[ ] denotes statistical expectation, or mean. Later in the paper to allow the model to reproduce oscillatory behaviour when a=0 we modify I_{net}, so that $I_{net} = I_0 + da$. The discrete stimuli are taken to have a lognormal distribution, *i.e.* stimulus size~$N(\mu, \sigma^2)$, median (stimulus size) =$\exp(\mu) = \Delta v$ and coefficient of variation (stimulus size)$\approx \sigma$. The proportion of the mean inter-stimulus interval, Δt, which is randomly varying throughout the simulation is $p_{stoch} = \tau/(r+\tau)$, also equal to the coefficient of variation of Δt_i; the proportion constituting the constant refractory period therefore being $1 - p_{stoch} = r/(r+\tau)$. The pattern of the stimuli ranges from *regular* ($p_{stoch} = 0$) to *Poissonian* ($p_{stoch} = 1$).

The motivation of the modelling exercise in this paper is as follows. In OVX rats, investigators have found that administration of either adrenergic receptor agonists or antagonists into the vicinity of the LHRH neurones results in a decrease to almost zero in LH pulse frequency[8,9]. Assuming that the antagonist reduces adrenergic input to near zero and that infusions of receptor agonists would greatly increase the frequency of receptor activation, this means that in our model system we require virtual abolition of LHRH pulses to result from both an increase and a decrease to zero in adrenergic input from some point over the range of possible adrenergic activity. This non-monotonic relationship between the frequency of LHRH pulses (f_p) and the frequency of stimulation (f_a) is reproduced in the mathematical model for regular and quasi-regular stimuli patterns.

Simulation experiments have been carried out on completely random (*i.e.* Poissonian) patterns of stimulation of excitable and oscillatory systems, starting with Perkel[10]: Poisson stimulation results in a much smoother, and usually monotonic, relationship between the expectation of the output pulse frequency and expectation of the stimulus frequency. Earlier work[5,11] deals with regular stimulation, but little appears to have been done on the range between these two extremes, with limited random variation in interstimulus timing. Relatively little has also been done on the effects of random variation in stimulus magnitude on the response of FHN/BVP models, except in the case of low amplitude, additive noise[12]. General work on the effects of random perturbations on nonlinear dynamical systems would lead us to believe that small departures from the deterministic case would result in only slight variation in the overall response pattern.

This study provides results of simulation using stimulation patterns involving these two types of randomness: in timing spanning the complete range from regular to Poisson, and (independently) in magnitude with coefficient of variation from zero to about 0.5. Generally speaking, increasing random variation has the eventual effect of removing the non-monotonicity of the f_p/f_a relationship. Slightly irregular timing and low amplitude random variation in stimulus size have the effect of smoothing out some of the minor gradual rise/steep fall fluctuations in the relationship, while still maintaining its overall substantially non-monotonic nature. Here, we attempt to determine the threshold in terms of variation in stimulus size and timing at which the larger scale non-monotonicity breaks down, and the interaction between the effects of the two types of variation on this threshold. This research perhaps gives a greater biological relevance to our previous studies. The LHRH system does not have absolutely regular inputs; neither do many of the applications considered previously, such as cardiac and invertebrate neural systems.

2 Relative importance of sources of variation

Previously we have only considered variation in the timing of the stimuli over a limited range of p_{stoch}[5]. We now consider, for the excitable system ($I_{net} = 0$), the effects of a wider range of random variation in the stimulus timing and introduce random variation in the stimulus magnitudes. We then consider possible interactions between p_{stoch} and σ. One of the main effects of increasing p_{stoch} is the collapse of the non-monotonic f_p/f_a relationship. Figure 1 shows the response of v (bottom trace in each window) to the stimuli (top trace) for (a) regular stimuli ($p_{stoch} = 0$) and (b) Poissonian stimuli ($p_{stoch} = 1$) over a range of f_a. In (a) f_p increases from (i) to (ii), is constant from (ii) to (iii) and then declines to zero in (iv) - a non-monotonic f_p/f_a relationship. In (b) f_p increases from (i) to (ii), but not as much as in (a), some of the stimuli are occurring during the refractory part of the pulse and so having no effect. From (ii) to (iv) the f_p continues rising - a monotonic f_p/f_a relationship.

Figure 2 shows f_p against f_a for a range of p_{stoch} with $\Delta v = 0.28$ for constant magnitude stimuli ($\sigma = 0$). For low values of f_a there is a strong sawtooth behaviour for $p_{stoch} = 0$. As p_{stoch} increases this behaviour is smoothed, although there is still an overall non-monotonic relationship. As p_{stoch} is increased further the f_p/f_a relationship becomes closer to monotonic, after allowing for the effects of sampling variation. Very similar results hold for $\sigma \neq 0$ with $p_{stoch} = 0$ and are not presented here. However it is difficult to discern the value of p_{stoch} (or σ) at which the non-monotonic behaviour ceases, due to sampling variation obscuring trends in these results. To overcome this we attempt to filter out these effects by the use of an empirically justified definition of monotonicity. We say that a series is 'monotonically increasing' if for any point in the series, the maximum drop that occurs between that point and any subsequent point is less than ζ, or less than ζ times the range of the series. More formally, a series $\{x_n\}$, $n = 1...t$, is 'monotonically increasing' if for any n, $(x_n - x_i) < \zeta \max[(x_{max} - x_{min}), 1]$ for all $i \in [n+1, t]$ where $x_{max} = \max(x_n)$, $x_{min} = \min(x_n)$, $n = 1...t$. $\zeta = 0.3$ was found to give intuitively acceptable results.

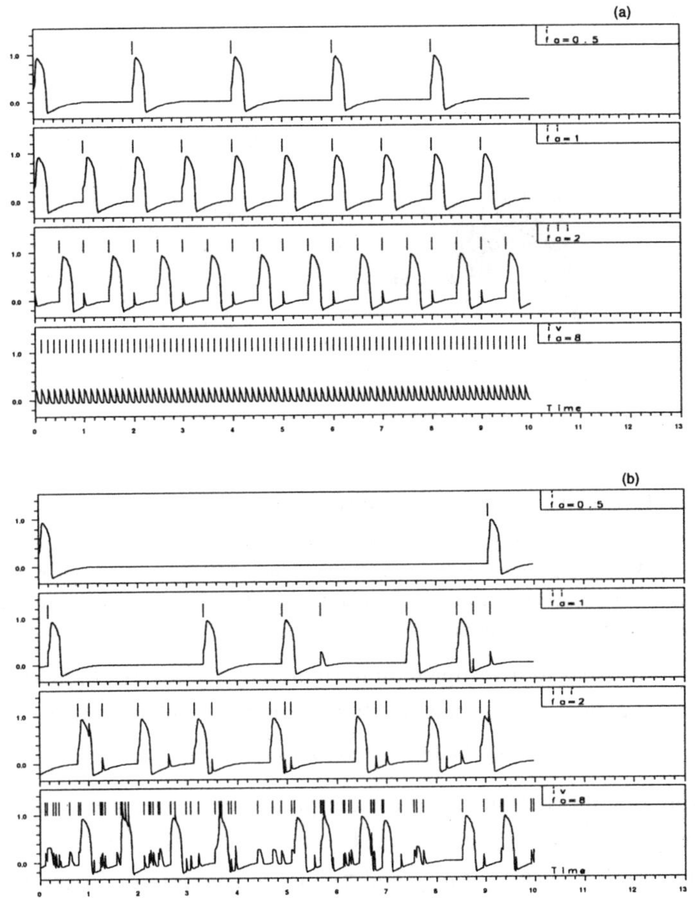

Figure 1: Plot of v and the discrete stimuli for (a) $p_{stoch} = 0$, & (b) $p_{stoch} = 1$ with (i) $f_a = 0.5$, (ii) $f_a = 1.0$, (iii) $f_a = 2$ & (iv) $f_a = 8$. In (b) these represent the mean (f_a). The differential equation parameters are $\alpha = 0.2$, $\beta = 2.5$, $\gamma = 200$ and $d = 0$, ($I_{net} = I_0 = 0$). The equations are solved by Gear's method using a custom written modelling package using NAG numerical routines and UNIRAS graphics on a DEC Alpha workstation. Note the non-monotonic f_p/f_a relationship in (a) but not in (b)

Applying this rule to Figure 2 results in the two-part Figure 3, where the lower part shows those traces which are not 'monotonic', and the top part those which are. This seems to be a usable definition for our purposes, with only those plots that have a clear drop in f_p as f_a increases not being declared 'monotonically increasing'. Applying this rule to similar graphs for σ is also effective. This gives

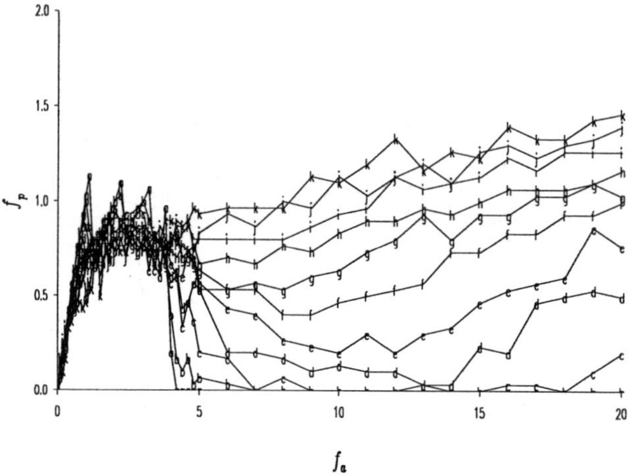

Figure 2: Plot of f_p against f_a for $\Delta\nu = 0.28$ with the letters in the plots denoting a-$p_{stoch} = 0$, b-$p_{stoch} = 0.1$, c-$p_{stoch} = 0.2$, d-$p_{stoch} = 0.3$, e-$p_{stoch} = 0.4$, f-$p_{stoch} = 0.5$, g-$p_{stoch} = 0.6$, h-$p_{stoch} = 0.7$, i-$p_{stoch} = 0.8$, j-$p_{stoch} = 0.9$ and k-$p_{stoch} = 1.0$. $\sigma = 0$. Other parameters as Figure 1. Data was collected over 30 time units (TU), after an initial transient period of 10 TU. It is difficult to distinguish when the traces are monotonic or not, allowing for experimental error.

us an empirical basis to investigate more rigorously the levels of random variation in the stimulus patterning that cause the 'non-monotonic' relationship to break down, and from this point we drop the quotation marks.

Applying this definition over wider parameter ranges gives Figure 4. Here we consider a grid of p_{stoch} and σ, for 4 values of $\Delta\nu$. The shaded regions of the graphs indicate those values of p_{stoch} and σ which are monotonic. These graphs show that the break-down point in the monotonicity decreases as $\Delta\nu$ increases. If the unshaded regions formed rectangles in the lower left corner, parallel to the axes, then the value of p_{stoch} bridging the transfer from monotonic to non-monotonic is constant, provided σ is less than a threshold σ_0, i.e. within this range of σ, it is *independent* of σ. If, on the other hand, the unshaded region is a triangle, the threshold in p_{stoch} is strongly correlated with σ, and there is a *linear trade-off* between p_{stoch} and σ. The observed results fall somewhere between these two extremes, but the underlying trends are obscured by random sampling variation. We therefore need some statistical test to determine which hypothesis between these two is tenable.

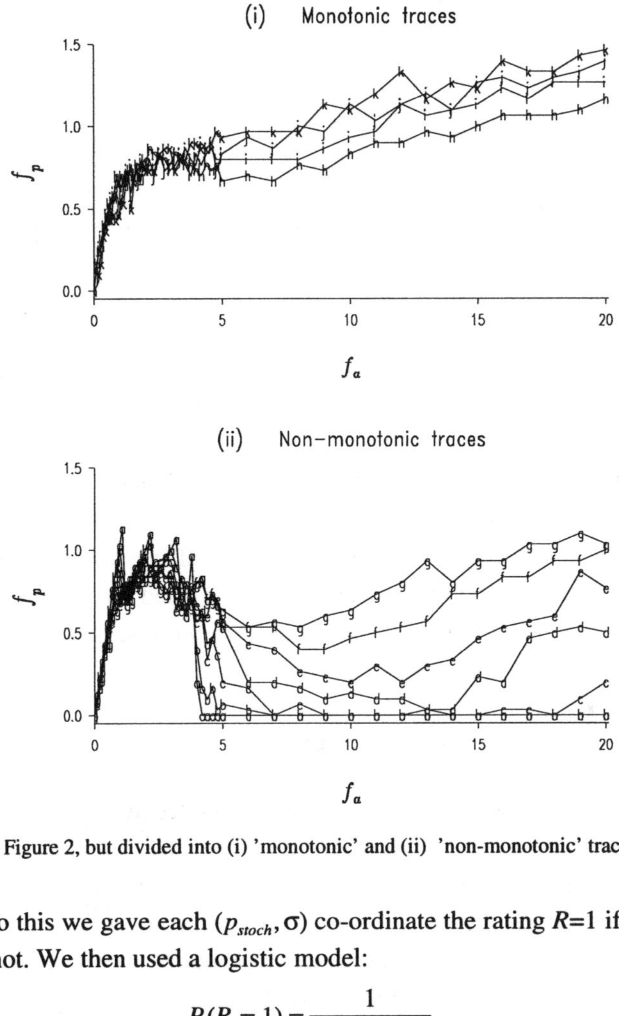

Figure 3: As in Figure 2, but divided into (i) 'monotonic' and (ii) 'non-monotonic' traces.

To do this we gave each (p_{stoch}, σ) co-ordinate the rating R=1 if monotonic, and R=0 if not. We then used a logistic model:

$$P(R=1)=\frac{1}{1+\exp(-S)}$$

where S is a score function involving σ and p_{stoch}. Various forms for S, involving parameters to be estimated from the data $(a, b, c, d, \sigma_0, p_{stoch0})$ were used, and the goodness of fit, as measured by the residual deviance of a generalised linear model with logit link and binomial error distribution, compared between models. The models and associated score functions we consider are:

M1 $S = a$	→	**M4** $S = a \qquad +e[(r_1\Psi(\sigma, p_{stoch}) + r_2\Psi(\sigma, p_{stoch})) - 1]$	
↓		↓	
M2 $S = a + b\,p_{stoch} + c\sigma$	→	**M5** $S = a + b\,p_{stoch} + c\sigma + e[(r_1\Psi(\sigma, p_{stoch}) + r_2\Psi(\sigma, p_{stoch})) - 1]$	
↓			
M3 $S = a + b\,p_{stoch} + c\sigma + d\,p_{stoch}\sigma$			

where

$$\Psi(\sigma, p_{stoch}) = \begin{cases} 1, & r_1 > r_2 \\ 0, & r_1 \le r_2 \end{cases}, \quad r_1 = \frac{\sigma}{\sigma_0}, \quad r_2 = \frac{p_{stoch}}{p_{stoch_0}}.$$

M1 is a baseline 'no-effects' model. M2 assumes that linear trade-off occurs, *i.e.* an increase in p_{stoch} is compensated for by a decrease in σ, or vice-versa. M3 is a extension of M2, allowing cross-product terms, and is hence more flexible. M4 allows independent thresholds in σ and p_{stoch}, so the fitted surface will be of a 'box' type. M5 is a composite of M2 and M4 giving a compromise between independence of p_{stoch} and σ and pure trade-off.

3 Statistical analysis and interpretation

The residual deviances and degrees of freedom generated by fitting the models given above to data given in Figure 4 are shown in Table 1. Differences between nested models for individual values of Δv can be tested by comparing residual deviances with percentage points of the chi-squared distribution on degrees of freedom equal to the difference in the number of parameters. M4 and M2 both have a lower residual deviance (and hence a better fit) than M1, however it is difficult to distinguish between them. Thus an independence of effect or a trade off hypothesis model are more tenable than the null-model. The composite M5 is a statistically better fit than both of these in all but one case, and so neither the hypotheses in M2 or M4 holds exactly; an intermediate hypothesis is more tenable. M3 is a simpler (2 fewer parameters) quite flexible model, which is frequently as good a fit as M5. The threshold boundary predicted by M3 is plotted on Figure 4 as a dotted line, showing that M3 gives a very good fit, and so a good indication of the interactions between p_{stoch} and σ in determining the boundary.

4 The effect of I_{net}

So far we have only considered the effects of randomness in timing and magnitude on the excitable form of the model. Some studies[13,14] have led to the proposal that the LHRH neurones are intrinsically oscillatory (previously discussed for regular stimulation[15]). We consider introducing random variation into the stimulation pattern and amplitude and look at the effects on the suitability of the oscillatory form of the model as a potential LHRH pulse generator.

Table 1 Summary of residual deviances and degrees of freedom from fitting models M1 - M5

	M1 null model	M2 trade-off	M3 trade-off and interaction	M4 independence	M5 composite
d.f.	98	96	95	95	93
No. parameters	1	3	4	4	6
Δv			DEVIANCES		
0.24	123.1	26.11	10.46	20.14	4.94
0.26	119.7	22.32	0.04	20.48	1.29
0.28	111.9	23.43	8.96	14.80	3.94
0.30	109.7	19.12	8.02	19.81	7.86
0.32	102.3	18.30	0.01	10.58	0.00
0.34	90.8	3.82	0.00	15.31	0.00
0.36	84.2	11.11	0.01	8.86	0.00
Total	741.70	124.21	27.50	109.98	18.03

Figures 5 and 6 show f_p as a contour diagrams of I_{net} against f_a. Figure 5(a), the regular case ($p_{stoch} = 0, \sigma = 0$), shows a non-monotonic relationship between f_p and f_a when the system is excitable ($I_{net} < 0.0495$), but much less variation when the system is oscillatory ($0.0495 \le I_{net} < 0.1745$). Figures 6(a) and 6(b) show the effects of $p_{stoch} = 0.2$ and $p_{stoch} = 0.6$. Here the slopes of f_p, as both f_a and I_{net} change, become lower, and the non-monotonic f_p/f_a relationship disappears in Figure 6(b) for $I_{net} = 0$. The f_p/f_a relationship for constant $I_{net} \in [0.0495, 0.1745]$ (system oscillatory) is more-or-less constant. Figure 6(c) shows f_p for Poissonian stimulation patterns ($p_{stoch} = 1$). As expected the f_p/f_a relationship is monotonic when the system is excitable. It is also approximately monotonic when the system is oscillatory, although f_p varies over a smaller range, but the maximum f_p has increased.

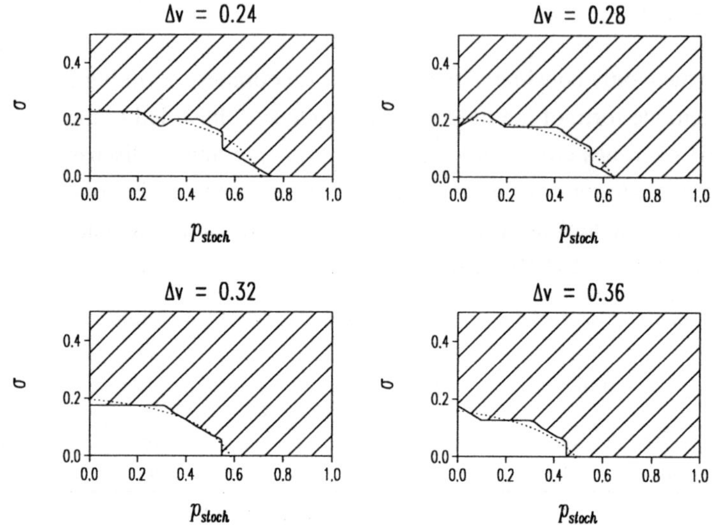

Figure 4: Plot of σ against p_{stoch} for $p_{stoch} \in [0,1]$, $\sigma \in [0,0.5]$ for $\Delta\nu = 0.24$, $\Delta\nu = 0.28$, $\Delta\nu = 0.32$ & $\Delta\nu = 0.36$. Other parameters and methods as Figures 1 and 2. The shaded area indicates those values which are 'monotonic'. Note the unshaded area decreases in size as $\Delta\nu$ increases. Dotted lines indicate the fitted model M3.

Figures 5(a) ($\sigma = 0$), 5(b) ($\sigma = 0.2$) and 5(c) ($\sigma = 0.4$) show the effects of increasing σ whilst keeping $p_{stoch} = 0$ over the range of I_{net}. In Figure 5(c) the strong non-monotonic f_p/f_a relationship has disappeared, as expected, for the excitable system. For the oscillator the features of the contour diagram for low f_a are remarkably robust to this level of variation in the stimulus magnitude, in contrast to the variation in timing. Randomness in stimulus timing results in some stimuli occurring during the refractory period of the oscillatory cycle, so having little effect on the ensuing time course, whereas when regular more stimuli are able to induce an entrainment. On the other hand, random variation in stimulus magnitude has little effect on the ability of the regularly occurring stimuli to entrain the oscillator. The changes in the other contour diagrams are similar to those for variation in p_{stoch}, with the exception of the contours for $I_{net} \in [0.15, 0.20]$ which are not smoothed to the same extent as in Figure 6(a) to (c). Here the regularly stimulated system is being trapped in a high dynamic equilibrium[11], which is more robust to randomness in stimulus magnitude than in stimulus timing.

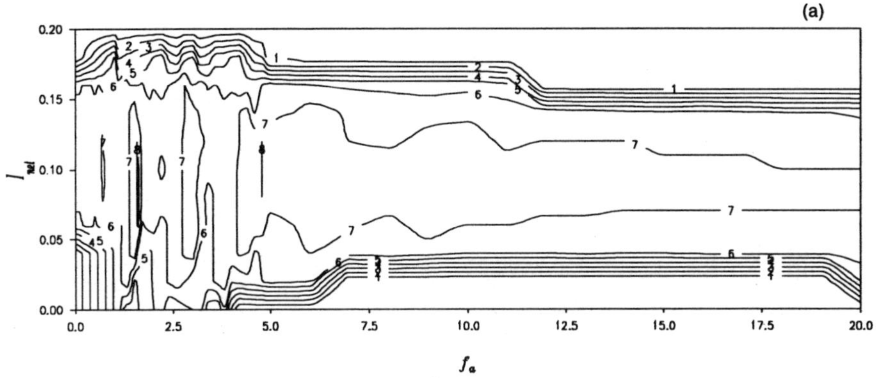

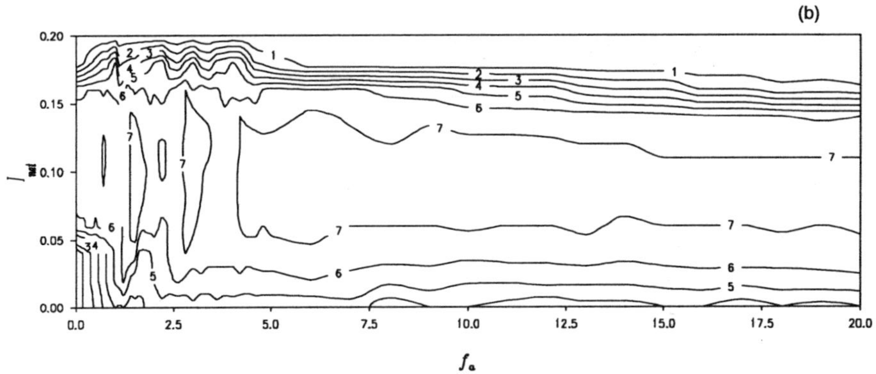

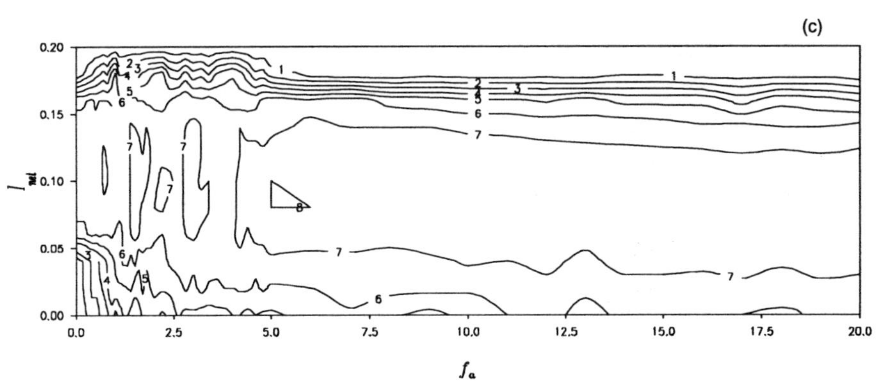

Figure 5: Plot of f_p as a contour diagram of f_a against I_{net} for (a) $\sigma = 0$ and (b)$\sigma = 0.2$, (c)$\sigma = 0.4$ (all with $p_{stoch} = 0$). $\Delta\nu = 0.33$, other parameters as Figure 1. Data was collected over 10 time units (TU), after an initial transient period of 10 TU. Contour values obtained by multiplying the contour labels by 0.2. Note that the plot changes more for low values of I_{net} than for intermediate values. Irregularities in the contours are caused by the numerical interpolation.

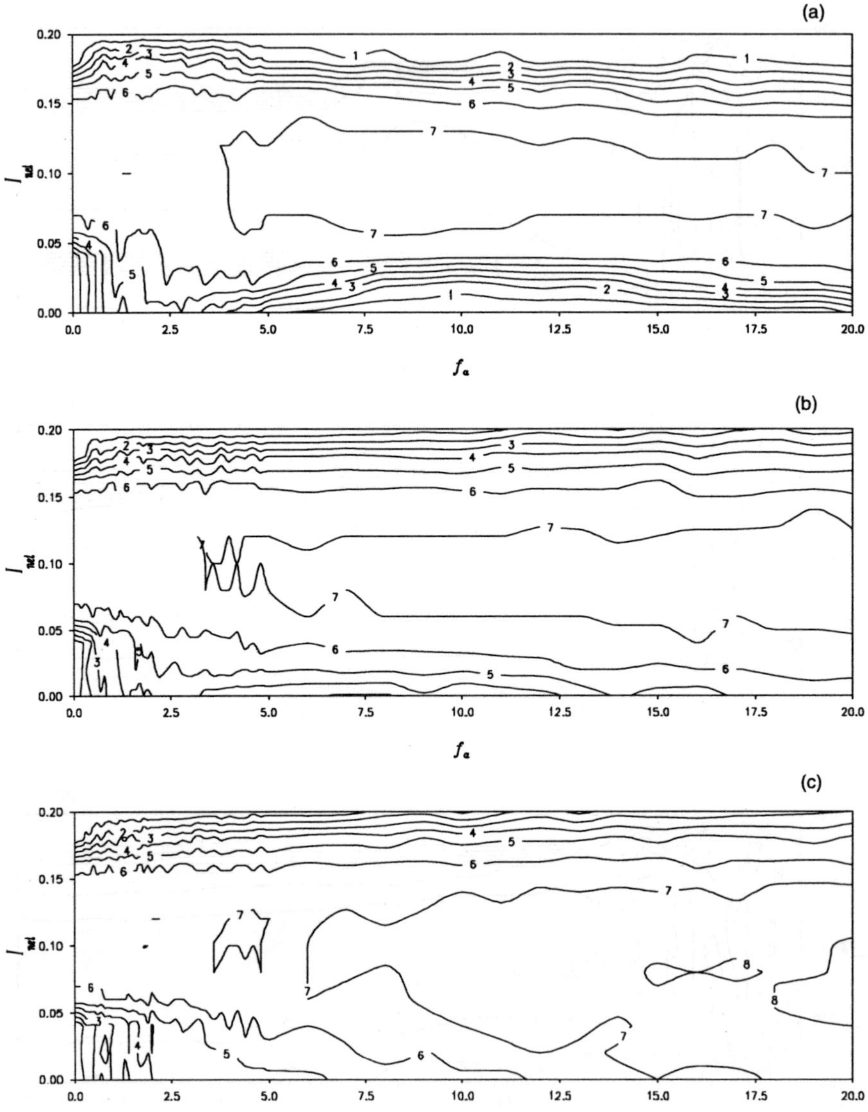

Figure 6: Plot of f_p as a contour diagram of f_a against I_{net} for (a) $p_{stoch} = 0.2$ (b) $p_{stoch} = 0.6$ and (c) $p_{stoch} = 1$ (all with $\sigma = 0$). $\Delta\nu = 0.33$, other parameters as Figure 1. Data was collected over 10 time units (TU), after an initial transient period of 10 TU. Contour values obtained by multiplying the contour labels by 0.2. Note that the plot changes more for low values of I_{net} than for intermediate values. Irregularities in the contours are caused by the numerical interpolation.

Summary

In modelling involving this type of pulsatile input it is necessary to balance the ease of analysis of the model with compatibility with the biology. Strictly regular inputs are not biologically plausible, although for the excitable system they give the appropriate model responses. We here provide a quantification of the relationship between the occurrence of a non-monotonic f_p/f_a relationship which is observed experimentally and two types of random variation. We have demonstrated that the non-monotonic relationship still occurs for a range of biologically plausible values of p_{stoch} and σ. These results form a useful basis for work involving more realistic models of adrenergic stimulation. The effects of random variation in the stimulus magnitude and timing on the oscillatory form of the system are not so great as for the excitable form, as the f_p/f_a relationship is dominated by the unstimulated pulse frequency. However for the values of Δv considered here the changes in σ have less effect than those in p_{stoch}.

1 D. Brown, A.E. Herbison, J.E. Robinson, R.W. Marrs & G. Leng, *Neuroscience* **63**, 869-879 (1994).
2 H. Jarry H, A. Perschl, & W. Wuttke, *Acta Endocr.* **118**, 573-578 (1988).
3 C. Leranth, N.J. MacLusky, H. Sakamoto, M. Shanbrough & F. Naftolin, *Neuroendocrinology* **40**, 536-539 (1985).
4 T. Funabashi, R. Hashimoto & F. Kimura, *Neuroendocrinology* **52** suppl. 1, 167 (1990).
5 J.P.A. Foweraker, D. Brown and R.W. Marrs, *Chaos* **5**, 200-209 (1995).
6 R. FitzHugh, *Biophys. J.* **1**, 445-466 (1961).
7 J.S. Nagumo, S. Arimoto & S. Yoshizawa, *Proc IRE.* **50**, 2061-2071 (1962).
8 H. Jarry, S. Leonhardt & W. Wuttke, *Neuroendocrinology* **51**, 337-344 (1990).
9 R.V. Gallo & R.E. Leipheimer, *Neuroendocrinology* **40**, 345-361 (1985).
10 D.H. Perkel, J.H. Schulman, T.H. Bullock, G.P. Moore and J.P. Segundo, *Science* **145**, 63-67 (1964).
11 D. Brown, J.P.A. Foweraker and R.W. Marrs, *Chaos, Solitons and Fractals* **5**, 359-369 (1995).
12 F. Moss, J.K. Douglass, L. Wilkens, D. Pierson and E. Pantazelou, *Annals N.Y. Acad. Sci.* **706**, 26-41 (1993).
13 G.J. Kokoris, N.Y. Lam, M. Ferin, A. Silvermans and M.J. Gibson, *Neuroendocrinology,* **48**, 45-52 (1988).
14 S. Leonhardt, H. Jarry, G. Falkenstein, J. Palmer and W. Wuttke, *Brain Research,* **562**, 105-110 (1991).
15 D. Brown, J.P.A. Foweraker, A.E, Herbison and R.W. Marrs, *Proceedings of the IMACS symposium on mathematical modelling* **2**, 311-314 (1994).

A MODEL OF PITUITARY RELEASE OF GROWTH HORMONE

E.A. STEPHENS, D. BROWN
Department of Neurobiology, Babraham Institute, Cambridge, CB2 4AT, UK

G. LENG
Department of Physiology, University of Edinburgh, Edinburgh, EH8 9AG, UK

R.G. SMITH
Merck Research Laboratories, Rahway, NJ, USA

Growth hormone-releasing hormone (GHRH) and somatostatin (SST) are released from secretory neurones of the hypothalamus into the median eminence where these neurones terminate. The hormones are transported in the hypophysial portal system to the anterior pituitary where they regulate the pituitary's release of growth hormone (GH). At the pituitary, GHRH and SST bind to specific, and different, receptors on the surface of the somatotroph cells. GHRH very quickly causes a dose-dependent release of GH, induces transcription of the GH gene, and differentiation, growth and proliferation of the somatotroph cells. SST has an adverse effect on the GH-releasing action of GHRH, and does so in a dose dependent manner. It also suppresses GH expression, and proliferation. For a general review see references 1-5. Synthetic growth hormone-releasing compounds, such as the GH-releasing peptides (GHRP), also stimulate release from the pituitary, and these have medical applications in reversing GH deficiency and in growth promotion. At the pituitary GHRP binds to a receptor different to that of GHRH, and potentiates a dose dependent release via a different mechanism. When applied together, GHRP and GHRH interact synergistically to elicit a greater GH response, both *in vitro* and to a greater extent *in vivo*.

Desensitization is the process by which prolonged exposure to a high concentration of a hormone or transmitter substance results in decreased responsiveness of the target tissue. Despite large pituitary stores of GH, desensitization of the somatotroph to both GHRH and GHRP occur. Paradoxically, application of SST resensitizes the pituitary to both ligands. Two possible explanations for desensitization which have received support, singly or in concert, are as follows. (i) Readily releasable reserves of GH in the pituitary are depleted. GH is synthesized in the golgi apparatus and stored in the cell in vesicles. Exocytosis can only occur when a vesicle is close enough to the plasma membrane for fusion to occur. The readily releasable reserves are made up of GH in those vesicles which are sufficiently close to the plasma membrane. It is interesting, however, that GHRH can still induce a large GH response during an infusion of GHRP. This suggests that a simple depletion of readily releasable reserves is not the sole cause of desensitization. (ii) Available binding sites on the somatotroph surface are used up. For each of the factors which bind on the somatotroph cell, there will be a maximum number of binding sites that can be available at a given time. Once binding of GHRH or GHRP with its receptor has occurred, the ligand-receptor complex appears to be desensitized and must pass through a recycling process before the receptor is reactivated. While the number of surface receptors is low, the somatotroph may be less sensitive to stimulation by the ligand.

The extent to which SST restores responsiveness to GHRH/GHRP depends on the dose of SST and history of GHRH, GHRP and SST encountered. Those supporting explanation (i) for desensitization have suggested that SST acts to resensitize the somatotroph by encouraging GH synthesis. The readily releasable reserves of GH are thus replenished, allowing further GHRH/GHRP stimulation to release GH. Explanation (ii) suggests that SST encourages free receptors on the somatotroph surface, so that maximal GHRH/GHRP binding can occur.

Modelling the pituitary release mechanism

We aim to model the pituitary release mechanism so that we may later incorporate this model into a larger model of the complete system. For simplicity we only choose certain aspects from the highly complex theory of the mechanism. Since so much is unknown, and we are interested in the system as a whole, it is more important that the GH response be realistic than the underlying mechanism correct in detail.

Of the two theories for desensitization described previously, we choose the second as an approach to modelling the system. Desensitization will be facilitated by the depletion of available receptors, and SST resensitization by its ability to return free receptors to the somatotroph surface and accumulate them there. We assume that depletion of readily releasable reserves plays no part in desensitization, i.e. that reserves are effectively infinite. The model therefore need not incorporate GH synthesis; neither does it account for proliferation of the somatotroph cells. This approach is readily extendable to also describe the influence of GHRP.

So far we have developed and investigated a number of models, each consisting of a system of non-linear differential equations. Problems were encountered with previous models. These reflected the biological processes more literally, and were therefore more complicated than our current model which has been successful in qualitatively mimicking the GH response.

The Model

The model is based on the Law of Mass Action, the central process being reversible binding of GHRH (r) and its receptor (f), $r \; + \; f \; \Leftrightarrow \; u$.

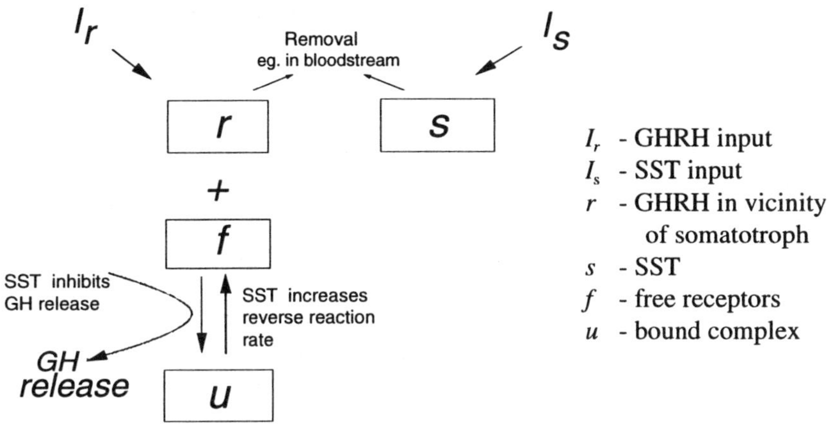

Figure 1. A schematic representation of the model.

The rate constants of the forward and back reactions are k_1 and $k_2 + k_3\phi(s)$ respectively:

$$\frac{dr}{dt} = I_r - k_1 r f + (k_2 + k_3\phi(s))u - k_6 r \tag{1.1}$$

$$\frac{df}{dt} = -k_1 r f + (k_2 + k_3\phi(s))u \tag{1.2}$$

$$\frac{du}{dt} = k_1 r f - (k_2 + k_3\phi(s))u \tag{1.3}$$

where $\phi(s) = 1/(1 + \exp(-(s - s_0)/\delta_0))$. This function switches between a high and a low value, dependent on the level of SST relative to its critical value s_0; the switch can be made more severe by decreasing δ_0. We assume that SST takes either a high or a low level, with intermediate values taken only when switching between the two, as is thought to be the case in the intact animal. The GHRH-receptor complex which results from binding (u) is thus recycled to release free receptors at a rate which is increased by the presence of SST. Desensitization results from a depletion in the proportion of free receptors, whilst resensitization depends on the return of free receptors to the surface which is encouraged by SST. I_r is the rate of input of GHRH, either from the hypothalamus or exogenously given in the *in vivo* case, or added to the medium surrounding isolated cells *in vitro*. GHRH is removed at a rate $k_6 r$ in the bloodstream or from the medium. Similarly, I_s and k_7s are the rates of input and removal of SST

$$\frac{ds}{dt} = I_s - k_7 s. \tag{1.4}$$

GH release occurs at a rate that is proportional to the amount of binding. SST decreases the constant of proportionality, inhibiting the GH response to GHRH

$$\text{release rate} \quad = \quad [k_4 + k_5(1-\phi(s))]rf. \tag{1.5}$$

Figures 2-7 show simulations of the model using a FORTRAN modelling program developed at the Babraham Institute. The parameter values used in these simulations were $k_1=10$, $k_2=-0.95$, $k_3=2$, $k_4=0.15$, $k_5=10$, $k_6=10$, $k_7=10$, $s_0=0$, $\delta_0 = 0.1$. The model demonstrates not only (i) the stimulatory effect of GHRH on GH release, (ii) the inhibitory effect of SST on GH release, but also, (iii) desensitization of the somatotroph to GHRH, and (iv) resensitization of the somatotroph under the influence of SST (Figure 2). The remaining behavioural requirements are also satisfied. The magnitude of the desensitization is dependent on the dose of GHRH (Figure 3) and the time interval between doses (Figure 4). Brief pulses of SST between GHRH pulses prevent desensitization (Figure 5). These features occur also when GHRH is infused rather than pulsed (Figures 6,7).

Qualitative analysis of behaviour

Since $df/dt + du/dt = 0$ we need not consider both f and u as one can be obtained given the other. Thus equations (1.1)-(1.3) can be reduced to 2 differential equations by substituting $u = f_0 - f$ into equations (1.1) and (1.2). Since equation (1.4) is uncoupled from equations (1.1)-(1.3), we can simplify a first consideration of the model's behaviour by assuming SST remains constant.

There is a steady state at

$$f = \frac{f_0}{1 + k_1 I_r / \{k_6[k_2 + k_3\phi(s)]\}} \qquad r = \left(\frac{k_1 I_r}{k_6[k_2+k_3\phi(s)] + k_1 I_r}\right)\left(\frac{k_2+k_3\phi(s)}{k_1} + \frac{I_r}{k_6}\right) \tag{2}$$

The f- and r-nullclines are given respectively by

$$f = \frac{f_0[k_2+k_3\phi(s)]}{k_2+k_3\phi(s)+k_1 r} \qquad f = \frac{f_0[k_2+k_3\phi(s)] + I_r - k_6 r}{k_2+k_3\phi(s)+k_1 r} \tag{3}$$

The nullclines and equilibria for I_r=0 in the absence of SST, and presence of high SST, are given in Figures 8(a) and 9(a) respectively. Trajectories for 0.25 time units from widely distributed points in the phase space represent recovery after an instantaneous pulse or perturbation in r. These reveal that the system quickly moves to the vicinity of the nullclines, before moving along the r-nullcline towards

Key for Figures 2-7. Each figure shows 7 plots, labelled (a)-(g). Plot (a) gives GHRH concentration around the somatotroph cells, and (b) the SST concentration. Proportions of free and bound receptor are given in (c) and (d). Plots (e), (f) and (g) display measures of the GH output. RATE is the actual rate of release, proportional to binding. CONC represents the concentration of GH at the pituitary given the rate at which it is released and that it is removed at a constant relative rate (k_8=10) in the bloodstream. RELEAS measures the total amount of GH released in equally spaced time intervals throughout the simulation, hence it is given as a series of vertical lines, a presentation which allows comparison of our results with experimental data.

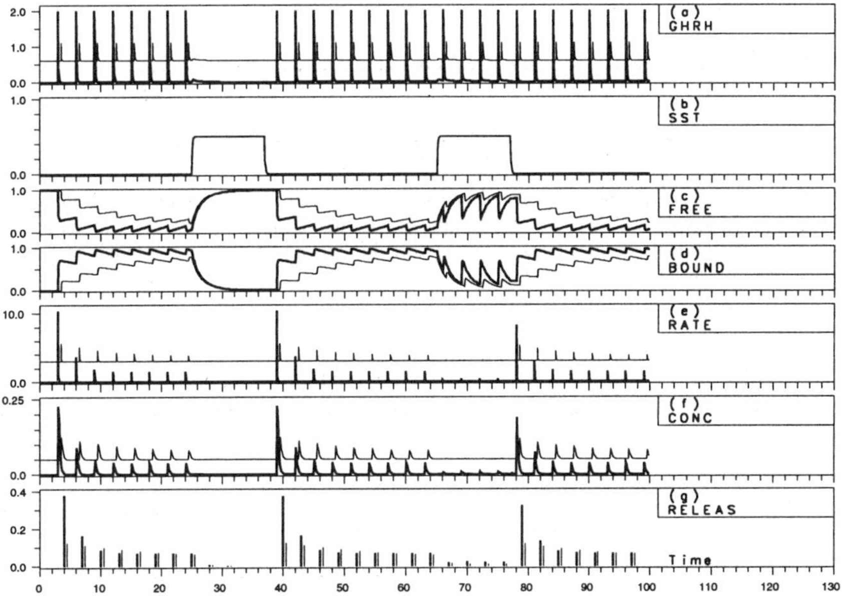

Figure 2. GHRH Pulses: desensitization; resensitization & inhibition by SST; rebound. Predicted GH response to a series of bolus injections of GHRH of 2 (thick line) and 0.5 units (thin). The thin trace is vertically displaced in plots (a), (e) and (f) for clarity. In plot (a) it is displaced by 0.6 units, in (e) by 3 units, and in (f) by 0.05 units. During the first series of injections, desensitization of the response at both GHRH levels occurs, but more so for the larger dose. An infusion of SST is given to resensitize the somatotroph. During the second series of pulses a SST infusion is applied, and release is almost completely inhibited. During the infusion the average level of free receptors is elevated, and when the infusion ends GH pulses occur again at almost as high a level as they did initially.

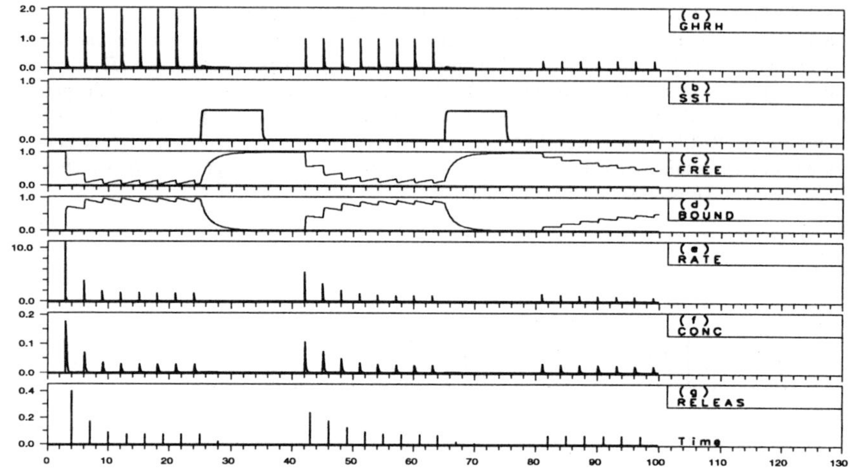

Figure 3. GHRH pulses: dose dependent desensitization. Predicted GH response to bolus injections of 2, 1 and 0.25 units of GHRH. Desensitization occurs slightly more rapidly for the larger pulses. Interestingly, in the desensitized state, there is little difference between the magnitude of the GH response to the different GHRH pulse sizes.

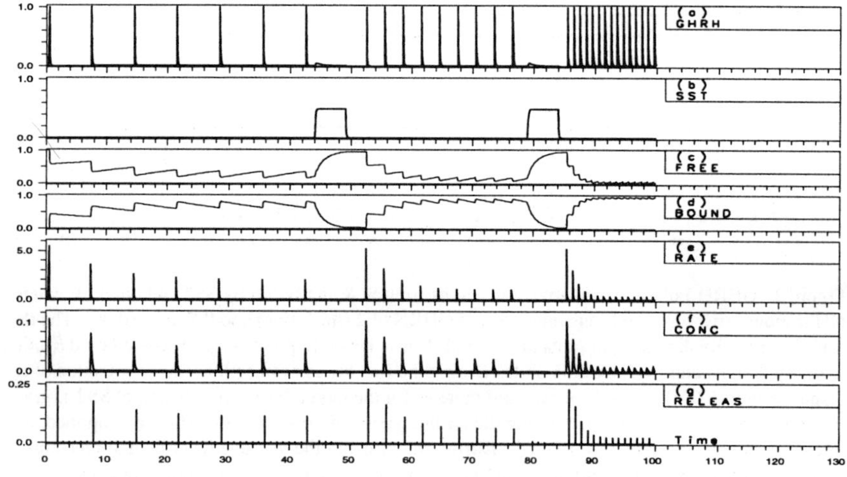

Figure 4. GHRH pulses: interval dependent desensitization. GH response to GHRH pulses at three different frequencies. Pulses of one unit are given at intervals of 7, 3 and 1 time units. Desensitization occurs over a shorter time period for more frequent pulses, although a similar number of pulses are required for desensitization in each case. More frequent pulses caused desensitization to a lower level pulse magnitude because less time for resensitization between pulses is allowed.

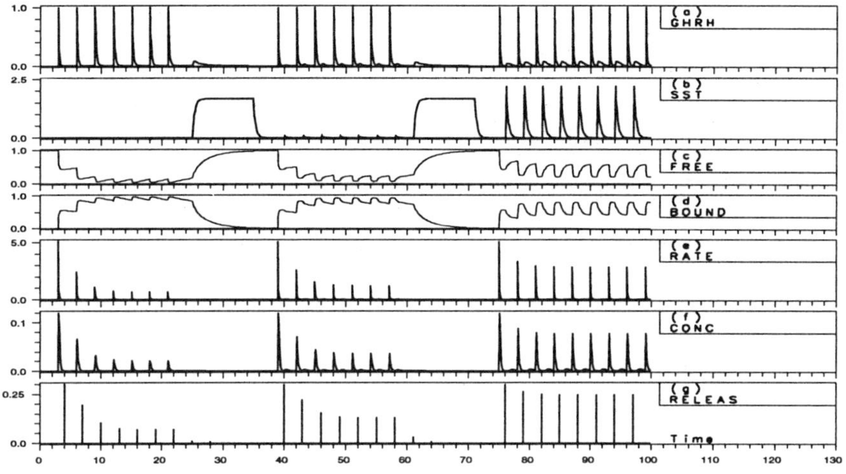

Figure 5. GHRH pulses: resensitization with SST pulses. No, small and large bolus injections (0, 0.1, 3 units respectively) of SST are given between GHRH pulses. Resensitizing effect of SST can be seen in preventing total desensitization. If infused for a longer period between pulses desensitization can be entirely prevented.

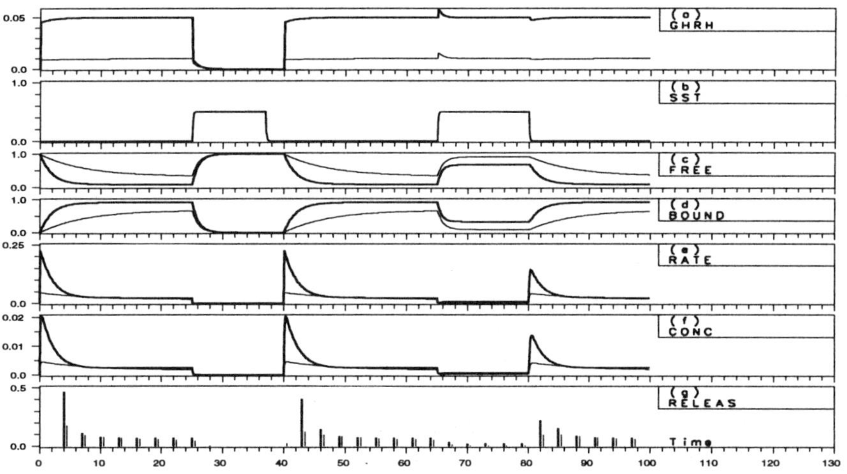

Figure 6. Infusion of GHRH: Desensitization, Resensitization, Inhibition and Rebound. Infusion analogue to Figure 2. GHRH is infused at a high and a low rate. As for Figure 2, desensitization, resensitization, almost complete inhibition and rebound are observed. Desensitization occurs more quickly and to a lower level, and the rebound is less noticeable.

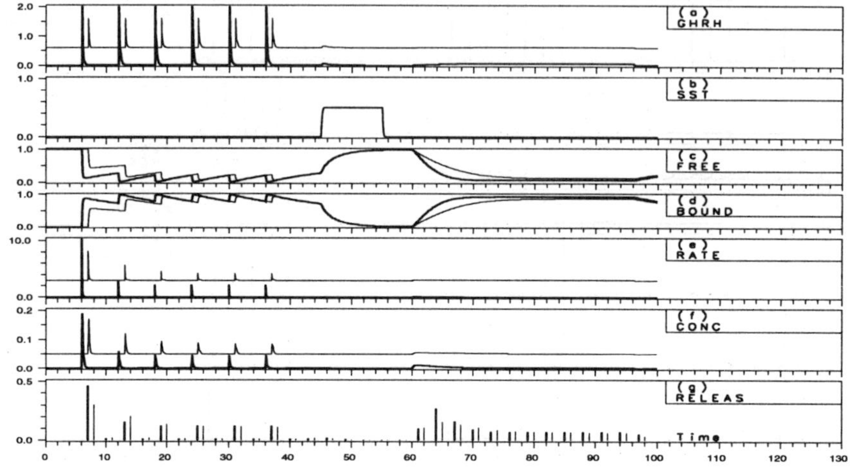

Figure 7. Comparative response to pulses and infusion of GHRH. Series of bolus injections of 2 (thick) and 1 (thin) units GHRH given at regular intervals, followed by resensitizing SST infusion. Equivalent amounts of GHRH are infused over the same time period. Vertical displacement as for Figure 2.

equilibrium - resulting in resensitization. In the absence of SST progress along the *r*-nullcline is very slow (Figure 8(b)), but the presence of SST dramatically speeds up recovery (Figure 9(b)). Given a constant infusion of GHRH the equation solutions will tend to the stable steady state ($r = I_r/k_6 = I_r/10, f = 1/(1+2I_r)$) in the absence of SST, and $f = 1/(1+I_r/10.5)$ if SST is very high.

Conclusions and further development

Despite the simplicity of the model, we found that it shows quite complex behaviour consistent with experimental findings on isolated pituitary cells *in vitro* subject to pulses and infusions of GHRH and SST. It incorporates the main features required for a model of the pituitary release mechanism. We do not make the claim that receptor desensitization is the mechanism by which desensitization in the GH response occurs, but it is sufficient to enable the model to reproduce observed behaviour. The two different biological mechanisms may be broadly equivalent when modelled; both involve depletion of an available resource during GHRH stimulated GH release, which is (possibly) replenished under the influence of SST. The receptor binding/desensitization approach will be used to also include the influence of GHRP in the model.

The model qualitatively reflects the behaviour seen in *in vitro* experiments. We aim next to produce a quantitative model with biologically plausible parameters, inputs and output. *In vivo* experimental data provide us with the GH output of the

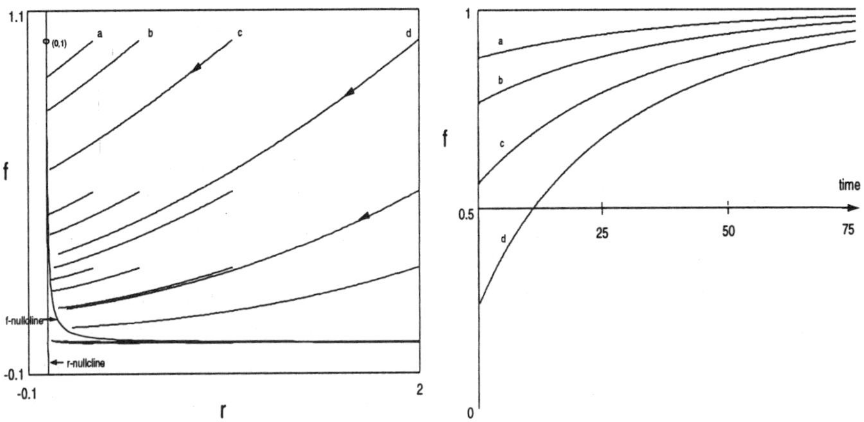

(a) Figure 8. SST absent (b)

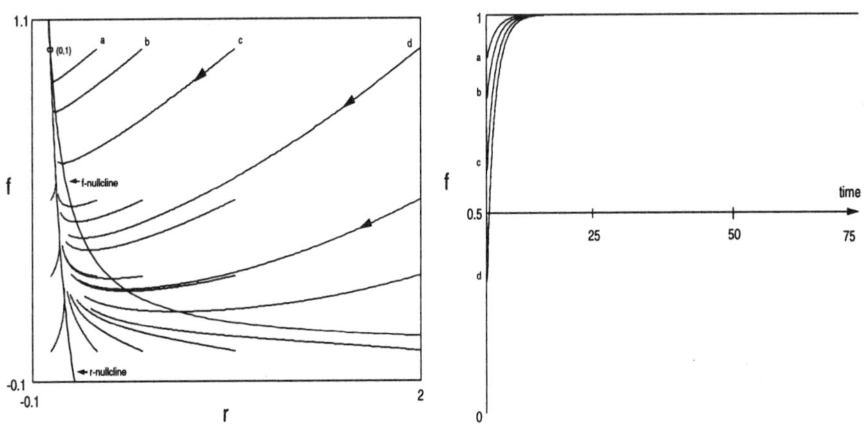

(a) Figure 9. SST very high (b)

Figures 8 & 9 (a) Phase plane diagrams for the case I_r=0, displaying nullclines, stable steady state at (r=0,f=1) and solution trajectories over 0.25 time units starting from a range of points in the phase space. (b) Plots of f against *time* starting from points in the phase space labelled a-d in (a). Figure 8: in the absence of SST ($\phi(s) = 0.5$), and Figure 9: SST very high ($\phi(s) = 1$).

pituitary, but the GHRH and SST inputs are more difficult to measure. We intend to exploit deconvolution methods to infer the inputs required for observed patterns of GH release, and this will be facilitated by the simplicity of the model.

References

1. Robinson ICAF (1991) REVIEW PAPER The growth hormone secretory pattern: a response to neuroendocrine signals. *Acta Paediatr Scand [Suppl]* **372:** 70-78.
2. Tannenbaum GS (1993) Genesis of episodic growth Hormone secretion. *J. Paediatric Endocrinol* **6**: 273-82
3. Mason WT, Dickson SL and Leng G (1993) REVIEW PAPER Control of growth hormone secretion at the single cell level. *Acta Paediatr Suppl* **388**: 84-92.
4. Dickerman ZVI, Guyda H and Tannenbaum GS (1993) Pretreatment with somatostatin analog SMS 201-995 potentiates growth hormone (GH) responsiveness to GH-releasing factor in short children. *J. Clin Endocrinol Metab* **77**: 652-657.
5. Tannenbaum GS, Painson J-C, Lengyel AMJ and Brazeau P (1989) Paradoxical enhancement of pituitary growth hormone (GH) responsiveness to GH-releasing factor in the face of high somtatostatin tone. *Endocrinology* **124**: 1380-88.

The Effect of the Embryonic Bottleneck on Vertical Microparasite Transmission

M.J. Hatcher, A.M. Dunn
Department of Pure and Applied Biology Leeds University,pab6mjh,pab6ad@gps.leeds.ac.uk

C. Tofts
Department of Computer Science,Manchester University, cmnt@cs.man.ac.uk

1 Introduction

For a parasitic organism one of the most important aspects of its life cycle is the manner in which it is transmitted to its host. In general, two methods of parasitic invasion can be distinguished:

> **Horizontal** in this case the parasite moves between unrelated hosts for example by ingestion, uptake in breathing and physical invasion - when an individual encouters infected food, air or an infected host;
>
> **Vertical** in this case the parasite [22] is transmitted from parents to offspring, usually in or on the eggs. Vertical transmission is a means employed by some intracellular parasites. Those parasites which are present in the cytoplasm of the gametes can be transmitted to the next generation of hosts. These parasites are necessarily smaller than eukaryotic cells they inhabit and usually transovarially transmitted (sperm are too small).

In many cases [1–2–14–13] the parasite may be horizontally and vertically transmitted at different stages of its life-cycle. Vertical transmission has benefits to the parasite in that it does not have to rely on random encounters between itself and potential hosts. There are many example systems of this nature.[17–16–1–2–15–6–12–21–14–7–23–4–8]

For a vertically transmitted microparasite there are two inherent bottlenecks which govern its transmission rate to the offspring of infected mothers. The first is the point of the zygote, at this point there is a single cell which contains all of the information for the production of the eventual animal. A vertically transmitted microparasite will have to be present in this cell if it is to be transmitted at all. Further, as the embryo cells divide eventually a level of differentiation amongst the cells is reached where primordial germ cells can be identified. For an ovarial microparasite transmission to future generations of the animal is only assured if it has a presence in this(ese) putative germ cells. A similar process of cell division may occur within the germline, that is the cells continue to divide and the parasite will only be successfully transmitted if it is present in the eventual gametes.

For a vertically transmitted parasite to invade a population it has to cause some change in the infected organism, for example manipluating its sex ratio [22] or enhancing host survival and/or fecundity [6] (improving its fitness). If the parasite *doubles* host fitness then to invade a population it still needs a transmission rate of at least 50% [3–26–11]. The presence of a parasite in a zygote of an animal is the result of it successfully surmounting the two barriers above to its transmission, which define the adult to offspring transmission rate. There are many potential methods whereby the parasite may surmount these two barriers. The first is by sheer force of numbers; given a large initial dose then there will be a very high probability that any eventual egg cells will contain parasites. A second is to reproduce within the host animals; as host cells divide, this will maintain relative numbers of parasite and host cells and result in a higher probability of successful transmission. In both of the above cases there is liable to be a high dosage of the parasite in the somatic tissue which could reduce host fitness. As the transmission rate of parasite in the population is the result of successful host reproduction selection will favour minimum pathogenicity [22]. A further possibility is that the parasite 'tracks' the tissue types that it is interested in (gonadal germline) and only reproduces when necessary. Clearly of the postulated mechanisms the last is the most complex, requiring that the parasite be capable of detecting and segregating towards potential germ cell tissue.

One important aspect of the interaction between the hosts and the parasites is the cost to the parasite of imposing too heavy a burden upon its hosts [5]. If the parasite reduces the fitness of the host then its transmission rate must be increased by a compensatory amount [26–11]. For a vertically transmitted parasite there are benefits to be gained by **only** being present in the germ tissue. However, such accurate targeting of the parasite could make it relatively easy for the host to resist its invasion.

The microsporidian parasite of *Gammarus duebeni* [7] shows a transmission rate from mother to offspring of approximately 70% [7]. We examine the transmission of the parasite under two scenarios; in the first a zygote has an initial load of parasite, which will not breed, and we examine the probability distribution of the parasite in subsequent generations of the cells; in the second, we assume that the parasite breeds at the same rate or less than that of the host cells and examine the distribution of parasite within the daughter cells.

Both models assume that parasites are physically incapable of crossing cell boundaries and are therefore confined within their current host cell lineage.

This model can account for within cell lineage transmission of the parasite both in the embryo before the differentiation of gonad tissue or germ cells, and the subsequent development of the gametes if they arise by division of the gonad or germ cells. The probability of successful transmission is a result of the relationship between the two given transmission rates.

2 Models

The most important property of the transmission to particular host cells of our microsporidian parasite is that **it is** dependent upon the transmission path. The parasite *cannot* move itself between cells, so once a cell line is free of the parasite it will remain so. Hence an analysis of the distribution of the parasite which ignores its transmission mechanism will greatly *underestimate* the number of parasite free cells. To give an accurate account of the parasite distribution we must take account of the path followed from the original infected cell to any particular generation of cells.

We present two different accounts of the transmission of the parasite to host cells from the original zygote. As a null model we assume that a dosage of parasite in the zygote is inherited and that the parasite will not divide until it reaches its target tissue. This approach will certainly achieve a minimal parasitic load for somatic tissue. However, to achieve the (necessary) transmission rates we observe an unfeasibly high initial loading may be required.

In the second model we assume that a dosage of parasite in the zygote is inherited and the parasite breeds with a rate less[a] than or equal to that of the host cells. In this case a lower intial dose of parasite will lead to a higher probability of successful transmission, but consequently the parasitic load in somatic tissue will be that much higher.

Throughout our models we shall assume that the host cells divide synchronously and that the parasites do likewise. Further, we assume that there is no difference in reproduction rates between cells or parasites in different cells. Given that an embryo is small and the timescale we shall be modelling it over (< 10 host cell divisions) is short both of these assumptions are at least (*ab initio*) reasonable.

---

[a]See the discussion for why it is unlikely that parasite breeding rate will exceed that of the host cells

2.1 The Null Model

We exploit probabilistic process algebra[10–24–25] developed from Milner's SCCS[19–20] to give a description of the evolution of the cell/parasite system. Consider a cell which contains a number m of the microparasite. If the division of the parasites to its daughter cells is unbiased then we should expect to see the following evolution of the process:

$$Cell(x) \stackrel{def}{=} \sum_{i=0}^{i=x} \begin{pmatrix} i \\ x \end{pmatrix} .\surd : (Cell(i) \times Cell(x-i))$$

So the evolution of the distribution of the parasite within an embryo given an intial dose of id is described by the following process:

$$Emb(id) \stackrel{def}{=} Cell(id)$$

For our purposes we are interested in the distribution of the parasites in individual cells, rather than in the complete system. By simplifying the above definition[b] we obtain the following description of the evolution of parasite number in the cells:

$$Cell'(x) \stackrel{def}{=} \sum_{i=0}^{i=x} \begin{pmatrix} i \\ x \end{pmatrix} .\surd : Cell'(i)$$

Consider an embryo with an intial dose of 2 parasites then the system will evolve as follows:

$$\begin{aligned} Emb(2) &= Cell(2) \\ &= \tfrac{1}{4}.\surd : Cell(0) + \tfrac{1}{2}.\surd : Cell(1) + \tfrac{1}{4}.\surd : Cell(2) \\ &= \tfrac{1}{4}.\surd : 1.\surd : Cell(0) + \tfrac{1}{2}.\surd : (\tfrac{1}{2}.\surd : Cell(0) + \tfrac{1}{2}.\surd : Cell(1)) \\ &\quad + \tfrac{1}{4}.\surd : (\tfrac{1}{4}.\surd : Cell(0) + \tfrac{1}{2}.\surd : Cell(1) + \tfrac{1}{4}.\surd : Cell(2)) \\ &\stackrel{p}{=} \tfrac{9}{16}.\surd : \surd : Cell(0) + \tfrac{6}{16}.\surd : \surd : Cell(1) + \tfrac{1}{16}.\surd : \surd : Cell(2) \end{aligned}$$

We could continue to expand the definitions above to obtain expressions for the probability of observing a particular dose of the parasite at subsequent generations of host cells. From the above structure it is possible to observe that the probability of a cell having a dose d of the parasite at generation g is given by the probability of there being a parental cell with d or more parasites and the probability of that cell giving d of its parasites to its daughters.

We generate the probability $P(g, d, id)$ that a cell in generation g will have d parasites given that the intial dosage of its ultimate parent cell was id

[b] This amounts to assuming that cells in an individual embryo are independent, this approximation is used to permit efficeint calculation and since we cannot measure all of the cells within a particular embryo.

(i.e., the original zygote). This probability can be described by the following recurrence relation.

$$P(0,k,id) = \begin{cases} 1 \text{ if } k = id \\ 0 \text{ otherwise} \end{cases}$$

$$P(g,d,id) = \sum_{k=id}^{k=d} \begin{pmatrix} d \\ k \end{pmatrix} \left(\frac{1}{2}\right)^k P(g-1,k,id)$$

In Figure 1 we present probability distributions for initial loads of 8 and 24 of the parasite, for generations 3 to 6 of the host.

2.2 *The Growth Model*

In this case we assume that the parasite can breed by cell doubling. We assume that the parasite breeds at no greater rate than the host does and shall generate the distribution of the parasite just before the cell next divides. For a sample system evolution see Figure 2.

In this case we shall assume that there is a signal (in real parasites this would be generated internally) which tells the parasites when to breed. This gives us a simple way of encoding different rates of parasite division with respect to that of the cells, since this is the only clock we can observe experimentally. To achieve this we redefine our cell process as follows:

$$\begin{aligned} BC(x) \quad &\stackrel{def}{=} \quad \textstyle\sum_{i=0}^{i=x} \begin{pmatrix} i \\ x \end{pmatrix} . \surd : BC(i) \\ &\quad + \textstyle\sum_{i=0}^{i=x} \begin{pmatrix} i \\ x \end{pmatrix} .breed : BC(2i) \end{aligned}$$

So in the above if a *breed* signal is input, then the cell will contain twice as many parasites as it started with prior to division. In this case we must supply a process to define when the parasite will breed with respect to the cells:

$$Br \stackrel{def}{=} \surd : \surd : \overline{breed} : \surd : \overline{breed} : 0$$

which defines the parasite as breeding at generations 2 and 4.

So the following process generates the probability distribution for the parasite over the first four cell divisions:

$$Bemb(id) \stackrel{def}{=} (BC(id) \times Br)\lceil\{\surd\}$$

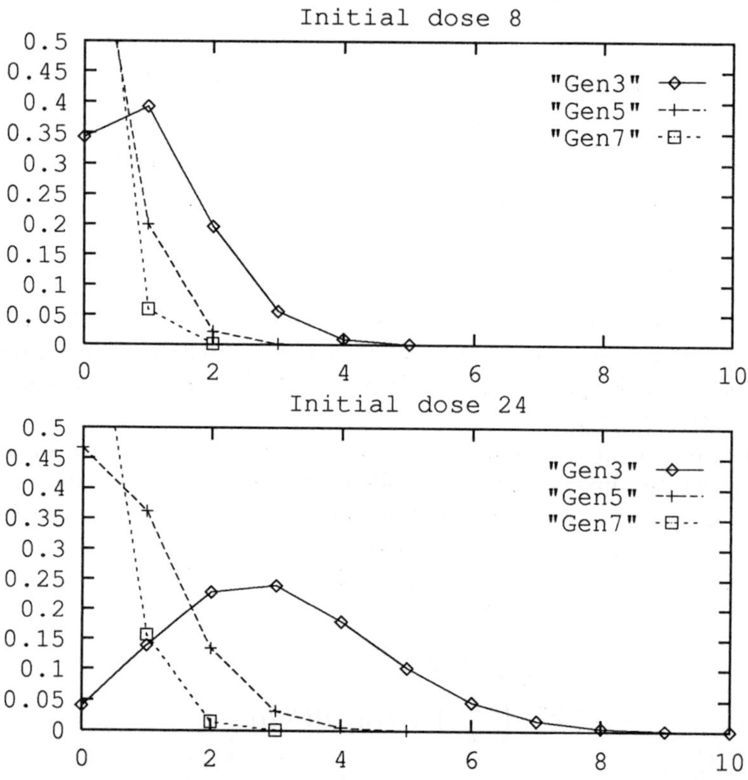

Figure 1: The probability distribution of microparasites given they do not breed for initial doses of 8 and 24 in generations 3, 5 and 7 inclusive of the embryo's cells. Horizontal axis is the number of parasites in a cell. Vertical axis is the probability of that number occuring.

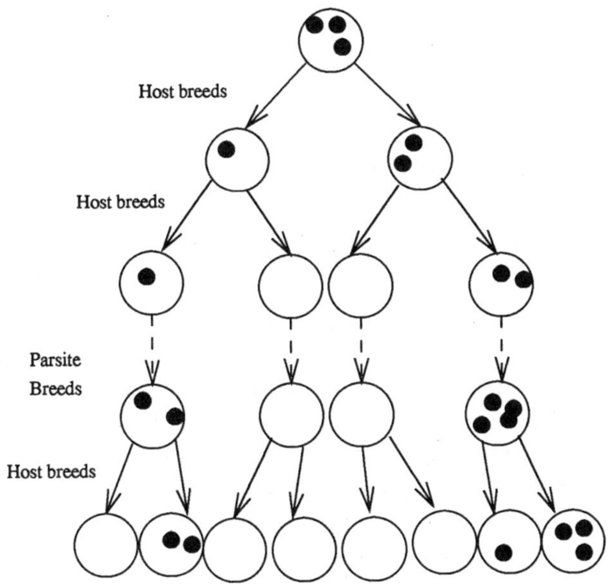

Figure 2: A possible development path for an embryo with 3 parasites initially which breed at half the rate of the cell division rate.

To give a recurrence account of the above description we define the predicate *breed*(g) which defines whether the parasite breeds at this generation, and *pgens*(g) which defines the number of parasite generations that have occurred before this cell generation. In the process theoretic account these predicates are given by the process Br.

We generate the probability $PG(g, i, d)$ that a cell in generation g will have i parasites given that the intial dosage of its ultimate parent cell was d, assuming that the division rate of the parasite is identical to that of the cell. This probability can be described by the following recurrence relation:

$$PG(0, i, d) = \begin{cases} 1 \text{ if } i = d \\ 0 \text{ otherwise} \end{cases}$$

$$PG(g, i, d) = \begin{cases} \sum_{k=i}^{k=d2^{pgens(g)}} \binom{i}{k} \left(\frac{1}{2}\right)^{2k} PG(g-1, k, d) \text{ not } breed(g) \\ \sum_{k=\frac{i}{2}}^{k=d2^{(pgens(g)-1)}} \binom{\frac{i}{2}}{k} \left(\frac{1}{2}\right)^{2k} PG(g-1, k, d) \text{ otherwise} \end{cases}$$

We evaluated the above recurrence in two circumstances, the first when the reproduction rate of the parasite was the same as that of the host $\forall g.breed(g)$ and $pgens(g) = g$. Parasite distributions for initial loads of 4 and 8 and generations 3 to 7 are presented in Figure 3. In the second case we assume that the parasite breeds at generations 2,4,6,8 (as this is the breeding pattern observed in our experimental system [9]) and probability distributions for the same initial loads and generations are given in Figure 4.

3 Conclusions and Discussion

Unsurprisingly if the parasite does not breed in the host embryo then to achieve the necessary level of infection the initial number of parasites will have to be very large. In the following table:

Generation	2	3	4	5	6	7
Initial Dose	3	6	11	22	45	89

the initial number of parasites required to achieve a 50% transmission rate to any particular generation of cells is presented, as this can effectively be regarded as a lower bound on transmission rate for the parasite population to be stable. To invade a population with this level of transmission the parasite will have to exert a considerable effect upon its host, for instance doubling its fitness or producing 100% feminization [3–26–11]. Even this may not be enough

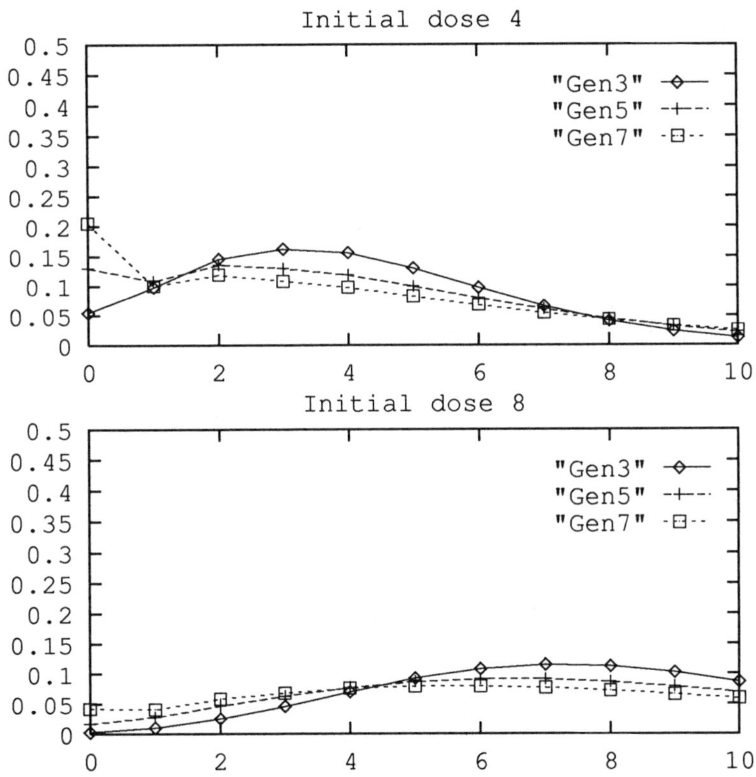

Figure 3: The probability distribution of microparasites given they breed at the same rate as the host for initial doses of 4 and 8 in generations 3, 5 and 7 of the host embryo's cells. Horizontal axis is the number of parasites in a cell. Vertical axis is the probability of that number occuring.

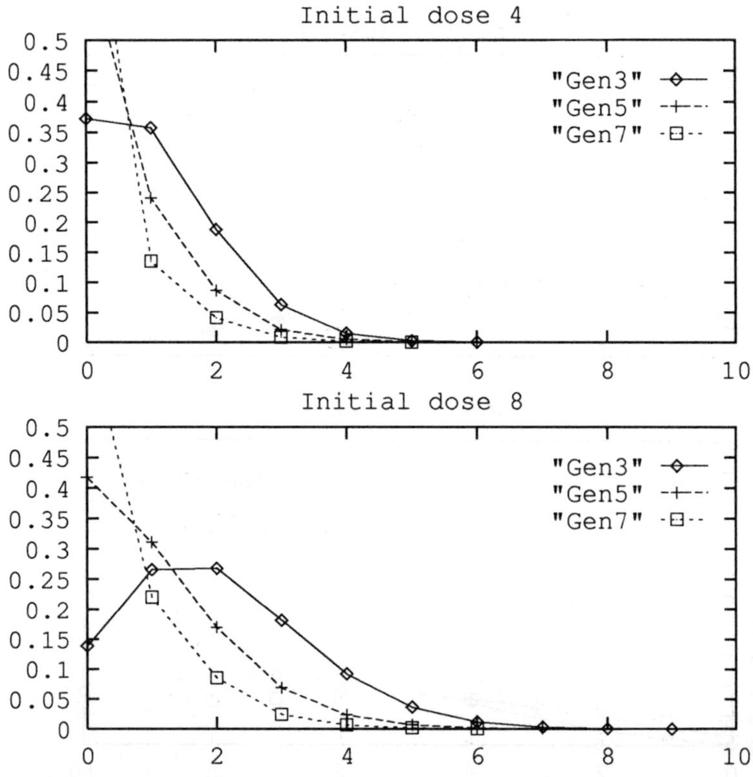

Figure 4: The probability distribution of microparasites given they breed at half the rate of the host for initial doses of 4 and 8 in generations 3, 5 and 7 of the host embryo's cells. Horizontal axis is the number of parasites in a cell. Vertical axis is the probability of that number occuring.

as there are other lossy processes involved, if for instance the gonadal or germ-cell tissue further divides to form gametes.

For our parasite transmission from adult to embryo is 96% [9], and from adult to young is 69% [7–22–8], hence a transmission rate of about 72% from embryo to gonad is the minimum consistent with these values. The following table:

Generation	2	3	4	5
Initial Dose	5	9	38	77

gives the initial parasite loads we would have to observe in order that this transmission rate (72%) could be achieved. These values should be compared with observed intial doses in the region of 1-19 [9] with a mean initial dose of 8.

In the case that the parasites breed at the same rate as the host cell (Figure 3), we observe that a far lower initial load of parasites can achieve a sufficiently high level of transmission. However, the distributions of the parasites observed in this case show an interesting property. The proportion of cells with no parasites *increases* monotonically (since once a cell is free of parasites all of its daughter cells will be, and new cells become parasite free as a result of stochastic interactions). Furthermore, since the host and parasite are dividing at the same rate, the average load of parasites per host is *constant*. Hence, over the generations, an increasing burden (Figure 3) of the parasite falls upon those cells that are infected. As the parasite is a metabolic burden upon the host cell then infected cells will be threatened by the presence of these high parasite numbers. In this case even if the parasite manages to reach the cells it is 'interested' in then it could kill the cell by overloading.

For our parasite we observe [9] that the parasite breeds at host generations 2, 4 and 6 (observations only go as far as generation 7). Whilst Figure 4 indicates that with an initial load of 8 this would give a reasonable transmission rate, we find that the distribution of the parasite within the cells is *not* consistent with this model. There is a greater prevelance of unifected cells than is consistent either with our model or even the basic transmission rate. Essentially, the observed rates of transmission and parasite distribution must arise from an ability on the part of the parasite to detect host cells which a likely to become germline. If the parasite can detect the parental cells of the germline then it can respond in two fashions, either by segregating preferentially to this cell line, or by breeding more rapidly within this cell line.

There is an interesting parallel between the behaviour of microparasites and that of organelles [18–11]. Organelles can be considered to be vertically transmitted parasites which confer a benefit upon their hosts[6–11]. The problem of distributing organelles within an organism is similar to that of parasite transmission (except that the organelle is probably targetting all of the cells).

In the case of organelles however, the maintenance of average load per cell is important, and therefore they *must* breed at a comparable rate to the host cells. In the absence of any mechanism to prevent overloading over time the organelles will tend to become concentrated in particular cell lines and this could prove fatal to these lineages.

By allowing for more classes of cell it would be possible to extend our model to examine the consequences of differential segregation or breeding rate of the parasite within these cells. Such models will be developed as suitable experimental techniques permit their resolution. An SML implementation of the calculation program for the models can be obtained by email from CT.

Acknowledgements

M.J. Hatcher is supported by NERC grant (GST/02/688), A.M. Dunn is supported by a NERC Fellowship (GT5/F/92/ALS/1), and C. Tofts is supported by an EPSRC Advanced Fellowship (B/93/AF/1556). The authors also acknowledge support from a grant from the Leverhulme Trust (F/122/AK). We thank R. Terry, D. Blakeley and S. Olbison for technical assistance, and English Nature for permission to collect animals.

References

1. Andreadis, T.G. & Hall, D.W., Significance of transovarial infections of *Amblyospora* sp. (Microspora: Thelohaniidae) in relation to parasite maintenance in the mosquito *Culex salinarius*. J. Invertebr. Pathol. 34, 152-157, 1979.
2. Andreadis, T.G., Experimental transmission of a microsporidian pathogen from mosquitos to alternate copepod host. Proceedings of the National Academy of Sciences, USA 82, 5574-5577, 1985.
3. Bull, J.J., The evolution of sex determining mechanisms, Benjamin Cummings, London, 1983.
4. Callaini, G., Riparbelli, M.G. & Dallai, R., The distribution of cytoplasmic bacteria in the early *Drosophila* embryos is mediated by astral microtubules. Journal of Cell Science 107, 673-682, 1994.
5. Dawkins, R., Parasite desiderata lists and the paradox of the organism. Parasitology 100, S63-S73, 1990.
6. Douglas, A.E., Mycetocyte symbiosis in insects. Biological Review 64, 409-434, 1989.
7. Dunn, A.M., Adams, J. and Smith, J.E. Transovarial transmission and sex ratio distortion by a microsporidian parasite in a shrimp. Journal of Invertebrate Pathology 61, 248-252, 1993.
8. Dunn, A.M., Hatcher, M.J. & Adams, J. Parasitic sex ratio distortion and host sexual strategies: factors underlying parasite prevalence in *Gammarus duebeni* populations. In submission to Journal of Evolutionary Biology, 1995.
9. Dunn, A.M., Hatcher, M.J., Terry, S. & Tofts, C., Host parasite sexual strategies: transovarial transmission of a microsporidian sex ratio distorter in *Gammarus duebeni*, in press, Parasitology, 1996.

10. van Glabbek R., S. A. Smolka, B. Steffen & Tofts, C., Reactive, Generative and Stratified Models of Probabilistic Processes, Proceedings LICS, 1990.
11. Hatcher, M.J. & Dunn, A.M., Evolutionary consequences of sex ratio distortion by cytoplasmically inherited feminizing factors. Philosophical Transactions of the Royal Society, London, B, 348, 445-456, 1995.
12. Houk, E.J. & Griffiths, G.W., Intracellular symbiotes of the Homoptera. Annual Review of Entomology. 25, 161-187, 1980.
13. Hurst L.D., The incidences and evolution of cytoplasmic male killers. Proceedings of the Royal Society of London B. 244, 91-99, 1991.
14. Hurst, G.D.D., Majerus, M.E.N. & Walker, L.E., Cytoplasmic male killing elements in *Adalia bipunctata* (Linnaeus)(Coloeptera: Coccinellidae). Heredity 69, 84-91, 1992.
15. Juchault, P., Moraga, D. & Legrand-Hamelin, E. Microorganismes symbiotiques et speciation. Bulletin de la Societe Zoologique de France 111, 135-147, 1987.
16. Kellen, W.R. & Wills, W., The transovarial transmission of *Thelohania californica* Kellen and Lipa in *Culex tarsalis* Coquillett. Journal of Insect Pathology 4, 321-326, 1962.
17. Malogolowkin, C., Maternally inherited 'sex ratio' conditions in *Drosophila willistoni* and *Drosophila paulistorum*. Genetics 43, 274-286, 1958.
18. Margulis, L., Origin of Eukaryotic Cells. New Haven: Yale University Press, 1970.
19. R. Milner, Calculi for Synchrony and Asynchrony, Theoretical Computer Science 25(3), pp 267-310, 1983.
20. R. Milner, Communication and Concurrency, Prentice Hall, 1990.
21. Rigaud, T. & Juchault, P. , Genetic control of the vertical transmission of a cytoplasmic sex factor in *Armadillidium vulgare* Latr. (Crustacea, Oniscidea). Heredity 68, 47-52, 1992.
22. Smith, J.E. & Dunn, A.M.,Transovarial transmission. Parasitology Today 7(6), 146-148, 1991.
23. Stouthamer, R. & Werren, J.H., Microbes associated with parthogenesis in wasps of the genus *Trichogramma*. Journal of Invertebrate Pathology 61, 6-9, 1993.
24. Tofts C., A Synchronous Calculus of Relative Frequency, CONCUR '90, Springer Verlag, LNCS 458, 1990.
25. C. Tofts, Processes with Probabilities, Priorities and Time, Formal Aspects of Computer Science 6(5):536-565, 1994.
26. Werren, J.H., The coevolution of autosomal and cytoplasmic sex ratio factors. Journal of Theoretical Biology 124, 313-334, 1987.

ON THE SIMULATION OF HEALTH WITH THE TISSUE AUTOMAT

G. ZAJICEK
H.H.Humphrey Center for Experimental Medicine and Cancer Research
Hebrew University-Hadassah Medical School, P.O.Box 12272
Jerusalem 91120, Israel.

1 Introduction

1.1 Structured turnover

Recent innovations in computer modeling, e.g., neural nets and artificial life [1,2], provide new tools for simulating complex non-linear medical systems. These programming tools have the following properties: Elements in the system operate in parallel, they interact in a non-linear fashion, the system evolves in an unpredictable way and its structure emerges. Despite their great flexibility for simulating complex phenomena, they lack two essential ingredients of the organism: Structured turnover, and negative feedback.

The turnover of ingredients in the organism proceeds in streams. Molecules are replaced according to the first in - first out rule, the first that are formed are also the first to leave, like proteins synthesized in the pancreas. The protein is born when its DNA code is translated into RNA that serves as a template for its assembly. As the molecule is being synthesized, it gradually streams from nucleus to cell surface, through the endoplasmic reticulum. Each gene may be regarded as a source of a tiny stream of matter in which proteins are gradually synthesized. Since the organism maintains steady state, for each molecule that leaves the cell, a new one is formed and starts streaming outward. Structured turnover is manifested also at the tissue level. The organism is made of cells that aggregate to tissues, and tissues to organs that make the human organism. Cells stream from tissue interior, outward, which will be illustrated on the crypt-villus unit of the inner gut lining [3,4].

1.2 Crypt-villus unit

All cells in the unit are the progeny of one cell, called the determined stem cell (DS) (Fig. 1). When the DS divides, one of its progeny replaces the parent and remains a DS, and the other starts streaming outward. It may differentiate along four cell lines; absorptive, goblet, endocrine, and Paneth. When entering a differentiation pathway it turns into a committed stem cell (CS), that differentiates while streaming outward, and when arriving at the unit periphery it dies. If for example the DS is scheduled to generate a mucus producing cell line, one of its progeny becomes a mucus-CS with a big nucleus and scanty cytoplasm. Gradually the cell prepares the necessary organelles, and then starts forming its product. This process is called differentiation. Fig. 1 depicts only one differentiation pathway. Differentiation is time dependent. The older the cell the more differentiated it becomes. In the crypt-villus unit the entire differentiation lasts several days.

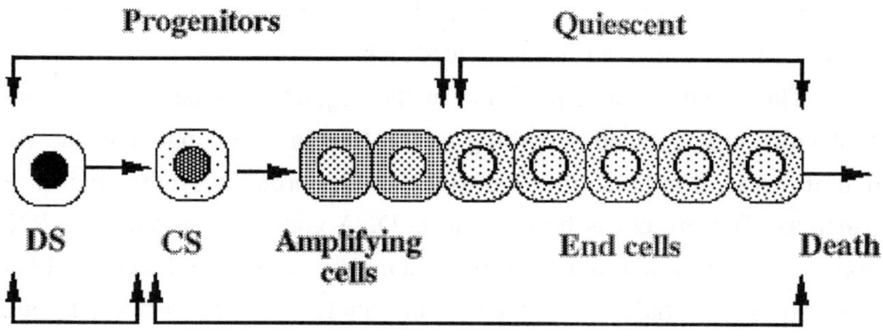

Cells in the inner part of the unit proliferate and are called amplifying progenitors (P-cells). Later on they lose the capacity to synthesize DNA and become non-dividing end cells (Q-cells). When reaching the unit border they die and slough off into the gut cavity. In parenchymatous organs like liver, cells die by apoptosis. In the healthy unit, cell death is a differentiation state. This differentiation dependent cell death has to be distinguished from random cell death, or necrosis, that occurs only in disease.

When the DS divides, one of its progeny replaces the parent and remains a DS and the other becomes a streaming transitional cell. A DS division generally results in two different cells, a stem and a transitional, and is called an asymmetric division. When a transitional cell divides both its descendants remain transitional. Since both cells are the same, the division is called symmetric. All transitional cells in a unit are descendants of one DS, they belong to one clone of cells, and the unit is mono-clonal. All cells except the DS are transitional and short lived. In the gut they live several days. Only the DS is permanent. It is the source of a cell stream from tissue interior and outward. When transitional cells are destroyed they are replenished by the DS. Yet when the DS dies, transitional cells continue streaming outward and are not replaced by new cells. After they die the unit disappears. The set of stem cells is invariant and exists as long as the organism does. It stores the information for generating tissues units.

1.3 Feedback

Cells in the organism interact by their products that can be either stimulatory or inhibitory. This interaction is called here feedback. Generally negative feedback stabilizes the system while positive feedback de stabilizes it. Cellular automata and genetic algorithms are open evolving systems without internal feedback. Their evolution may be constrained [5] appearing as if having a feedback yet this feedback is too general while in the organism it operates at all levels.

1.4 Extensive and intensive magnitudes

Magnitudes are divided into two sorts, extensive and intensive. Extensive magnitudes allow both comparison and measurement. Intensive magnitudes allow comparison but not measurement. This is because they do not admit of units which can be added together [6]. Some tissue properties are clearly extensive, e.g., size and cell count, yet many properties that concern medicine, are intensive, e.g., tissue pathology or health. Traditionally, models of medical systems simulated extensive magnitudes, e.g., a heart-lung model made of differential equations, describing the exchange of gases and energy. These models generally simulate physiological mechanisms, yet medicine is more concerned with health and disease that cannot be simulated by them. The heart-lung system is dynamic and complex. It traverses through attractor states known in medicine as diseases. Each attractor is endowed with a property known as health, and the organism drives this heart-lung system through the healthiest attractors. How does one simulate such a multidimensional trajectory that attempts to maximize health? Only with models of intensive magnitudes.

These ideas are illustrated here on the Tissue automat (TA), that simulates medical systems with intensive and extensive magnitudes.

2 Tissue automat (TA)

The tissue unit model or tissue automat (TA) (Fig 2) consists of a cell column with a stem cell at its bottom. Then follow progenitor (P) cells and finally non proliferating (Q) cells. Cell division in P creates a cell stream outward whose velocity depends on cell production rate. Before dividing, the cell progresses through several preparatory phases, known as cell cycle, during which it doubles its DNA content (S-phase) and ultimately divides (M-phase). Cell production is controlled by two negative feedback loops: Cell turnover and P-compartment size are inversely proportional to TA size. P-compartment size is also inversely proportional to Q-compartment size. The arrow represents a product made by Q-cells that inhibits P-cell division.

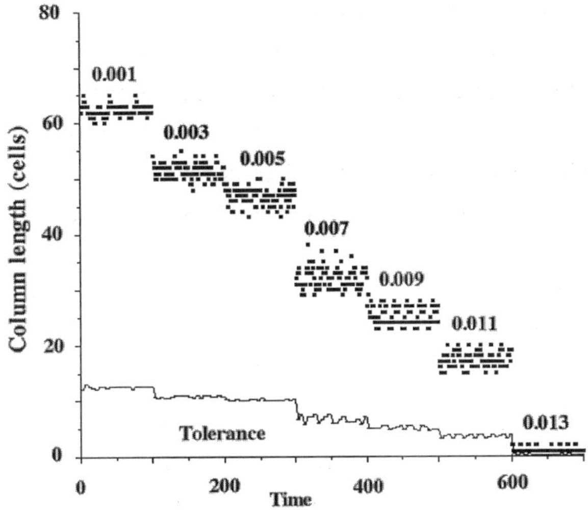

Fig. 2

The environment interacts with the TA by two death processes: 1. Differentiation dependent death eliminating a fraction of mature cells. 2. Random death represented here by x-irradiation that inhibits cell division in the P-compartment. The program was written in Mathematica [7]. It simulates cell kinetic experiments, e.g., labeling with tritiated thymidine (autoradiography), x-irradiation, regeneration, and first stages of neoplastic progression [3,4].

2.1 Steady state configurations with rising death rate

The experiment starts with a single stem cell that divides and the unit elongates. When birth rate equals death rate, the TA reaches its steady state. Figure 2 depicts TA response to a rising death rate. Only steady state values are depicted. The transitions between them are not shown. At each level the points oscillate around a mean. Time is given in arbitrary units. The lower curve depicts TA tolerance, a measure of TA "health" that was computed as follows: Tolerance = 0.1 (TA length + P-compartment size + Q-compartment size). Its magnitude is given in cells or cell units. The TA continually attempts to maximize its tolerance, adjusting the three components accordingly. Q-cells are the first to be affected since they are the oldest. In order to replenish the lost cells, more and more progenitors are recruited and the P-compartment elongates (Fig. 3). At a death rate of 0.007 all cells are progenitors. Higher death rates eliminate also P-cells, cell production declines and TA shrinks. When death rate rises above 0.013 the TA dies.

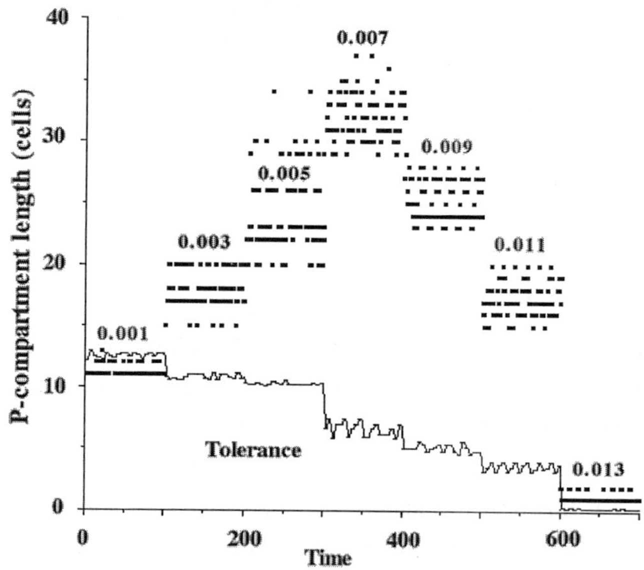

Fig. 3

2.3 TA resection

In the following experiments 95% of TA cells were resected (Fig. 4). At death rate of 0.003 TA replenishes all its cells and its length is restored (Fig. 4, A). In the next experiment TA was resected at death rate 0.003 (Fig. 4, B). As it regenerated death rate rose to 0.005 and the TA could not regain its initial length. When resected at death rate of 0.007 the TA could not regenerate at all and died (Fig. 4, C). The TA could still regenerate when less than 10% of its cells were removed.

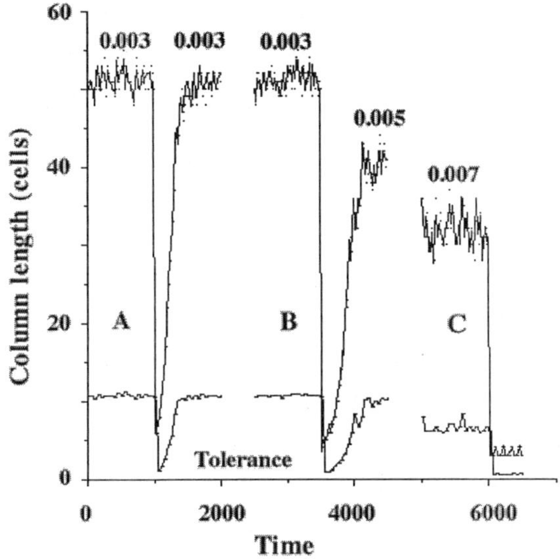

Fig. 4

2.4 Tolerance

Tolerance is an indicator of TA reserves. It depends on the environmental death rate. The higher the death rate the lower tolerance (Fig. 2). Tolerance can be estimated by challenging the TA, e.g., by resection. At high tolerances, the TA restores its initial length. When tolerance declines below a certain value, the TA does not regenerate and dies. The TA microcosm is represented by a tolerance surface (Fig. 5). Tolerance depends on three factors; TA length, P- and Q-compartment sizes, and is four dimensional. Three dimensions represent TA states (TA length, P- and Q-compartment size) and the fourth, its tolerance. Since a four dimensional surface is difficult to depict we shall restrict our discussion to a three dimensional surface that depicts the relationship between TA length (x-coordinate), P-compartment size (y-coordinate), and tolerance (z-coordinate) (Fig. 5). The x and y coordinates are death rates extending from 0.001 to 0.013. The linear relationship between TA length and death rate is evident along a line that extends from (0.001, 0.001) to (0.013,

0.001) (compare with Fig. 2). P-compartment change is best observed between the coordinates (0.013, 0.001) and (0.013, 0.013) (compare with Fig. 3). Tolerance trajectory of Fig. 2, extends from (0.001, 0.001) to (0.013, 0.013).

2.5 Aging

The white diagonal line represents the tolerance change with rising death rate, and was depicted also in Fig. 2. It illustrates a salient feature of aging. Suppose that in this microcosm, a TA is born when death rate is 0.001. From then and on, death rate continually rises until at 0.013 the TA dies. The diagonal represents the declining tolerance during aging, known also as vitality. The white diagonal is the TA aging trajectory extending from birth to death. Age is represented here by distance. A TA at point B (Fig. 5) is older than that at point A. We may even specify how much older it is : = SQRT[(0.005-0.003)**2+(0.005-0.003)**2].

Age may thus be represented by two measures, chronological, known also as physical, and biological [8,9]. Actually the tolerance set is a set of attractors, and does not represent inter-attractor transitions. The (chronological) time necessary for a TA to settle at an attractor, varies and is not represented by the tolerance set. A TA may attain an attractor quickly or slowly. The set depicts tolerance change, and not the rate of change, and this change is expressed by biological time. The difference between the two age measures is real. A chronologically old man may still be biologically young, and vice versa. Vitality is proportional to biological age and not to the chronological.

2.6 Acute and chronic pathology

The three resection experiments (Fig. 4) are depicted as three trajectories connecting points A, B and C (Fig. 5). Each trajectory starts on the aging trajectory. Resection drives the TA away from the aging trajectory and returns during regeneration. The first resection returns to its initial state (Fig. 5-A5). Following the second resection death rate rose from 0.003 to 0.005. The TA starts at A and joins the aging trajectory at B (Fig. 5 A-B).

The third resection was fatal and the trajectory starting at C ended in the death region (Fig. 5-C). The trajectories represent three pathologies, acute, chronic and fatal. A pathology is cured when the trajectory joins the aging trajectory .

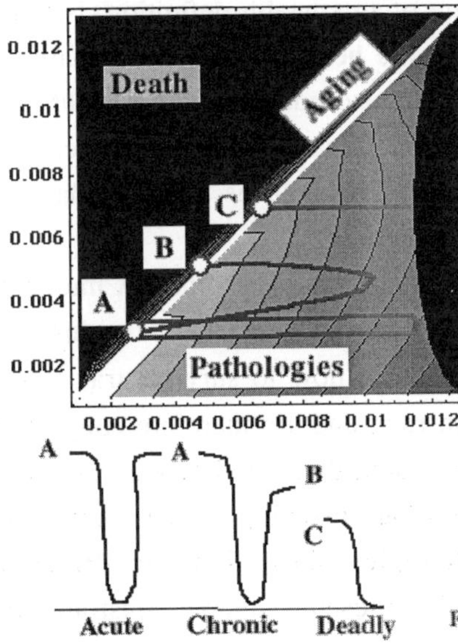

Fig. 5

Biological time determines whether a pathology is acute or chronic. If the trajectory returns to its starting point (Fig. 5-A) it is acute since the biological time or the distance between the beginning and of the trajectory (A-A) is zero. The chronic pathology advanced from A to B (Fig. 5), whose distance is 0.0028 and lasted therefore 0.0028 biological time units.

3 Health

The model provides a simple definition of health. Suppose a plague kills 30% of the population. Retrospectively we may conclude that before the epidemic started, 70% were healthier. When a poison given to mice kills 50% of them it is called LD-50 (lethal dose of 50%). The same experiment indicates also that before challenging the group with this poison, 50% could tolerate it better, and were therefore healthier than the others. This example illustrates Canguilhem's' concept of health: "Health is a margin of tolerance for the inconsistencies of the environment" [10]. Suppose that an old timer continues smoking daily four packs of cigarettes and remains unharmed, while his younger companions die from lung cancer. He is healthier, than the others.

The disadvantage of this health definition, is that it is destructive. One knows which mice are healthier only after challenging them with a poison. However there are also non-destructive health measures. One may take components that are linearly related with health, or vitality, e.g., TA length (Fig. 2). Since TA-length is linearly related with tolerance, it is a good estimator of tolerance.

The computer program that drives the TA through the tolerance set may be regarded as TA chromosome, and the instructions, its genes. While genes instruct the TA how to survive, trajectories in the disease set are driven by death rate, and are therefore epigenetic. A trajectory is carved out by an independent threat and dependent strategies that emerge from genes. Even a thorough mapping of genes cannot explain or predict the outcome of pathologies since they are driven by external and independent factors. The present example illustrates that genes do not suffice for explaining emerging diseases.

References

1. CG Langton, (Addison-Wessley Publ. Co. Inc. Reading MA, 1989)
2. G Zajicek,. *Meth. Inform. Med.* **31**, 167 (1992).
3. G Zajicek, *ACM Sigbio, Newsletter* **12**, 18 (1992) .
4. G Zajicek, ed. R Paton (Chapman & Hall London, 1994).
5. JR Koza, (MIT Press. Cambridge MA,1992).
6. AR Lacey, (Routledge, London, 1989).
7. S Wolfram (Addison Wesley Publ. Co. Inc. New York 1991).
8. G Zajicek, et al. *Appl. Optics* **26**,3408 (1987).
9. G Zajicek, *Path. Res. Pract.* **188**, 410 (1992).
10. G Canguilhem (Zone Books, New York 1991).